Modern Electronic Instrumentation and Measurement Techniques

Modern Electronic Instrumentation and Measurement Techniques

Albert D. Helfrick
William D. Cooper

PRENTICE HALL, *Englewood Cliffs, New Jersey 07632*

Library of Congress Cataloging-in-Publication Data

Helfrick, Albert D.
  Modern electronic instrumentation and measurement techniques /
Albert D. Helfrick, William D. Cooper.

  Rev. ed. of: Electronic instrumentation and measurement techniques
/ William David Cooper, Albert D. Helfrick. 3rd ed. c1985.
  Includes bibliographies and index.
  ISBN 0-13-593294-7
  1. Electric measurements. 2. Electric meters. 3. Electronic
measurements. 4. Electronic instruments. I Cooper, William
David. II. Cooper, William David. Electronic instrumentation and
measurement techniques. III. Title.
TK275.H45   1990
621.3815′4—dc19   88-34209

Editorial/production supervision and
  interior design: *Arthur Hamparian*
Cover design: *Lundgren Graphics, Ltd.*
Manufacturing buyer: *Mike Woerner*

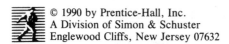 © 1990 by Prentice-Hall, Inc.
A Division of Simon & Schuster
Englewood Cliffs, New Jersey 07632

This book was previously published as
*Electronic Instrumentation and Measurement
Techniques.*

Printed in the United States of America

10 9 8 7 6 5 4 3 2 1

ISBN 0-13-593294-7

PRENTICE-HALL INTERNATIONAL (UK) LIMITED, *London*
PRENTICE-HALL OF AUSTRALIA PTY. LIMITED, *Sydney*
PRENTICE-HALL CANADA INC., *Toronto*
PRENTICE-HALL HISPANOAMERICANA, S.A., *Mexico*
PRENTICE-HALL OF INDIA PRIVATE LIMITED, *New Delhi*
PRENTICE-HALL OF JAPAN, INC., *Tokyo*
SIMON & SCHUSTER ASIA PTE. LTD., *Singapore*
EDITORA PRENTICE-HALL DO BRASIL, LTDA., *Rio de Janeiro*

Contents

8 *Signal Generation*   246

9 *Signal Analysis*   282

10 *Frequency Counters and Time-Interval Measurements*   315

Contents

Contents

Preface

This new edition of *Electronic Instrumentation and Measurement Techniques* is a modernization of an old and effective text. The characteristics that has made this book successful over the years have been retained while every effort was taken to ensure a modern text that covers all aspects of instrumentation. To enforce this concept, the title has been changed to *Modern Electronic Instrumentation and Measurement Techniques*.

Basic measurement techniques such as accuracy, precision, standards, and so on, are retained, with some clarification and modernization to include new standards. Understanding these basics is an absolute prerequisite for the discussion of more sophisticated systems.

Some information concerning moving-coil meters was removed and modified, as these instruments find fewer applications in modern electronics. Some of the material is retained as an introduction to the general problems of measurement without bogging the reader down with excessively complex measuring systems.

The digital storage oscilloscope is a new subject, as its use has become more commonplace in recent years. The Fourier transform or digital spectrum analyzer is also included in this edition. These two digital instruments are gaining wide acceptance in electronic instrumentation.

Chapters 11 and 12 on transducers and data acquisition have received considerable overhaul to include more modern transducers and to include such important subjects as instrumentation and isolation amplifiers, and data transmission. An important inclusion in Chapter 12 is fiber optics data transmission, which is gaining rapid acceptance in the industrial environment.

Chapter 14 is totally new and covers fiber optics measurements. There is very little material available to the student on the subject of optical measurements relative to fiber optics, and this chapter makes this edition unique.

Above all, those items that make a book a textbook, such as worked-out examples, references, and review problems at the end of the chapters have been retained and expanded.

ACKNOWLEDGMENTS

Grateful appreciation is offered to the following individuals for reviewing our work: David G. Delker, Kansas State University; Val Feldkircher, Electronic Technology Institute; Dewey J. Gray, Nashville State College; Richard A. Hultin, Rochester Institute of Technology; Earl C. Iselin, Jr., University of Dayton; O. M. Kuritza; William Middendorf, University of Cincinnati; and Donald J. Poulin, Northeastern University.

Modern Electronic Instrumentation and Measurement Techniques

1

Measurement and Error

1-1 DEFINITIONS

Measurement generally involves using an instrument as a physical means of determining a quantity or variable. The instrument serves as an extension of human faculties and in many cases enables a person to determine the value of an unknown quantity which his unaided human faculties could not measure. An instrument, then, may be defined as *a device for determining the value or magnitude of a quantity or variable*. The *electronic* instrument, as its name implies, is based on electrical or electronic principles for its measurement function. An electronic instrument may be a relatively uncomplicated device of simple construction such as a basic dc current meter (see Chapter 4). As technology expands, however, the demand for more elaborate and more accurate instruments increases and produces new developments in instrument design and application. To use these instruments intelligently, one needs to understand their operating principles and to appraise their suitability for the intended application.

Measurement work employs a number of terms which should be defined here.

> *Instrument:* a device for determining the value or magnitude of a quantity or variable.

Accuracy: closeness with which an instrument reading approaches the true value of the variable being measured.

Precision: a measure of the reproducibility of the measurements; i.e., given a fixed value of a variable, precision is a measure of the degree to which successive measurements differ from one another.

Sensitivity: the ratio of output signal or response of the instrument to a change of input or measured variable.

Resolution: the smallest change in measured value to which the instrument will respond.

Error: deviation from the true value of the measured variable.

Several techniques may be used to minimize the effects of errors. For example, in making precision measurements, it is advisable to record a series of observations rather than rely on one observation. Alternate methods of measurement, as well as the use of different instruments to perform the same experiment, provide a good technique for increasing accuracy. Although these techniques tend to increase the *precision* of measurement by reducing environmental or random error, they cannot account for instrumental error.*

This chapter provides an introduction to different types of error in measurement and to the methods generally used to express errors, in terms of the most reliable value of the measured variable.

1-2 ACCURACY AND PRECISION

Accuracy refers to the degree of closeness or conformity to the true value of the quantity under measurement. *Precision* refers to the degree of agreement within a group of measurements or instruments.

To illustrate the distinction between accuracy and precision, two voltmeters of the same make and model may be compared. Both meters have knife-edged pointers and mirror-backed scales to avoid parallax, and they have carefully calibrated scales. They may therefore be read to the same *precision*. If the value of the series resistance in one meter changes considerably, its readings may be in error by a fairly large amount. Therefore the *accuracy* of the two meters may be quite different. (To determine which meter is in error, a comparison measurement with a standard meter should be made.)

Precision is composed of two characteristics: *conformity* and the number of *significant figures* to which a measurement may be made. Consider, for example, that a resistor, whose true resistance is 1,384,572 Ω, is measured by an ohmmeter which consistently and repeatedly indicates 1.4 MΩ. But can the observer "read" the true value from the scale? His estimates from the scale reading consistently yield a value of 1.4 MΩ. This is as close to the true value as he can read

---

* Melville B. Stout, *Basic Electrical Measurements,* 2nd ed. (Englewood Cliffs, N.J.: Prentice-Hall, Inc., 1960), pp. 21–26.

the scale by estimation. Although there are no deviations from the observed value, the error created by the limitation of the scale reading is a *precision* error. The example illustrates that conformity is a necessary, but not sufficient, condition for precision because of the lack of significant figures obtained. Similarly, precision is a necessary, but not sufficient, condition for accuracy.

Too often the beginning student is inclined to accept instrument readings at face value. He is not aware that the accuracy of a reading is *not* necessarily guaranteed by its precision. In fact, good measurement technique demands *continuous skepticism* as to the accuracy of the results.

In critical work, good practice dictates that the observer make an independent set of measurements, using different instruments or different measurement techniques, not subject to the same systematic errors. He must also make sure that the instruments function properly and are calibrated against a known standard, and that no outside influence affects the accuracy of his measurements.

1-3 SIGNIFICANT FIGURES

An indication of the precision of the measurement is obtained from the number of *significant figures* in which the result is expressed. Significant figures convey actual information regarding the magnitude and the measurement precision of a quantity. The more significant figures, the greater the precision of measurement.

For example, if a resistor is specified as having a resistance of 68 Ω, its resistance should be closer to 68 Ω than to 67 Ω or 69 Ω. If the value of the resistor is described as 68.0 Ω, it means that its resistance is closer to 68.0 Ω than it is to 67.9 Ω or 68.1 Ω. In 68 Ω there are two significant figures; in 68.0 Ω there are three. The latter, with more significant figures, expresses a measurement of greater precision than the former.

Often, however, the total number of digits may not represent measurement precision. Frequently, large numbers with zeros before a decimal point are used for approximate populations or amounts of money. For example, the population of a city is reported in six figures as 380,000. This may imply that the true value of the population lies between 379,999 and 380,001, which is six significant figures. What is meant, however, is that the population is closer to 380,000 than to 370,000 or 390,000. Since in this case the population can be reported only to two significant figures, how can large numbers be expressed?

A more technically correct notation uses *powers of ten,* 38×10^4 or 3.8×10^5. This indicates that the population figure is only accurate to two significant figures. Uncertainty caused by zeros to the *left* of the decimal point is therefore usually resolved by *scientific notation* using powers of ten. Reference to the velocity of light as 186,000 mi/s, for example, would cause no misunderstanding to anyone with a technical background. But 1.86×10^5 mi/s leaves no confusion.

It is customary to record a measurement with all the digits of which we are sure nearest to the true value. For example, in reading a voltmeter, the voltage may be read as 117.1 V. This simply indicates that the voltage, read by the observer to best estimation, is closer to 117.1 V than to 117.0 V or 117.2 V.

Another way of expressing this result indicates the *range of possible error*. The voltage may be expressed as 117.1 ± 0.05 V, indicating that the value of the voltage lies between 117.05 V and 117.15 V.

When a number of independent measurements are taken in an effort to obtain the best possible answer (closest to the true value), the result is usually expressed as the arithmetic *mean* of all the readings, with the range of possible error as the *largest deviation* from that mean. This is illustrated in Example 1-1.

EXAMPLE 1-1

A set of independent voltage measurements taken by four observers was recorded as 117.02 V, 117.11 V, 117.08 V, and 117.03 V. Calculate (a) the average voltage; (b) the range of error.

SOLUTION

(a) $E_{av} = \dfrac{E_1 + E_2 + E_3 + E_4}{N}$

$= \dfrac{117.02 + 117.11 + 117.08 + 117.03}{4} = 117.06$ V

(b) Range $= E_{max} - E_{av} = 117.11 - 117.06 = 0.05$ V

but also

$E_{av} - E_{min} = 117.06 - 117.02 = 0.04$ V

The average range of error therefore equals

$\dfrac{0.05 + 0.04}{2} = \pm 0.045 = \pm 0.05$ V

When two or more measurements with different degrees of accuracy are added, *the result is only as accurate as the least accurate measurement*. Suppose that two resistances are *added* in series as in Example 1-2.

EXAMPLE 1-2

Two resistors, R_1 and R_2, are connected in series. Individual resistance measurements, using a digital multimeter, give $R_1 = 18.7\ \Omega$ and $R_2 = 3.624\ \Omega$. Calculate the total resistance to the appropriate number of significant figures.

SOLUTION

$R_1 = 18.7\ \Omega$ (three significant figures)

$R_2 = 3.624\ \Omega$ (four significant figures)

$R_T = R_1 + R_2 = 22.324\ \Omega$ (five significant figures) $= 22.3\ \Omega$

The doubtful figures are written in *italics* to indicate that in the addition of R_1 and R_2 the last three digits of the sum are doubtful figures. There is no value whatsoever in retaining the last two digits (the *2* and the *4*) because one of the

---

Measurement and Error   Chap. 1

resistances is accurate only to three significant figures or tenths of an ohm. The result should therefore also be reduced to three significant figures or the nearest tenth, i.e., *22.3* Ω.

The number of significant figures in *multiplication* may increase rapidly, but again only the appropriate figures are retained in the answer, as shown in Example 1-3.

EXAMPLE 1-3

In calculating voltage drop, a current of 3.18 A is recorded in a resistance of 35.68 Ω. Calculate the voltage drop across the resistor to the appropriate number of significant figures.

SOLUTION

$$E = IR = (35.68) \times (3.18) = 113.4624 = 113 \text{ V}$$

Since there are three significant figures involved in the multiplication, the answer can be written only to a maximum of three significant figures.

In Example 1-3, the current, I, has three significant figures and R has four; and the result of the multiplication has only three significant figures. This illustrates that the answer cannot be known to any accuracy greater than the *least* poorly defined of the factors. Note also that if extra digits accumulate in the answer, they should be discarded or rounded off. In the usual practice, if the (least significant) digit in the first place to be discarded is less than five, it and the following digits are dropped from the answer. This was done in Example 1-3. If the digit in the first place to be discarded is five or greater, the previous digit is increased by one. For three-digit precision, therefore, 113.46 should be rounded off to 113; and 113.74 to 114.

Addition of figures with a range of doubt is illustrated in Example 1-4.

EXAMPLE 1-4

Add 826 ± 5 to 628 ± 3.

SOLUTION

$$N_1 = 826 \pm 5 \ (= \pm 0.605\%)$$
$$N_2 = 628 \pm 3 \ (= \pm 0.477\%)$$
$$\text{Sum} = 1,454 \pm 8 \ (= \pm 0.55\%)$$

Note in Example 1-4 that the doubtful parts are *added*, since the ± sign means that one number may be high and the other low. The worst possible combination of range of doubt should be taken in the answer. The percentage doubt in the original figure N_1 and N_2 does not differ greatly from the percentage doubt in the final result.

If the same two numbers are *subtracted,* as in Example 1-5, there is an interesting comparison between addition and subtraction with respect to the range of doubt.

EXAMPLE 1-5

Subtract 628 ± 3 from 826 ± 5 and express the range of doubt in the answer as a percentage.

SOLUTION

$$N_1 = 826 \pm 5 \ (= \pm 0.605\%)$$

$$N_2 = 628 \pm 3 \ (= \pm 0.477\%)$$

$$\text{Difference} = 198 \pm 8 \ (= \pm 4.04\%)$$

Again, in Example 1-5, the doubtful parts are added for the same reason as in Example 1-4. Comparing the results of addition and subtraction of the same numbers in Examples 1-4 and 1-5, note that the precision of the results, when expressed in *percentages,* differs greatly. The final result after subtraction shows a large increase in percentage doubt compared to the percentage doubt after addition. The percentage doubt increases even more when the difference between the numbers is relatively small. Consider the case illustrated in Example 1-6.

EXAMPLE 1-6

Subtract 437 ± 4 from 462 ± 4 and express the range of doubt in the answer as a percentage.

SOLUTION

$$N_1 = 462 \pm 4 \ (= \pm 0.87\%)$$

$$N_2 = 437 \pm 4 \ (= \pm 0.92\%)$$

$$\text{Difference} = 25 \pm 8 \ (= \pm 32\%)$$

Example 1-6 illustrates clearly that one should avoid measurement techniques depending on subtraction of experimental results because the range of doubt in the final result may be greatly increased.

1-4 TYPES OF ERROR

No measurement can be made with perfect accuracy, but it is important to find out what the accuracy actually is and how different errors have entered into the measurement. A study of errors is a first step in finding ways to reduce them. Such a study also allows us to determine the accuracy of the final test result.

Errors may come from different sources and are usually classified under three main headings:

Gross errors: largely human errors, among them misreading of instruments, incorrect adjustment and improper application of instruments, and computational mistakes.

Systematic errors: shortcomings of the instruments, such as defective or worn parts, and effects of the environment on the equipment or the user.

Random errors: those due to causes that cannot be directly established because of random variations in the parameter or the system of measurement.

Each of these classes of errors will be discussed briefly and some methods will be suggested for their reduction or elimination.

1-4.1 Gross Errors

This class of errors mainly covers *human* mistakes in reading or using instruments and in recording and calculating measurement results. As long as human beings are involved, some gross errors will inevitably be committed. Although complete elimination of gross errors is probably impossible, one should try to anticipate and correct them. Some gross errors are easily detected; others may be very elusive. One common gross error, frequently committed by beginners in measurement work, involves the improper use of an instrument. In general, indicating instruments change conditions to some extent when connected into a complete circuit, so that the measured quantity is altered by the method employed. For example, a well-calibrated voltmeter may give a misleading reading when connected across two points in a high-resistance circuit (Example 1-7). The same voltmeter, when connected in a low-resistance circuit, may give a more dependable reading (Example 1-8). These examples illustrate that the voltmeter has a "loading effect" on the circuit, altering the original situation by the measurement process.

EXAMPLE 1-7

A voltmeter, having a sensitivity of 1,000 Ω/V, reads 100 V on its 150-V scale when connected across an unknown resistor in series with a milliammeter. When the milliammeter reads 5 mA, calculate (a) the apparent resistance of the unknown resistor; (b) the actual resistance of the unknown resistor; (c) the error due to the loading effect of the voltmeter.

SOLUTION

(a) The total circuit resistance equals

$$R_T = \frac{V_T}{I_T} = \frac{100 \text{ V}}{5 \text{ mA}} = 20 \text{ k}\Omega$$

Neglecting the resistance of the milliammeter, the value of the unknown resistor is $R_X = 20$ kΩ.

(b) The voltmeter resistance equals

$$R_V = 1,000\,\frac{\Omega}{V} \times 150\ V = 150\ k\Omega$$

Since the voltmeter is in parallel with the unknown resistance, we can write

$$R_X = \frac{R_T R_V}{R_V - R_T} = \frac{20 \times 150}{130} \times 23.05\ k\Omega$$

(c) % Error $= \dfrac{\text{actual} - \text{apparent}}{\text{actual}} \times 100\% = \dfrac{23.05 - 20}{23.05} \times 100\%$

$$= 13.23\%$$

EXAMPLE 1-8

Repeat Example 1-7 if the milliammeter reads 800 mA and the voltmeter reads 40 V on its 150-V scale.

SOLUTION

(a) $R_T = \dfrac{V_T}{I_T} = \dfrac{40\ V}{0.8\ A} = 50\ \Omega$

(b) $R_V = 1,000\,\dfrac{\Omega}{V} \times 150\ V = 150\ k\Omega$

$$R_X = \frac{R_T R_V}{R_V - R_T} = \frac{50 \times 150}{149.95} = 50.1\ \Omega$$

(c) % Error $= \dfrac{50.1 - 50}{50.1} \times 100\% = 0.2\%$

Errors caused by the loading effect of the voltmeter can be avoided by using it intelligently. For example, a low-resistance voltmeter should not be used to measure voltages in a vacuum tube amplifier. In this particular measurement, a high-input impedance voltmeter (such as a VTVM or TVM) is required.

A large number of gross errors can be attributed to carelessness or bad habits, such as improper reading of an instrument, recording the result differently from the actual reading taken, or adjusting the instrument incorrectly. Consider the case in which a multirange voltmeter uses a single set of scale markings with different number designations for the various voltage ranges. It is easy to use a scale which does not correspond to the setting of the range selector of the voltmeter. A gross error may also occur when the instrument is not set to zero before the measurement is taken; then all the readings are off.

Errors like these cannot be treated mathematically. They can be avoided only by taking care in reading and recording the measurement data. Good practice requires making more than one reading of the same quantity, preferably by a different observer. Never place complete dependence on one reading but take at least three separate readings, preferably under conditions in which instruments are switched off-on.

1-4.2 Systematic Errors

This type of error is usually divided into two different categories: (1) instrumental errors, defined as shortcomings of the instrument; (2) environmental errors, due to external conditions affecting the measurement.

Instrumental errors are errors inherent in measuring instruments because of their mechanical structure. For example, in the d'Arsonval movement friction in bearings of various moving components may cause incorrect readings. Irregular spring tension, stretching of the spring, or reduction in tension due to improper handling or overloading of the instrument will result in errors. Other instrumental errors are calibration errors, causing the instrument to read high or low along its entire scale. (Failure to set the instrument to zero before making a measurement has a similar effect.)

There are many kinds of instrumental errors, depending on the type of instrument used. The experimenter should always take precautions to insure that the instrument he is using is operating properly and does not contribute excessive errors for the purpose at hand. Faults in instruments may be detected by checking for erratic behavior, and stability and reproducibility of results. A quick and easy way to check an instrument is to compare it to another with the same characteristics or to one that is known to be more accurate.

Instrumental errors may be avoided by (1) selecting a suitable instrument for the particular measurement application; (2) applying correction factors after determining the amount of instrumental error; (3) calibrating the instrument against a standard.

Environmental errors are due to conditions external to the measuring device, including conditions in the area surrounding the instrument, such as the effects of changes in temperature, humidity, barometric pressure, or of magnetic or electrostatic fields. Thus a change in ambient temperature at which the instrument is used causes a change in the elastic properties of the spring in a moving-coil mechanism and so affects the reading of the instrument. Corrective measures to reduce these effects include air conditioning, hermetically sealing certain components in the instrument, use of magnetic shields, and the like.

Systematic errors can also be subdivided into *static* or *dynamic* errors. Static errors are caused by limitations of the measuring device or the physical laws governing its behavior. A static error is introduced in a micrometer when excessive pressure is applied in torquing the shaft. Dynamic errors are caused by the instrument's not responding fast enough to follow the changes in a measured variable.

1-4.3 Random Errors

These errors are due to unknown causes and occur even when all systematic errors have been accounted for. In well-designed experiments, few random errors usually occur, but they become important in high-accuracy work. Suppose a voltage is being monitored by a voltmeter which is read at half-hour intervals. Although the instrument is operated under ideal environmental conditions and has been accurately calibrated before the measurement, it will be found that the readings vary slightly over the period of observation. This variation cannot be corrected by any method of calibration or other known method of control and it cannot be explained without minute investigation. The only way to offset these errors is by increasing the number of readings and using statistical means to obtain the best approximation of the true value of the quantity under measurement.

1-5 STATISTICAL ANALYSIS

A statistical analysis of measurement data is common practice because it allows an analytical determination of the uncertainty of the final test result. The outcome of a certain measurement method may be predicted on the basis of sample data without having detailed information on all the disturbing factors. To make statistical methods and interpretations meaningful, a large number of measurements is usually required. Also, systematic errors should be small compared with residual or random errors, because statistical treatment of data cannot remove a fixed bias contained in all the measurements.

1-5.1 Arithmetic Mean

The most probable value of a measured variable is the arithmetic mean of the number of readings taken. The best approximation will be made when the number of readings of the same quantity is very large. Theoretically, an infinite number of readings would give the best result, although in practice, only a finite number of measurements can be made. The arithmetic mean is given by the following expression:

$$\bar{x} = \frac{x_1 + x_2 + x_3 + x_4 + \cdots + x_n}{n} = \frac{\Sigma x}{n} \tag{1-1}$$

where
$$\bar{x} = \text{arithmetic mean}$$
$$x_1, x_2, x_n = \text{readings taken}$$
$$n = \text{number of readings}$$

Example 1-1 showed how the arithmetic mean is used.

1-5.2 Deviation from the Mean

Deviation is the departure of a given reading from the arithmetic mean of the group of readings. If the deviation of the first reading, x_1, is called d_1, and that of the second reading, x_2, is called d_2, and so on, then the deviations from the mean can be expressed as

$$d_1 = x_1 - \bar{x} \qquad d_2 = x_2 - \bar{x} \qquad d_n = x_n - \bar{x} \tag{1-2}$$

Note that the deviation from the mean may have a positive or a negative value and that the algebraic sum of all the deviations must be zero.

Example 1-9 illustrates the computation of deviations.

EXAMPLE 1-9

A set of independent current measurements was taken by six observers and recorded as 12.8 mA, 12.2 mA, 12.5 mA, 13.1 mA, 12.9 mA, and 12.4 mA. Calculate (a) the arithmetic mean; (b) the deviations from the mean.

SOLUTION (a) Using Eq. (1-1), we see that the arithmetic mean equals

$$\bar{x} = \frac{12.8 + 12.2 + 12.5 + 13.1 + 12.9 + 12.4}{6} = 12.65 \text{ mA}$$

(b) Using Eq. (1-2), we see that the deviations are

$$d_1 = 12.8 - 12.65 = 0.15 \text{ mA}$$

$$d_2 = 12.2 - 12.65 = -0.45 \text{ mA}$$

$$d_3 = 12.5 - 12.65 = -0.15 \text{ mA}$$

$$d_4 = 13.1 - 12.65 = 0.45 \text{ mA}$$

$$d_5 = 12.9 - 12.65 = 0.25 \text{ mA}$$

$$d_6 = 12.4 - 12.65 = -0.25 \text{ mA}$$

Note that the algebraic sum of all the deviations equals zero.

1-5.3 Average Deviation

The average deviation is an indication of the precision of the instruments used in making the measurements. Highly precise instruments will yield a low average deviation between readings. By definition, average deviation is the sum of the *absolute* values of the deviations divided by the number of readings. The absolute value of the deviation is the value without respect to sign. Average deviation may be expressed as

$$D = \frac{|d_1| + |d_2| + |d_3| + \cdots + |d_n|}{n} = \frac{\Sigma |d|}{n} \tag{1-3}$$

Example 1-10 shows how average deviation is calculated.

EXAMPLE 1-10

Calculate the average deviation for the data given in Example 1-9.

SOLUTION

$$D = \frac{0.15 + 0.45 + 0.15 + 0.45 + 0.25 + 0.25}{6} = 0.283 \text{ mA}$$

1-5.4 Standard Deviation

In statistical analysis of random errors, the root-mean-square deviation or *standard deviation* is a very valuable aid. By definition, the standard deviation σ of an infinite number of data is the square root of the sum of *all* the individual deviations squared, divided by the number of readings. Expressed mathematically:

$$\sigma = \sqrt{\frac{d_1^2 + d_2^2 + d_3^2 + \cdots + d_n^2}{n}} = \sqrt{\frac{\Sigma\, d_i^2}{n}} \qquad (1\text{-}4)$$

In practice, of course, the possible number of observations is finite. The standard deviation of a *finite* number of data is given by

$$\sigma = \sqrt{\frac{d_1^2 + d_2^2 + d_3^2 + \cdots + d_n^2}{n-1}} = \sqrt{\frac{\Sigma\, d_i^2}{n-1}} \qquad (1\text{-}5)$$

Equation (1-5) will be used in Example 1-11.

Another expression for essentially the same quantity is the *variance* or *mean square deviation,* which is the same as the standard deviation except that the square root is not extracted. Therefore

$$\text{variance } (V) = \text{mean square deviation} = \sigma^2$$

The variance is a convenient quantity to use in many computations because variances are additive. The standard deviation, however, has the advantage of being of the same units as the variable, making it easy to compare magnitudes. Most scientific results are now stated in terms of standard deviation.

1-6 PROBABILITY OF ERRORS

1-6.1 Normal Distribution of Errors

Table 1-1 shows a tabulation of 50 voltage readings that were taken at small time intervals and recorded to the nearest 0.1 V. The nominal value of the measured voltage was 100.0 V. The result of this series of measurements can be presented

TABLE 1-1 Tabulation of Voltage Readings

Voltage reading (volts)	Number of readings
99.7	1
99.8	4
99.9	12
100.0	19
100.1	10
100.2	3
100.3	1
	50

Measurement and Error   Chap. 1

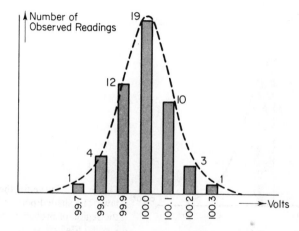

Figure 1-1 Histogram showing the frequency of occurrence of the 50 voltage readings of Table 1-1. The dashed curve represents the limiting case of the histogram when a large number of readings at small increments are taken.

graphically in the form of a block diagram or *histogram* in which the number of observations is plotted against each observed voltage reading. The histogram of Fig. 1-1 represents the data of Table 1-1.

Figure 1-1 shows that the largest number of readings (19) occurs at the central value of 100.0 V, while the other readings are placed more or less symmetrically on either side of the central value. If more readings were taken at smaller increments, say 200 readings at 0.05-V intervals, the distribution of observations would remain approximately symmetrical about the central value and the shape of the histogram would be about the same as before. With more and more data, taken at smaller and smaller increments, the contour of the histogram would finally become a smooth curve, as indicated by the dashed line in Fig. 1-1. This bell-shaped curve is known as a Gaussian curve. The sharper and narrower the curve, the more definitely an observer may state that the most probable value of the true reading is the central value or mean reading.

The Gaussian or Normal law of error forms the basis of the analytical study of random effects. Although the mathematical treatment of this subject is beyond the scope of this text, the following qualitative statements are based on the Normal law:

(a) All observations include small disturbing effects, called random errors.

(b) Random errors can be positive or negative.

(c) There is an equal probability of positive and negative random errors.

We can therefore expect that measurement observations include plus and minus errors in more or less equal amounts, so that the total error will be small and the mean value will be the true value of the measured variable.

The possibilities as to the form of the error distribution curve can be stated as follows:

(a) Small errors are more probable than large errors.

(b) Large errors are very improbable.

Sec. 1-6   Probability of Errors

13

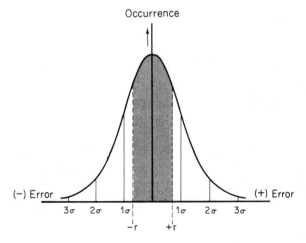

Occurrence

(−) Error   (+) Error

3σ   2σ   1σ   1σ   2σ   3σ

−r   +r

Figure 1-2 Curve for the Normal law. The shaded portion indicates the region of probable error, where $r = \pm 0.6745\sigma$.

(c) There is an equal probability of plus and minus errors so that the probability of a given error will be symmetrical about the zero value.

The error distribution curve of Fig. 1-2 is based on the Normal law and shows a symmetrical distribution of errors. This normal curve may be regarded as the limiting form of the histogram of Fig. 1-1 in which the most probable value of the true voltage is the mean value of 100.0 V.

1-6.2 Probable Error

The area under the Gaussian probability curve of Fig. 1-2, between the limits $+\infty$ and $-\infty$, represents the entire number of observations. The area under the curve between the $+\sigma$ and $-\sigma$ limits represents the cases that differ from the mean by no more than the standard deviation. Integration of the area under the curve within the $\pm\sigma$ limits gives the total number of cases within these limits. For normally dispersed data, following the Gaussian distribution, approximately 68 per cent of all the cases lie between the limits of $+\sigma$ and $-\sigma$ from the mean. Corresponding values of other deviations, expressed in terms of σ, are given in Table 1-2.

If, for example, a large number of nominally 100-Ω resistors is measured and the mean value is found to be 100.00 Ω, with a standard deviation (S.D.) of 0.20 Ω,

TABLE 1-2 Area Under the Probability Curve

Deviation ($\pm$), σ	Fraction of total area included
0.6745	0.5000
1.0	0.6828
2.0	0.9546
3.0	0.9972

we know that on the average 68 per cent (or roughly two-thirds) of all the resistors have values which lie between limits of $\pm 0.20\ \Omega$ of the mean. There is then approximately a two to one chance that any resistor, selected from the lot at random, will lie within these limits. If larger odds are required, the deviation may be extended to a limit of $\pm 2\sigma$, in this case $\pm 0.40\ \Omega$. According to Table 1-2, this now includes 95 per cent of all the cases, giving ten to one odds that any resistor selected at random lies within $\pm 0.40\ \Omega$ of the mean value of $100.00\ \Omega$.

Table 1-2 also shows that half of the cases are included in the deviation limits of $\pm 0.6745\sigma$. The quantity r is called the *probable error* and is defined as

$$\text{probable error } r = \pm 0.6745\sigma \qquad (1\text{-}6)$$

This value is *probable* in the sense that there is an even chance that any one observation will have a random error no greater than $\pm r$. Probable error has been used in experimental work to some extent in the past, but standard deviation is more convenient in statistical work and is given preference.

EXAMPLE 1-11

Ten measurements of the resistance of a resistor gave 101.2 Ω, 101.7 Ω, 101.3 Ω, 101.0 Ω, 101.5 Ω, 101.3 Ω, 101.2 Ω, 101.4 Ω, 101.3 Ω, and 101.1 Ω. Assume that only random errors are present. Calculate (a) the arithmetic mean; (b) the standard deviation of the readings; (c) the probable error.

SOLUTION With a large number of readings a simple tabulation of data is very convenient and avoids confusion and mistakes.

		Deviation	
Reading, x		d	d^2
	101.2	−0.1	0.01
	101.7	0.4	0.16
	101.3	0.0	0.00
	101.0	−0.3	0.09
	101.5	0.2	0.04
	101.3	0.0	0.00
	101.2	−0.1	0.01
	101.4	0.1	0.01
	101.3	0.0	0.00
	101.1	−0.2	0.04
$\Sigma x = 1{,}013.0$		$\Sigma |d| = 1.4$	$\Sigma d^2 = 0.36$

(a) Arithmetic mean, $\bar{x} = \dfrac{\Sigma x}{n} = \dfrac{1{,}013.0}{10} = 101.3\ \Omega$

(b) Standard deviation, $\sigma = \sqrt{\dfrac{d^2}{n-1}} = \sqrt{\dfrac{0.36}{9}} = 0.2\ \Omega$

(c) Probable error $= 0.6745\sigma = 0.6745 \times 0.2 = 0.1349\ \Omega$

1-7 LIMITING ERRORS

In most indicating instruments the accuracy is guaranteed to a certain percentage of full-scale reading. Circuit components (such as capacitors, resistors, etc.) are guaranteed within a certain percentage of their rated value. The limits of these deviations from the specified values are known as *limiting errors* or *guarantee errors*. For example, if the resistance of a resistor is given as 500 Ω ± 10 per cent, the manufacturer guarantees that the resistance falls between the limits 450 Ω and 550 Ω. The maker is not specifying a standard deviation or a probable error, but promises that the error is no greater than the limits set.

EXAMPLE 1-12

A 0–150-V voltmeter has a guaranteed accuracy of 1 per cent full-scale reading. The voltage measured by this instrument is 83 V. Calculate the limiting error in per cent.

SOLUTION   The magnitude of the limiting error is

$$0.01 \times 150 \text{ V} = 1.5 \text{ V}$$

The percentage error at a meter indication of 83 V is

$$\frac{1.5}{83} \times 100 \text{ per cent} = 1.81 \text{ per cent}$$

It is important to note in Example 1-12 that a meter is guaranteed to have an accuracy of better than 1 per cent of the full-scale reading, but when the meter reads 83 V the limiting error increases to 1.81 per cent. Correspondingly, when a smaller voltage is measured, the limiting error will increase further. If the meter reads 60 V, the per cent limiting error is $1.5/60 \times 100 = 2.5$ per cent; if the meter reads 30 V, the limiting error is $1.5/30 \times 100 = 5$ per cent. The increase in per cent limiting error, as smaller voltages are measured, occurs because the magnitude of the limiting error is a fixed quantity based on the full-scale reading of the meter. Example 1-12 shows the importance of taking measurements *as close to full scale as possible*.

Measurements or computations, *combining* guarantee errors, are often made. Example 1-13 illustrates such a computation.

EXAMPLE 1-13

The voltage generated by a circuit is equally dependent on the value of three resistors and is given by the following equation:

$$V_{\text{out}} = \frac{R_1 R_2}{R_3}$$

If the tolerance of each resistor is 0.1 per cent, what is the maximum error of the generated voltage?

SOLUTION The highest resulting voltage occurs when R_1 and R_2 are at the maximum value allowed by the tolerance, while R_3 is at the lowest value allowed by the tolerance. The actual value need not be known but only the relative value. For a variation of 0.1 per cent the highest value of a resistor is 1.001 times the nominal value, while the lowest value is 0.999 times the nominal value. Using the maximum value of R_1 and R_2 and the minimum value for R_3 results in the greatest value for V_{out} of

$$V_{out} = \frac{(1.001R_1)(1.001R_2)}{0.999R_3} = 1.003$$

The lowest resulting voltage occurs when the value of R_3 is highest and R_1 and R_2 are the lowest. The resulting voltage is

$$V_{out} = \frac{(0.999R_1)(0.999R_2)}{1.003R_3} = 0.997$$

The total variation of the resultant voltage is ± 0.3 per cent, which is the algebraic sum of the three tolerances. This is true in the first approximation. The maximum error is slightly different from the sum of the individual tolerances. On the other hand, it is highly unlikely that all three components of this example would have the maximum error and in such a fashion to produce the maximum or minimum voltage. Therefore, the statistical methods outlined in the previous sections must be used.

EXAMPLE 1-14

The current passing through a resistor of 100 ± 0.2 Ω is 2.00 ± 0.01 A. Using the relationship $P = I^2 R$, calculate the limiting error in the computed value of power dissipation.

SOLUTION Expressing the guaranteed limits of both current and resistance in percentages instead of units, we obtain

$$I = 2.00 \pm 0.01 \text{ A} = 2.00 \pm 0.5\%$$

$$R = 100 \pm 0.2\% = 100 \pm 0.2\%$$

If the worst possible combination of errors for the calculation of power, that is, the highest value of resistance and the highest value of current, is used, the power dissipation becomes

$$P = I^2(1 + 0.005)^2 R(1.002) = 1.012I^2R$$

For the lowest power dissipation,

$$P = I^2(1 - 0.005)^2 R(1 - 0.002) = 0.988I^2R$$

The error is ± 1.2 per cent, which is two times the 0.5 per cent error of the current plus the 0.2 per cent error of the resistor. This is because the I term of the equation essentially appears twice in the equation. This can be seen by rewriting the equation

$$P = I \times I \times R = I^2R$$

REFERENCES

1-1. Bartholomew, Davis, *Electrical Measurements and Instrumentation*, chaps. 1, 2. Boston: Allyn and Bacon, Inc., 1963.

1-2. Maloney, Timothy J., *Electric Circuits: Principles and Applications*, chap. 1. Englewood Cliffs, N.J.: Prentice-Hall, Inc., 1984.

1-3. Young, Hugh D., *Statistical Treatment of Experimental Data*. New York: McGraw-Hill Book Company, 1962.

PROBLEMS

1-1. What is the difference between accuracy and precision?

1-2. List four sources of possible errors in instruments.

1-3. What are the three general classes of errors?

1-4. Define **(a)** instrumental error; **(b)** limiting error; **(c)** calibration error; **(d)** environmental error; **(e)** random error; **(f)** probable error.

1-5. A 0–1-mA milliammeter has 100 divisions which can easily be read to the nearest division. What is the resolution of the meter?

1-6. A digital voltmeter has a read-out range from 0 to 9,999 counts. Determine the resolution of the instrument in volts when the full-scale reading is 9.999 V.

1-7. State the number of significant figures in each of the following: **(a)** 542; **(b)** 0.65; **(c)** 27.25; **(d)** 0.00005; **(e)** 40×10^6; **(f)** 20,000.

1-8. Four capacitors are placed in parallel. The capacitor values are 36.3 μF, 3.85 μF, 34.002 μF, and 850 nF, with an uncertainty of one digit in the last place. What is the total capacitance? Give only the significant figures in the answer.

1-9. A voltage drop of 112.5 V is measured across a resistor passing a current of 1.62 A. Calculate the power dissipation of the resistor. Give only significant figures in the answer.

1-10. What voltage would a 20,000-Ω/V meter on a 0–1-V scale show in the circuit of Fig. P1-10?

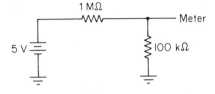

Figure P1-10

1-11. The voltage across a resistor is 200 V, with a probable error of ±2 per cent, and the resistance is 42 Ω with a probable error of ±1.5 per cent. Calculate **(a)** the power dissipated in the resistor; **(b)** the percentage error in the answer.

1-12. The following values were obtained from the measurements of the value of a resistor: 147.2 Ω, 147.4 Ω, 147.9 Ω, 148.1 Ω, 147.1 Ω, 147.5 Ω, 147.6 Ω, 147.4 Ω, 147.6 Ω, and

147.5 Ω. Calculate (a) the arithmetic mean; (b) the average deviation; (c) the standard deviation; (d) the probable error of the average of the ten readings.

1-13. Six determinations of a quantity, as entered on the data sheet and presented to you for analysis, are 12.35, 12.71, 12.48, 10.24, 12.63, and 12.58. Examine the data and on the basis of your conclusions calculate (a) the arithmetic mean; (b) the standard deviation; (c) the probable error in per cent of the average of the readings.

1-14. Two resistors have the following ratings:

$$R_1 = 36 \ \Omega \pm 5\% \quad \text{and} \quad R_2 = 75 \ \Omega \pm 5\%$$

Calculate (a) the magnitude of error in each resistor; (b) the limiting error in ohms and in per cent when the resistors are connected in series; (c) the limiting error in ohms and in per cent when the resistors are connected in parallel.

1-15. The resistance of an unknown resistor is determined by the Wheatstone bridge method. The solution for the unknown resistance is stated as $R_x = R_1 R_2 / R_3$, where

$$R_1 = 500 \ \Omega \pm 1\%$$
$$R_2 = 615 \ \Omega \pm 1\%$$
$$R_3 = 100 \ \Omega \pm 0.5\%$$

Calculate (a) the nominal value of the unknown resistor; (b) the limiting error in ohms of the unknown resistor; (c) the limiting error in per cent of the unknown resistor.

1-16. A resistor is measured by the voltmeter-ammeter method. The voltmeter reading is 123.4 V on the 250-V scale and the ammeter reading is 283.5 mA on the 500-mA scale. Both meters are guaranteed to be accurate within ±1 per cent of full-scale reading. Calculate (a) the indicated value of the resistance; (b) the limits within which you can guarantee the result.

1-17. In a dc circuit, the voltage across a component is 64.3 V and the current is 2.53 A. Both current and voltage are given with an uncertainty of one unit in the last place. Calculate the power dissipation to the appropriate number of significant figures.

1-18. A power transformer was tested to determine losses and efficiency. The input power was measured as 3,650 W and the delivered output power was 3,385 W, with each reading in doubt by ±10 W. Calculate (a) the percentage uncertainty in the losses of the transformer; (b) the percentage uncertainty in the efficiency of the transformer, as determined by the difference in input and output power readings.

1-19. The power factor and phase angle in a circuit carrying a sinusoidal current are determined by measurements of current, voltage, and power. The current is read as 2.50 A on a 5-A ammeter, the voltage as 115 V on a 250-V voltmeter, and the power as 220 W on a 500-W wattmeter. The ammeter and voltmeter are guaranteed accurate to within ±0.5 per cent of full-scale indication and the wattmeter to within ±1 per cent of full-scale reading. Calculate (a) the percentage accuracy to which the power factor can be guaranteed; (b) the possible error in the phase angle.

2

Systems of Units of Measurement

2-1 FUNDAMENTAL AND DERIVED UNITS

To specify and perform calculations with physical quantities, the physical quantities must be defined both in *kind* and *magnitude*. The standard measure of each kind of physical quantity is the *unit*; the number of times the unit occurs in any given amount of the same quantity is the *number of measure*. For example, when we speak of a distance of 100 meters, we know that the meter is the unit of length and that the number of units of length is one hundred. The physical quantity, *length,* is therefore defined by the unit, *meter*. Without the unit, the number of measure has no physical meaning.

In science and engineering, two kinds of units are used: *fundamental units* and *derived units*. The fundamental units in mechanics are measures of *length, mass,* and *time*. The sizes of the fundamental units, whether foot or meter, pound or kilogram, second or hour, are arbitrary and can be selected to fit a certain set of circumstances. Since length, mass, and time are fundamental to most other physical quantities besides those in mechanics, they are called the *primary* fundamental units. Measures of certain physical quantities in the thermal, electrical, and illumination disciplines are also represented by fundamental units. These units are used only when these particular classes are involved, and they may therefore be defined as *auxiliary* fundamental units.

20

All other units which can be expressed in terms of the fundamental units are called *derived* units. Every derived unit originates from some physical law defining that unit. For example, the area (A) of a rectangle is proportional to its length (l) and breadth (b), or $A = lb$. If the meter has been chosen as the unit of length, then the area of a rectangle of 3 m by 4 m is 12 m². Note that the numbers of measure are multiplied ($3 \times 4 = 12$) as well as the units (m $\times$ m $=$ m²). The derived unit for area (A) is then the square meter (m²).

A derived unit is recognized by its *dimensions,* which can be defined as the complete algebraic formula for the derived unit. The dimensional *symbols* for the fundamental units of length, mass, and time are L, M, and T, respectively. The dimensional symbol for the derived unit of area is L^2 and that for volume, L^3. The dimensional symbol for the unit of force is LMT^{-2}, which follows from the defining equation for force. The dimensional formulas of the derived units are particularly useful for converting units from one system to another, as is shown in Sec. 2-6.

For convenience, some derived units have been given new names. For example, the derived unit of force in the SI system is called the newton (N), instead of the dimensionally correct name kg m/s².

2-2 SYSTEMS OF UNITS

In 1790 the French government issued a directive to the French Academy of Sciences to study and to submit proposals for a single system of weights and measures to replace all other existing systems. The French scientists decided, as a first principle, that a *universal system* of weights and measures should not depend on man-made reference standards, but instead be based on permanent measures provided by nature. As the *unit of length,* therefore, they chose the *meter,* defined as the ten-millionth part of the distance from the pole to the equator along the meridian passing through Paris. As the *unit of mass* they chose the mass of a cubic centimeter of distilled water at 4°C and normal atmospheric pressure (760 mm Hg) and gave it the name *gram.* As the third unit, the *unit of time,* they decided to retain the traditional second, defining it as 1/86,400 of the mean solar day.

As a second principle, they decided that all other units should be *derived* from the aforementioned three *fundamental units* of length, mass, and time. Next—the third principle—they proposed that all multiples and submultiples of basic units be in the *decimal system,* and they devised the system of prefixes in use today. Table 2-1 lists the decimal multiples and submultiples.

The proposals of the French Academy were approved and introduced as the *metric system* of units in France in 1795. The metric system aroused considerable interest elsewhere and finally, in 1875, 17 countries signed the so-called Metre Convention, making the metric system of units the legal system. Britain and the United States, although signatories of the convention, recognized its legality only in international transactions but did not accept the metric system for their own domestic use.

TABLE 2-1 Decimal Multiples and Submultiples

Name	Symbol	Equivalent
tera	T	10^{12}
giga	G	10^{9}
mega	M	10^{6}
kilo	k	10^{3}
hecto	h	10^{2}
deca	da	10
deci	d	10^{-1}
centi	c	10^{-2}
milli	m	10^{-3}
micro	μ	10^{-6}
nano	n	10^{-9}
pico	p	10^{-12}
femto	f	10^{-15}
atto	a	10^{-18}

Britain, in the meantime, had been working on a system of electrical units, and the British Association for the Advancement of Science decided on the centimeter and the gram as the fundamental units of length and mass. From this developed the *centimeter-gram-second* or *CGS absolute system* of units, used by physicists all over the world. Complications arose when the CGS system was extended to electric and magnetic measurements because of the need to introduce at least one more unit in the system. In fact, two parallel systems were established. In the CGS *electrostatic system,* the unit of electric charge was derived from the centimeter, gram, and second by assigning the value 1 to the permittivity of free space in Coulomb's law for the force between electric charges. In the CGS *electromagnetic system,* the basic units are the same and the unit of magnetic pole strength is derived from them by assigning the value 1 to the permeability of free space in the inverse square formula for the force between magnetic poles.

The *derived* units for electric current and electric potential in the electromagnetic system, the ampere and the volt, are used in practical measurements. These two units, and the corresponding ones, such as the coulomb, ohm, henry, farad, etc., were incorporated in a third system, called the *practical system.* Further simplification in the establishment of a truly universal system came as the result of pioneer work by the Italian engineer Giorgi, who pointed out that the practical units of current, voltage, energy, and power, used by electrical engineers, were compatible with the meter-kilogram-second system. He suggested that the metric system be expanded into a *coherent* system of units by including the practical electrical units. The Giorgi system, adopted by many countries in 1935, came to be known as the MKSA system of units in which the ampere was selected as the fourth basic unit.

A more comprehensive system was adopted in 1954 and designated in 1960 by international agreement as the Système International d'Unités (SI). In the SI system, six basic units are used, namely, the meter, kilogram, second, and ampere

TABLE 2-2 Basic SI Quantities, Units, and Symbols

Quantity	Unit	Symbol
Length	meter	m
Mass	kilogram	kg
Time	second	s
Electric current	ampere	A
Thermodynamic temperature	kelvin	K
Luminous intensity	candela	cd

of the MKSA system and, in addition, the kelvin and the candela as the units of temperature and luminous intensity, respectively. The SI units are replacing other systems in science and technology; they have been adopted as the legal units in France, and will become obligatory in other metric countries.

The six basic SI quantities and units of measurement, with their unit symbols, are listed in Table 2-2.

2-3 ELECTRIC AND MAGNETIC UNITS

Before listing the SI units (sometimes called the *International* MKS system of units), a brief look at the origin of the electrical and magnetic units seems appropriate. The practical electrical and magnetic units with which we are familiar, such as the volt, ampere, ohm, henry, etc., were first derived in the CGS systems of units.

The *CGS electrostatic system* (CGSe) is based on Coulomb's experimentally derived law for the force between two electric charges. Coulomb's law states that

$$F = k \frac{Q_1 Q_2}{r^2} \tag{2-1}$$

where
F = force between the charges, expressed in CGSe units of force ($\text{g cm/s}^2 = \text{dyne}$)

k = proportionality constant

$Q_{1,2}$ = electric charges, expressed in (derived) CGSe units of electric charge (statcoulomb)

r = separation between the charges, expressed in the fundamental CGSe unit of length (centimeter)

Coulomb also found that the proportionality factor k depended on the medium, varying inversely as its permittivity ε. (Faraday called permittivity the *dielectric constant*.) Coulomb's law then takes the form

$$F = \frac{Q_1 Q_2}{\varepsilon r^2} \tag{2-2}$$

Since ε is a numerical value depending only on the medium, a value of 1 was assigned to the permittivity of free space, ε_0, thereby defining ε_0 as the *fourth fundamental unit* of the CGSe system. Coulomb's law then allowed the unit of electric charge Q to be determined in terms of these four fundamental units by the relation

$$\text{dyne} = \frac{\text{g cm}}{\text{s}^2} = \frac{Q^2}{(\varepsilon_0 = 1)\ \text{cm}^2}$$

and therefore, dimensionally,

$$Q = \text{cm}^{3/2}\text{g}^{1/2}\text{s}^{-1} \tag{2-3}$$

The CGSe unit of electric charge was given the name *statcoulomb*.

The derived unit of electric charge in the CGSe system of units allowed other electrical units to be determined by their defining equations. For example, *electric current* (symbol I) is defined as the rate of flow of electric charge and is expressed as

$$I = \frac{Q}{t} \qquad \text{(statcoulomb/sec)} \tag{2-4}$$

The unit for electric current in the CGSe system was given the name *statampere*. Electric *fieldstrength, E, potential difference, V*, and *capacitance, C*, can similarly be derived from their defining equations.

The basis of the *CGS electromagnetic system* of units (CGSm) is Coulomb's experimentally determined law for the force between two magnetic poles, which states that

$$F = k\,\frac{m_1 m_2}{r^2} \tag{2-5}$$

The proportionality factor, k, was found to depend on the medium in which the poles were placed, varying inversely with the magnetic *permeability* μ of the medium. The factor k was assigned the value 1 for the permeability of free space, μ_0, so that $k = 1/\mu_0 = 1$. This established the permeability of free space, μ_0, as the *fourth fundamental unit* of the CGSm system. The derived electromagnetic unit of polestrength was then defined in terms of these four fundamental units by the relation:

$$\text{dyne} = \frac{\text{g cm}}{\text{s}^2} = \frac{m^2}{(\mu_0 = 1)\ \text{cm}^2}$$

and therefore, dimensionally,

$$m = \text{cm}^{3/2}\text{g}^{1/2}\text{s}^{-1} \tag{2-6}$$

The derived unit of magnetic polestrength in the CGSm system led to the determination of other magnetic units, again by their defining equations. *Magnetic flux density* (symbol B), for example, is defined as the magnetic force per unit polestrength, where both force and polestrength are derived units in the

Systems of Units of Measurement   Chap. 2

CGSm system. Dimensionally, B is found to be equal to $cm^{-1/2}g^{1/2}s^{-1}$ (dyne-second/abcoulomb-centimeter) and is given the name *gauss*. Similarly, other magnetic units can be derived from defining equations and we find that the unit for *magnetic flux* (symbol Φ) is given the name *maxwell*; the unit for *magnetic field-strength* (symbol H), the name *oersted*; and the unit for *magnetic potential difference* or *magnetomotive force* (symbol U), the name *gilbert*.

The two CGS systems were linked together by Faraday's discovery that a moving magnet could induce an electric current in a conductor, and conversely, that electricity in motion could produce magnetic effects. Ampere's law of the magnetic field relates electric current (I) to magnetic fieldstrength (H),* quantitatively connecting the magnetic units in the CGSm system to the electric units in the CGSe system. The dimensions of the two systems did not agree exactly, and numerical conversion factors were introduced. The two systems finally formed one *practical system of electrical units* which was officially adopted by the International Electrical Congress.

These practical electrical units, derived from the CGSm system, were later defined in terms of so-called international units. It was thought at the time (1908) that the establishment of the practical units from the definitions of the CGS system would be too difficult for most laboratories and it was therefore decided (unfortunately) to define the practical units in a way which would make it fairly simple to establish them. The *ampere,* therefore, was defined in terms of the rate of deposition of silver from a silver nitrate solution by passing a current through that solution and the *ohm* as the resistance of a specified column of mercury. These units and those derived from them were called *international units*. As measurement techniques improved, it was found that small differences existed between CGSm derived practical units and the international units, which were then specified as follows:

$$1 \text{ int. ohm} = 1.00049 \ \Omega \text{ (practical CGSm unit)}$$

$$1 \text{ int. ampere} = 0.99985 \text{ A}$$

$$1 \text{ int. volt} = 1.00034 \text{ V}$$

$$1 \text{ int. coulomb} = 0.99985 \text{ C}$$

$$1 \text{ int. farad} = 0.99951 \text{ F}$$

$$1 \text{ int. henry} = 1.00049 \text{ H}$$

$$1 \text{ int. watt} = 1.00019 \text{ W}$$

$$1 \text{ int. joule} = 1.00019 \text{ J}$$

Particulars of the electric and magnetic units, and their defining relationships, are given in Table 2-3. Multiplication factors for conversion into SI units are given in the columns headed CGSm and CGSe.

---

* See a textbook on electromagnetic theory.

TABLE 2-3 Electric and Magnetic Units

Quantity and symbol	SI unit Name and symbol		Defining equation[a]	Conversion factors CGSm	Conversion factors CGSe[b]
Electric current, I	ampere	A	$F_z = 10^{-7} I^2 \dfrac{dN}{dz}$	10	$10/c$
Electromotive force, E	volt	V	$p = IE$	10^{-8}	$10^{-8}c$
Potential, V	volt	V	$p = IV$	10^{-8}	$10^{-8}c$
Resistance, R	ohm	Ω	$R = V/I$	10^{-9}	$10^{-9}c$
Electric charge, Q	coulomb	C	$Q = It$	10	$10/c$
Capacitance, C	farad	F	$C = Q/V$	10^9	$10^9/c^2$
Electric fieldstrength, E	—	V/m	$E = V/l$	10^{-6}	$10^{-6}c$
Electric flux density, D	—	C/m²	$D = Q/l^2$	10^5	$10^5/c$
Permittivity, ε	—	F/m	$\varepsilon = D/E$	—	$10^{11}/4\pi c^2$
Magnetic fieldstrength, H	—	A/m	$\oint H\,dl = nI$	$10^{3/4}$	—
Magnetic flux, Φ	weber	Wb	$E = d\Phi/dt$	10^{-8}	—
Magnetic flux density, B	tesla	T	$B = \Phi/l^2$	10^{-4}	—
Inductance, L, M	henry	H	$M = \Phi/I$	10^{-9}	—
Permeability, μ	—	H/m	$\mu = B/H$	$4\pi \times 10^{-7}$	—

[a] N denotes Neumann's integral for two linear circuits each carrying the current I; F_z is the force between the two circuits in the direction defined by coordinate z, the circuits being in a vacuum; p denotes power; l^2 denotes area.

[b] c = velocity of light in free space in cm/s = 2.997925×10^{10}.

2-4 INTERNATIONAL SYSTEM OF UNITS

The international MKSA system of units was adopted in 1960 by the Eleventh General Conference of Weights and Measures under the name *Système International d'Unités* (SI). The SI system is replacing all other systems in the metric countries and its widespread acceptance dooms other systems to eventual obsolescence.

The six fundamental SI quantities are listed in Table 2-2. The derived units are expressed in terms of these six basic units by defining equations. Some examples of defining equations are given in Table 2-3 for the electric and magnetic quantities. Table 2-4 lists, together with the fundamental quantities which are repeated in this table, the supplementary and derived units in the SI which are recommended for use by the General Conference.

The first column in Table 2-4 shows the *quantities* (fundamental, supplementary, and derived). The second column gives the *equation symbol* for each quantity. The third column lists the *dimension* of each derived unit in terms of the six fundamental dimensions. The fourth column gives the name of each *unit*; the fifth, the unit *symbol*. The *unit* symbol should not be confused with the *equation* symbol; i.e., the equation symbol for resistance is R, but the unit abbreviation (symbol) for ohm is Ω.

TABLE 2-4 Fundamental, Supplementary, and Derived Units

Quantity	Equation symbol	Dimension	Unit	Unit symbol
Fundamental				
Length	l	L	meter	m
Mass	m	M	kilogram	kg
Time	t	T	second	s
Electric current	I	I	ampere	A
Thermodynamic temperature	T	Θ	kelvin	K
Luminous intensity			candela	cd
Supplementary[a]				
Plane angle	α, β, γ	$[L]^\circ$	radian	rad
Solid angle	Ω	$[L^2]^\circ$	steradian	sr
Derived				
Area	A	L^2	square meter	m^2
Volume	V	L^3	cubic meter	m^3
Frequency	f	T^{-1}	hertz	Hz (1/s)
Density	ρ	$L^{-3}M$	kilogram per cubic meter	kg/m^3
Velocity	v	LT^{-1}	meter per second	m/s
Angular velocity	ω	$[L]^\circ T$	radian per second	rad/s
Acceleration	a	LT^{-2}	meter per second squared	m/s^2
Angular acceleration	α	$[L]^\circ T^{-2}$	radian per second squared	rad/s^2
Force	F	LMT^{-2}	newton	N (kg m/s^2)
Pressure, stress	p	$L^{-1}MT^{-2}$	newton per square meter	N/m^2
Work, energy	W	L^2MT^{-2}	joule	J (N m)
Power	P	L^2MT^{-3}	watt	W (J/s)
Quantity of electricity	Q	TI	coulomb	C (A s)
Potential difference, electromotive force	V	$L^2MT^{-3}I^{-1}$	volt	V (W/A)
Electric fieldstrength	E, ε	$LMT^{-3}I^{-1}$	volt per meter	V/m
Electric resistance	R	$L^2MT^{-3}I^2$	ohm	Ω (V/A)
Electric capacitance	C	$L^{-2}M^{-1}T^4I^2$	farad	F (A s/V)
Magnetic flux	Φ	$L^2MT^{-2}I^{-1}$	weber	Wb (v s)
Magnetic fieldstrength	H	$L^{-1}I$	ampere per meter	A/m
Magnetic flux density	B	$MT^{-2}I^{-1}$	tesla	T (Wb/m^2)
Inductance	L	$L^2MT^{-2}I^2$	henry	H (V s/A)
Magnetomotive force	U	I	ampere	A
Luminous flux			lumen	lm (cd sr)
Luminance			candela per square meter	cd/m^2
Illumination			lux	lx (lm/m^2)

[a] The Eleventh General Conference designated these units as *supplementary*, although it could be argued that they are derived units.

2-5 OTHER SYSTEMS OF UNITS

The English system of units uses the *foot* (ft), the *pound-mass* (lb), and the *second* (s) as the three fundamental units of length, mass, and time, respectively. Although the measures of length and weight are legacies of the Roman occupation of Britian and therefore rather poorly defined, the *inch* (defined as one-twelfth of the foot) has since been fixed at *exactly* 25.4 mm. Similarly, the measure for the pound (lb) has been determined as *exactly* 0.45359237 kg. These two figures allow all units in the English system to be converted into SI units.

Starting with the fundamental units, foot, pound, and second, the mechanical units may be derived simply by substitution into the dimensional equations of Table 2-4. For example, the unit of density will be expressed in lb/ft^3 and the unit of acceleration in ft/s^2. The derived unit of force in the ft-lb-s system is called the *poundal* and is the force required to accelerate 1 pound-mass at the rate of 1 ft/s^2. As a result, the unit for work or energy becomes the foot-poundal (ft pdl).

Various other systems have been devised and were used in various parts of the world. The *MTS* (meter-tonne-second) system was especially designed for engineering purposes in France and provided a replica of the CGS system except that the length and mass units (meter and tonne, respectively) were more suitable in practical engineering applications. *Gravitational* systems define the second fundamental unit as the *weight* of a mass measure; i.e., as the force by which that mass is attracted to the earth by gravity. In contrast to the gravitational systems, the so-called absolute systems, as the CGS and SI, use the mass measure as the second fundamental unit, but its value is independent of gravitational attraction.

Since English measures are still extensively used, both in Britain and on the North American continent, conversion into the SI becomes necessary if we wish to work in that system. Table 2-5 lists some of the common conversion factors for English into SI units.

TABLE 2-5 English into SI Conversion

Quantity	English unit	Symbol	Metric equivalent	Reciprocal
Length	1 foot	ft	30.48 cm	0.0328084
	1 inch	in.	25.4 mm	0.0393701
Area	1 square foot	ft^2	9.29030×10^2 cm^2	0.0107639×10^{-2}
	1 square inch	in.^{2}	6.4516×10^2 mm^2	0.155000×10^{-2}
Volume	1 cubic foot	ft^3	0.0283168 m^3	35.3147
Mass	1 pound (avdp)	lb	0.45359237 kg	2.20462
Density	1 pound per cubic foot	lb/ft^3	16.0185 kg/m^3	0.062428
Velocity	1 foot per second	ft/s	0.3048 m/s	3.28084
Force	1 poundal	pdl	0.138255 N	7.23301
Work, energy	1 foot-poundal	ft pdl	0.0421401 J	23.7304
Power	1 horsepower	hp	745.7 W	0.00134102
Temperature	degree F	°F	$5(t - 32)/9$°C	—

2-6 CONVERSION OF UNITS

It is often necessary to convert physical quantities from one system of units into another. Section 2-1 stated that a physical quantity is expressed in both unit and number of measure: it is the unit that must be converted, not the number of measure. Dimensional equations are very convenient for converting the numerical value of a dimensional quantity, when the units are transformed from one system to the other. The technique requires a knowledge of the numerical relation between the fundamental units and some dexterity in the manipulation of multiples and submultiples of the units.

The method used in converting from one system into the other is illustrated by a number of examples of progressively increasing difficulty.

EXAMPLE 2-1

The floor area of an office building is 5,000 m². Calculate the floor area in ft².

SOLUTION To convert the unit m² into the new unit ft², we must know the relation between them. In Table 2-5 the metric equivalent of 1 ft is 30.48 cm, or 1 ft = 0.3048 m. Therefore

$$A = 5{,}000 \text{ m}^2 \times \left(\frac{1 \text{ ft}}{0.3048 \text{ m}}\right)^2 = 53{,}820 \text{ ft}^2$$

EXAMPLE 2-2

A flux density in the CGS system is expressed as 20 maxwells/cm². Calculate the flux density in lines/in². (NOTE: 1 maxwell = 1 line.)

SOLUTION

$$B = \frac{20 \text{ maxwells}}{\text{cm}^2} \times \left(\frac{2.54 \text{ cm}}{\text{in.}}\right)^2 \times \frac{1 \text{ line}}{1 \text{ maxwell}} = 129 \text{ lines/in.}^2$$

EXAMPLE 2-3

The velocity of light in free space is given as 2.997925×10^8 m/s. Express the velocity of light in km/hr.

SOLUTION

$$c = 2.997925 \times 10^8 \frac{\text{m}}{\text{s}} \times \frac{1 \text{ km}}{10^3 \text{ m}} \times \frac{3.6 \times 10^3 \text{ s}}{1 \text{ hr}} = 10.79 \times 10^8 \text{ km/hr}$$

EXAMPLE 2-4

Express the density of water, 62.5 lb/ft^3, in (a) lb/in.^{3}; (b) g/cm^3.

SOLUTION

(a) Density $= \dfrac{62.5 \text{ lb}}{\text{ft}^3} \times \left(\dfrac{1 \text{ ft}}{12 \text{ in.}}\right)^3 = 3.62 \times 10^{-2} \text{ lb/in.}^3$

(b) Density $= 3.62 \times 10^{-2} \dfrac{\text{lb}}{\text{in.}^3} \times \dfrac{453.6 \text{ g}}{1 \text{ lb}} \times \left(\dfrac{1 \text{ in.}}{2.54 \text{ cm}}\right)^3 = 1 \text{ g/cm}^3$

EXAMPLE 2-5

The speed limit on a highway is 60 km/hr. Calculate the limit in (a) mi/hr; (b) ft/s.

SOLUTION

(a) Speed limit $= \dfrac{60 \text{ km}}{\text{hr}} \times \dfrac{10^3 \text{ m}}{1 \text{ km}} \times \dfrac{10^2 \text{ cm}}{1 \text{ m}} \times \dfrac{1 \text{ in.}}{2.54 \text{ cm}} \times \dfrac{1 \text{ ft}}{12 \text{ in.}}$

$\times \dfrac{1 \text{ mi}}{5,280 \text{ ft}} = 37.3 \text{ mi/hr}$

(b) Speed limit $= \dfrac{37.3 \text{ mi}}{\text{hr}} \times \dfrac{5,280 \text{ ft}}{1 \text{ mi}} \times \dfrac{1 \text{ hr}}{3.6 \times 10^3 \text{ s}} = 54.9 \text{ ft/s}$

REFERENCES

2-1. Geczy, Steven, *Basic Electrical Measurements,* chap. 1 and Appendix. Englewood Cliffs, N.J.: Prentice-Hall, Inc., 1984.

2-2. ITT Staff, *Reference Data for Radio Engineers,* 7th ed., chap. 3. Indianapolis, Ind.: Howard W. Sams & Company, Inc., 1985.

PROBLEMS

2-1. Complete the following conversions:

$$1,500 \text{ MHz} = \text{GHz}$$

$$12.5 \text{ kHz} = \text{Hz}$$

$$125 \text{ nH} = \mu\text{H}$$

$$346.4 \text{ kV} = \text{V}$$

$$5.3 \text{ mA} = \text{A}$$

$$5\ \text{H} = \text{mH}$$

$$4.6\ \text{pJ} = \text{J}$$

$$1.4\ \mu\text{s} = \text{ms}$$

$$3.2\ \text{ns} = \text{hr}$$

$$14\ \text{fs} = \mu\text{s}$$

2-2. What is the velocity of light in free space in feet per second?

2-3. The charge of an electron is 1.6×10^{-19} C. How many electrons pass by a point each microsecond if the current at that point is 4.56 A?

2-4. Typical "room" temperature is 25°C. What is this temperature in degrees Fahrenheit and kelvin?

2-5. Calculate the height in cm of a man 5 ft 11 in. tall.

2-6. Calculate the mass in kg of 1 yd^3 of iron when the density of iron is 7.86 g/cm^3.

2-7. Calculate the conversion factor to change mi/hr to ft/s.

2-8. An electrically charged body has an excess of 10^{15} electrons. Calculate its charge in C.

2-9. A train covers a distance of 220 mi in 2 hr and 45 min. Calculate the average speed of the train in m/s.

2-10. Two electric charges are separated by a distance of 1 m. If one charge is $+10$ C and the other charge -6 C, calculate the force of attraction between the charges in N and in lb. Assume that the charges are placed in a vacuum.

2-11. The practical unit of electrical energy is the kWh. The unit of energy in the SI is the joule (J). Calculate the number of joules in 1 kWh.

2-12. A crane lifts a 100-kg mass a height of 20 m in 5 s. Calculate **(a)** the work done by the crane, in SI units; **(b)** the increase of potential energy of the mass, in SI units; **(c)** the power, or rate of doing the work, in SI units.

2-13. Calculate the voltage of a battery if a charge of 3×10^{-4} C residing on the positive battery terminal possesses 6×10^{-2} J of energy.

2-14. An electric charge of 0.035 C flows through a copper conductor in 5 min. Calculate the average current in mA.

2-15. An average current of 25 μA is passed through a wire for 30 s. Calculate the number of electrons transferred through the conductor.

2-16. The speed limit on a four-lane highway is 70 mi/hr. Calculate the speed limit in **(a)** km/hr; **(b)** ft/s.

2-17. The density of copper is 8.93 g/cm^3. Express the density in **(a)** kg/m^3; **(b)** lb/ft^3.

3

Standards of Measurement

3-1 CLASSIFICATION OF STANDARDS

A standard of measurement is a physical representation of a unit of measurement. A unit is realized by reference to an arbitrary material standard or to natural phenomena including physical and atomic constants. For example, the fundamental unit of mass in the international system (SI) is the *kilogram,* defined as the mass of a cubic decimeter of water as its temperature of maximum density of 4°C (see Sec. 2-2). This unit of mass is represented by a material standard: the mass of the International Prototype Kilogram, consisting of a platinum-iridium alloy cylinder. This cylinder is preserved at the International Bureau of Weights and Measures at Sèvres, near Paris, and is the *material representation* of the kilogram. Similar standards have been developed for other units of measurement, including standards for the fundamental units as well as for some of the derived mechanical and electrical units.

Just as there are fundamental and derived units of measurement, we find different types of *standards of measurement,* classified by their function and application in the following categories:

(a) International standards

(b) Primary standards

(c) Secondary standards

(d) Working standards

The *international standards* are defined by international agreement. They represent certain units of measurement to the closest possibly accuracy that production and measurement technology allow. International standards are periodically evaluated and checked by *absolute measurements* in terms of the fundamental units (see Table 2-2). These standards are maintained at the International Bureau of Weights and Measures and are not available to the ordinary user of measuring instruments for purposes of comparison or calibration.

The *primary* (basic) *standards* are maintained by national standards laboratories in different parts of the world. The National Bureau of Standards (NBS) in Washington is responsible for maintenance of the primary standards in North America. Other national laboratories include the National Physical Laboratory (NPL) in Great Britain and, the oldest in the world, the Physikalisch-Technische Reichsanstalt in Germany. The primary standards, again *representing* the fundamental units and some of the derived mechanical and electrical units, are independently calibrated by absolute measurements at each of the national laboratories. The results of these measurements are compared against each other, leading to a world average figure for the primary standard. Primary standards are not available for use outside the national laboratories. One of the main functions of primary standards is the verification and calibration of secondary standards.

Secondary standards are the basic *reference* standards used in industrial measurement laboratories. These standards are maintained by the particular involved industry and are checked locally against other reference standards in the area. The responsibility for maintenance and calibration of secondary standards rests entirely with the industrial laboratory itself. Secondary standards are generally sent to the national standards laboratories on a periodic basis for calibration and comparison against the primary standards. They are then returned to the industrial user with a *certification* of their measured value in terms of the primary standard.

Working standards are the principal tools of a measurement laboratory. They are used to check and calibrate general laboratory instruments for accuracy and performance or to perform comparison measurements in industrial applications. A manufacturer of precision resistances, for example, may use a *standard resistor* (a *working* standard) in the quality control department of his plant to check his testing equipment. In this case, he *verifies* that his measurement setup performs within the required limits of accuracy.

In electrical and electronic measurement we are concerned with the electrical and magnetic standards of measurement. These are discussed in the following sections. We have seen, however, that electrical units can be traced back to the basic units of length, mass, and time (in fact, the national laboratories perform *measurements* to relate derived electrical units to fundamental units) and they deserve some investigation here.

3-2 STANDARDS FOR MASS, LENGTH, AND VOLUME

The metric *unit of mass* was originally defined as the mass of a cubic decimeter of water at its temperature of maximum density. The *material representation* of this unit is the International Prototype Kilogram, preserved at the International Bureau of Weights and Measures near Paris. The *primary standard* of mass in North America is the United States Prototype Kilogram, preserved by the NBS to an accuracy of 1 part in 10^8 and occasionally verified against the standard at the International Bureau. *Secondary standards* of mass, kept by the industrial laboratories, generally have an accuracy of 1 ppm (part per million) and may be verified against the NBS primary standard. Commercial *working standards* are available in a wide range of values to suit almost any application. Their accuracy is in the order of 5 ppm. The working standards, in turn, are checked against the secondary laboratory standards.

The *pound* (lb), established by the Weights and Measures Act of 1963 (which actually came into effect on January 31, 1964), is defined as equal to 0.45359237 kg *exactly*. All countries which retain the pound as the basic unit of measurement have now adopted the new definition, which supersedes the former imperial standard pound made of platinum.

The metric unit of length, the meter, was initially defined as $1/10^4$ part of the meridional quadrant through Paris (Sec. 2-2). This was an outgrowth of a suggestion in 1790 by the well-known French astronomer Pierre-Simon Laplace that the right angle be divided into 100 degrees, rather than 90, and each degree into 100 minutes, rather than 60. The measure of one meter would be the distance on the surface of the earth covered by one second of arc, which would be one ten-thousandth of the meridional quadrant, or the line from the equator to the north geographical pole. This was materially represented by the distance between two lines engraved on a platinum-iridium bar preserved at the International Bureau of Weights and Measures near Paris. In 1960 the meter was redefined more accurately in terms of a number of wavelengths of light emitted from the krypton-86 atom. For over 20 years the international standard meter was 1,650,763.73 wavelengths of the orange-red radiation from a carefully specified and observed krypton discharge lamp. Because this standard did not prove as precise as originally thought, in 1983 a new standard for the meter was adopted. This standard is simply that one meter is the distance light that propagates in a vacuum in 1/299,792,458 seconds.

The yard is defined as 0.9144 meter exactly, or 1 inch is 25.4 mm exactly. This is because the standards for the English units of measurement are based on the metric standards. This definition of a yard and inch superseded the former definition in terms of a standard imperial yard. The few countries that have retained the yard and other English units of measurement have adopted this metric-based definition.

The most widely used industrial *working standards* of length are precision *gage blocks,* made of steel. These steel blocks have two plane parallel surfaces, a specified distance apart, with accuracy tolerances in the 0.5–0.25-micron range (1 micron = one millionth of 1 m). The development and use of precision gage blocks, low in cost and of high accuracy, have made it possible to manufacture interchangeable industrial components in a very economical application of precision measurement.

The unit of *volume* is a derived quantity and is not represented by an international standard. The NBS, however, has constructed a number of primary standards of volume, calibrated in terms of the absolute dimensions of length and mass. Secondary derived standards of volume are available and may be calibrated in terms of the NBS primary standards.

As the need for more accurate standards arises and the technology is developed to create and preserve these standards, the basis for international weights and measures will change to fill the needs of the scientific and commerce community. Additions and improvements will be added to the international standards to keep in pace with the needs of the world.

3-3 TIME AND FREQUENCY STANDARDS*

Since early times men have sought a reference standard for a uniform time scale together with means to interpolate from it a small time interval. For many centuries the time reference used was the rotation of the earth about its axis with respect to the sun. Precise astronomical observations have shown that the rotation of the earth about the sun is very irregular, owing to secular and irregular variations in the rotational speed of the earth. Since the time scale based on this apparent *solar time* does not represent a uniform time scale, other avenues were explored. *Mean solar time* was thought to give a more accurate time scale. A mean solar day is the average of all the apparent days in the year. A *mean solar second* is then equal to 1/86,400 of the mean solar day. The mean solar second, thus defined, is still inadequate as the fundamental unit of time, since it is tied to the rotation of the earth, which is now known to be nonuniform.

The system of *universal time* (UT), or mean solar time, is also based on the rotation of the earth about its axis. This system is known as UT_0 and is subject to periodic, long-term, and irregular variations. Correction of UT_0 has led to two subsequent universal time scales: UT_1 and UT_2. UT_1 recognizes the fact that the earth is subject to polar motion, and the UT_1 time scale is based on the true angular rotation of the earth, corrected for polar motion. The UT_2 time scale is UT_1 with an additional correction for *seasonal* variations in the rotation of the earth. These variations are apparently caused by seasonal displacement of matter over the earth's surface, such as changes in the amount of ice in the polar regions as the sun moves from the southern hemisphere to the northern and back again through the year. This *cyclic* redistribution of mass acts on the earth's rotation since it produces changes in its moment of inertia. The *epoch,* or *instant of time,* of UT_2 can be established to an accuracy of a few milliseconds, but it is not usually

---

* *Frequency and Time Standards,* Application Note AN 52, published by Hewlett-Packard, Palo Alto, Calif., describes methods of frequency comparisons, time scales, and worldwide time standards broadcasts.

distributed to this accuracy. The epoch indicated by the standard radio time signals may differ from the epoch of UT_2 by as much as 100 ms. The actual values of the differences are given in bulletins published by the national time services (NBS) and by the Bureau International de l'Heure (Paris Observatory).

The search for a truly universal time unit has led astronomers to define a time unit called *ephemeris time* (ET). ET is based on astronomical observations of the motion of the moon about the earth. Since 1956 the *ephemeris second* has been defined by the International Bureau of Weights and Measures as the fraction 1/31,556,925.9747 of the tropical year for 1900 January 0 at 12 h ET, and adopted as the *fundamental invariable unit of time*. A disadvantage of the use of the ephemeris second is that it can be determined only several years in arrears and then only indirectly, by observations of the positions of the sun and the moon. For *physical measurements*, the unit of time interval has now been defined in terms of an *atomic standard*. The universal second and the ephemeris second, however, will continue to be used for navigation, geodetic surveys, and celestial mechanics.

Development and refinement of *atomic resonators* have made possible control of the frequency of an oscillator and, hence, by frequency conversion, *atomic clocks*. The transition between two energy levels, E_1 and E_2, of an atom is accompanied by the emission (or absorption) of radiation having a frequency given by $h\eta = E_2 - E_1$, where h is Planck's constant. Provided that the energy states are not affected by external conditions, such as magnetic fields, the frequency η is a *physical constant*, depending *only* on the internal structure of the atom. Since frequency is the inverse of time interval, such an atom provides a *constant time interval*. Atomic transitions of various metals were investigated, and the first atomic clock, based on the cesium atom, was put into operation in 1955. The time interval, provided by the cesium clock, is more accurate than that provided by a clock calibrated by astronomical measurements. The *atomic unit of time* was first related to UT but was later expressed in terms of ET. The International Committee of Weights and Measures has now defined the second in terms of the frequency of the cesium transition, assigning a value of 9,192,631,770 Hz to the hyperfine transition of the cesium atom unperturbed by external fields.

The *atomic definition* of the second realizes an accuracy much greater than that achieved by astronomical observations, resulting in a more uniform and much more convenient time base. Determinations of time intervals can now be made in a few minutes to greater accuracy than was possible before in astronomical measurements that took many years to complete. An atomic clock with a precision exceeding 1 μs per day is in operation as a primary frequency standard at the NBS. An atomic time scale, designated NBS-A, is maintained with this clock.

Time and frequency standards are unique in that they may be transmitted from the primary standard at NBS to other locations via radio or television transmissions. Early standard time and frequency transmissions were in the high-frequency (HF) portion of the radio spectrum, but these transmissions suffered from Doppler shifts due to the fact that radio propagation was primarily ionospheric. Transmission of time and frequency standards via low-frequency and very low frequency radio reduces this Doppler shift because the propagation is strictly ground wave. Two NBS-operated stations, WWVL and WWVB, operate at 20 and 60 kHz, respectively, providing precision time and frequency transmissions.

Another source of precision time and frequency information is the low-frequency navigation system called LORAN-C. This navigation system transmits shaped pulses at a carrier frequency of 100 kHz with a bandwidth of 20 kHz. The LORAN-C transmitters are controlled by cesium beam clocks and provide strong signals within most of the United States and in other parts of the world. Because LORAN-C is primarily a marine navigation system, coverage is not provided away from significant bodies of water.

Another source of accurate time and frequency dissemination is via television transmissions. The color burst frequency, which is nominally 3.579545 MHz, is phase locked to a cesium clock and is distributed over the television networks. Because television programming is distributed via terrestrial and satellite microwave links, there is no significant Doppler shift, and the color burst frequency can be transmitted accurately and is readily available for use as a precision standard.

3-4 ELECTRICAL STANDARDS

3-4.1 The Absolute Ampere

The international system of units (SI) defines the *ampere* (the fundamental unit of electric current) as the constant current which, if maintained in two straight parallel conductors of infinite length and negligible circular cross section placed 1 m apart in a vacuum, will produce between these conductors a force equal to 2×10^{-7} newton per meter length. Early measurements of the absolute value of the ampere were made with a *current balance* which measured the force between two parallel conductors. These measurements were rather crude and the need was felt to produce a more practical and reproducible standard for the national laboratories. By international agreement, the value of the *International Ampere* was based on the electrolytic deposition of silver from a silver nitrate solution. The International Ampere was then defined as that current which deposits silver at the rate of 1.118 mg/s from a standard silver nitrate solution. Difficulties were encountered in the exact measurement of the deposited silver and slight discrepancies existed between measurements made independently by the various national standards laboratories.

In 1948 the International Ampere was superseded by the *Absolute Ampere*. The determination of the Absolute Ampere is again made by means of a current balance, which *weighs* the *force* exerted between two current-carrying coils. Improvement in the techniques of force measurement yields a value for the ampere far superior to the early measurements. The relationship between the force and the current which produces the force can be calculated from fundamental electromagnetic theory concepts and reduces to a simple computation involving the geometric dimensions of the coils. The Absolute Ampere is now the *fundamental unit of electric current* in the SI and is universally accepted by international agreement.

Instruments manufactured before 1948 are calibrated in terms of the International Ampere but newer instruments are using the Absolute Ampere as the basis for calibration. Since both types of instruments may be found side by side in one laboratory, the NBS has established conversion factors to relate both units. These factors are given in Sec. 2-3.

Voltage, current, and resistance are related by Ohm's law of constant proportionality ($E = IR$). The specification of any two quantities automatically sets the third. Two types of material standards form a combination which conveniently serves to maintain the ampere with high precision over long periods of time: the *standard resistor* and the *standard cell* (for voltage). Each of these is described below.

3-4.2 Resistance Standards

The absolute value of the ohm in the SI system is defined in terms of the fundamental units of length, mass, and time. The *absolute measurement* of the ohm is carried out by the International Bureau of Weights and Measures in Sèvres and also by the national standards laboratories, which preserve a group of *primary* resistance standards. The NBS maintains a group of those primary standards (1-Ω standard resistors) which are periodically checked against each other and are occasionally verified by absolute measurements. The standard resistor is a coil of wire of some alloy like *manganin* which has a high electrical resistivity and a low temperature coefficient of resistance (almost constant temperature-resistance relationship). The resistance coil is mounted in a double-walled sealed container (Fig. 3-1) to prevent changes in resistance due to moisture conditions in the atmosphere. With a set of four or five 1-Ω resistors of this type, the unit of

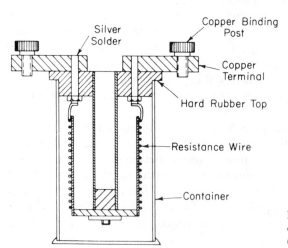

Figure 3-1 Cross-sectional view of a double-walled resistance standard. (Courtesy of Hewlett-Packard Co.)

Standards of Measurement   Chap. 3

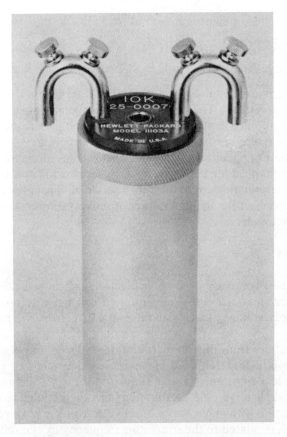

Figure 3-2 Ten-kilohm resistance standard. (Courtesy of Hewlett-Packard Company.)

resistance can be represented with a precision of a few parts in 10^7 over several years.

Secondary standards and *working* standards are available from some instrument manufacturers in a wide range of values, usually in multiples of 10 Ω. These standard resistors are made of alloy resistance wire, such as manganin or Evanohm. Figure 3-2 is a photograph of a laboratory secondary standard, sometimes referred to as a *transfer resistor*. The resistance coil of the transfer resistor is supported between polyester film to reduce stresses on the wire and to improve the stability of the resistor. The coil is immersed in moisture-free oil and placed in a sealed can. The connections to the coil are silver soldered, and the terminal hooks are made of nickel-plated oxygen-free copper. The transfer resistor is checked for stability and temperature characteristics at its rated power and a specified operating temperature (usually 25°C). A *calibration report* accompanying the resistor specifies its traceability to NBS standards and includes the α and β temperature coefficients. Although the selected resistance wire provides almost constant resistance over a fairly wide temperature range, the exact value of the resistance at any temperature can be calculated from the formula

$$R_t = R_{25°C} + \alpha(t - 25) + \beta(t - 25)^2 \qquad (3\text{-}1)$$

where R_t = resistance at the ambient temperature t

     $R_{25°C}$ = resistance at 25°C

     α, β = temperature coefficients

Temperature coefficient α is usually less than 10×10^{-6}, and coefficient β lies between -3×10^{-7} and -6×10^{-7}. This means that a change in temperature of 10°C from the specified reference temperature of 25°C may cause a change in resistance of 30 to 60 ppm (parts per million) from the nominal value.

    Transfer resistors find application in industrial, research, standards, and calibration laboratories. In typical applications, the transfer resistor may be used for resistance and ratio determinations or in the construction of ultralinear decade dividers which can then be used for the calibration of universal ratio sets, voltboxes, and Kelvin-Varley dividers.

3-4.3 Voltage Standards

For many years the standard volt was based on an electrochemical cell called the *saturated standard cell* or *standard cell*. The saturated cell has a temperature dependence, and the output voltage changes about -40 μV/°C from the nominal of 1.01858 V.

    The standard cell suffers from this temperature dependence and also from the fact that the voltage is a function of a chemical reaction and not related directly to any other physical constants. A new standard for the volt came about from the work of Brian Josephson in 1962. A thin-film junction is cooled to nearly absolute zero and irradiated with microwave energy. A voltage is developed across the junction, which is related to the irradiating frequency by the following relationship:

$$v = \frac{hf}{2e} \tag{3-2}$$

where h = Planck's constant (6.63×10^{-34} J-s)

    e = charge of an electron (1.602×10^{-19} C)

    f = frequency of the microwave irradiation

Because only the irradiating frequency is a variable in the equation, the standard volt is related to the standard of time/frequency. When the microwave irradiating frequency is locked to an atomic clock or a broadcast frequency standard such as WWVB, the accuracy of the standard volt, including all of the system inaccuracies, is one part in 10^8.

    The major method of transferring the volt from the standard based on the Josephson junction to secondary standards used for calibration is the standard cell. This device is called the *normal* or *saturated Weston cell*. The Weston cell has a positive electrode of mercury and a negative electrode of cadmium amalgam (10% cadmium). The electrolyte is a solution of cadmium sulfate. These components are placed in an H-shaped glass container, as shown in Fig. 3-3.

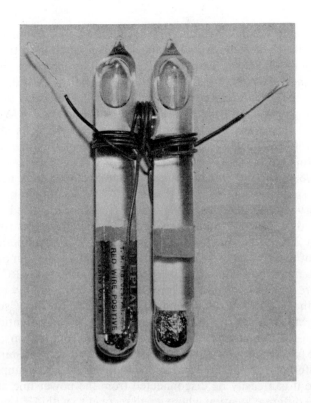

Figure 3-3 Weston cadmium cell: emf of 1.0193, accuracy of 0.1%. (Courtesy of The Eppley Laboratory, Inc.)

There are two types of Weston cell: the *saturated* cell, in which the electrolyte is saturated at all temperatures by cadmium sulfate crystals covering the electrodes, and the *unsaturated* cell, in which the concentration of cadmium sulfate is such that it produces saturation at 4°C. The unsaturated cell has a negligible temperature coefficient of voltage at normal room temperatures. The saturated cell has a voltage variation of approximately $-40\ \mu V$ per 1°C rise, but is better reproducible and more stable than the unsaturated cell.

National standards laboratories, such as the NBS, maintain a number of *saturated* cells as the *primary standard* for voltage. The cells are kept in an oil bath to control their temperature to within 0.01°C. The voltage of the Weston saturated cell at 20°C is 1.01858 V (absolute), and the emf at other temperatures is given by the formula

$$e_t = e_{20°C} - 0.000046(t - 20) - 0.00000095(t - 20)^2$$
$$+ 0.00000001(t - 20)^3 \qquad (3\text{-}3)$$

Saturated Weston cells remain satisfactory as voltage standards for periods of 10 to 20 years, provided that they are carefully treated. Their drift in voltage is on the order of 1 μV per year. Since saturated cells are temperature sensitive, they are unsuited for general laboratory use as secondary or working standards.

More rugged portable *secondary* and *working standards* are found in the *unsaturated* Weston cell. These cells are very similar in construction to the normal cell but they do not require exact temperature control. The emf of an unsaturated cell lies in the range of 1.0180 V to 1.0200 V and varies less than 0.01 per cent from 10°C to 40°C. The voltage of the cell is usually indicated on the cell housing, as shown in Fig. 3-3 (1.0193 abs. V). The internal resistance of Weston cells range from 500 Ω to 800 Ω. The current drawn from these cells should therefore not exceed 100 μA, because the nominal voltage would be affected by the internal voltage drop.

Versatile *laboratory working* standards have been developed with accuracies on the order of standard cell accuracy. Figure 3-4 is a photograph of a multipurpose laboratory voltage standard, called a *transfer standard,* based on the operation of a Zener diode as the voltage reference element. The instrument basically consists of a Zener-controlled voltage source placed in a temperature-controlled environment to improve its long-term stability, and a precision output voltage divider. The temperature-controlled oven is held to within ±0.03°C over an ambient temperature range of 0°C to 50°C, providing an output stability on the order of 10 ppm/month. The four available outputs are (a) a 0–1,000-μV source with 1-μV resolution, called (Δ); (b) a 1.000-V reference for voltbox potentiometric measurements; (c) a 1.018 + (Δ) reference for saturated cell comparisons; (d) a 1.0190 + (Δ) reference for unsaturated cell comparisons. The dc transfer standard can be used as a transfer instrument and can be moved to the piece of equipment to be calibrated, since it can easily be disconnected from the power line at one location and set up at a different location where it will recover to within ±1 ppm in approximately 30 minutes' warm-up time.

Figure 3-4 Dc transfer standard that can be used as a 1.000-V reference source, a standard cell comparison instrument, and a 0 to 1,000-μV dc source. (Courtesy of Hewlett-Packard Company.)

3-4.4 Capacitance Standards

Since the unit of resistance is represented by the standard resistor, and the unit of voltage by the standard Weston cell, many electrical and magnetic units may be expressed in terms of these standards. The unit of *capacitance* (the farad) can be measured with a Maxwell dc commutated bridge, where the capacitance is computed from the resistive bridge arms and the frequency of the dc commutation. This bridge is shown in Fig. 3-5. Although the exact derivation of the expression for capacitance in terms of the resistances and the frequency is rather involved, it may be seen that the capacitor could be measured by this method. Since both resistance and frequency can be determined very accurately, the value of the capacitance can be measured with great accuracy. *Standard capacitors* are usually constructed from interleaved metal plates with air as the dielectric material. The area of the plates and the distance between them must be known very accurately, and the capacitance of the air capacitor can be determined from these basic dimensions. The NBS maintains a bank of air capacitors as standards and uses them to calibrate the secondary and working standards of measurement laboratories and industrial users.

Capacitance working standards can be obtained in a range of suitable values. Smaller values are usually air capacitors, whereas the larger capacitors use solid dielectric materials. The high dielectric constant and the very thin dielectric layer account for the compactness of these standards. Silver-mica capacitors make excellent working standards; they are very stable and have a very low dissipation factor (Sec. 5-8), a very small temperature coefficient, and little or no aging effect. Mica capacitors are available in decade mounting, but decade capacitors are usually not guaranteed better than 1 per cent. Fixed standards are generally used where accuracy is important.

3-4.5 Inductance Standards

The *primary inductance standard* is derived from the ohm and the farad, rather than from the large geometrically constructed inductors used in the determination of the absolute value of the ohm. The NBS selected a *Campbell standard* of

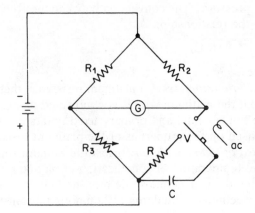

Figure 3-5 Commutated dc method for measuring capacitance. Capacitor C is alternately charged and discharged through the commutating contact and resistor R. Bridge balance is obtained by adjustment of R_3, allowing exact determination of the capacitance value in terms of bridge-arm constants and frequency commutation.

mutual inductance as the primary standard for both mutual and self-inductance. Inductance *working standards* are commercially available in a wide range of practical values, both fixed and variable. A typical set of fixed inductance standards includes values from approximately 100 μH to 10 H, with a guaranteed accuracy of 0.1 per cent at a specified operating frequency. Variable inductors are also available. Typical mutual inductance accuracy is on the order of 2.5 per cent and inductance values range from 0 to 200 mH. *Distributed capacitance* exists between the windings of these inductors, and the errors they introduce must be taken into account. These considerations are usually specified with commercial equipment.

3-5 STANDARDS OF TEMPERATURE AND LUMINOUS INTENSITY

Thermodynamic temperature is one of the basic SI quantities and its unit is the Kelvin (Sec. 2-2). The thermodynamic Kelvin scale is recognized as the *fundamental scale* to which all temperatures should be referred. The temperatures on this scale are designated as K and denoted by the symbol T. The magnitude of the Kelvin has been fixed by defining the thermodynamic temperature of the *triple point* of water at *exactly* 273.16 K. The triple point of water is the temperature of equilibrium between ice, liquid water, and its vapor.

Since temperature measurements on the thermodynamic scale are inherently difficult, the Seventh General Conference of Weights and Measures adopted in 1927 a *practical scale* which has been modified several times and is now called the *International Practical Scale of Temperature*. The temperatures on this scale are designated as °C (degree Celsius) and denoted by the symbol t. The Celsius scale has two *fundamental* fixed points: the boiling point of water as 100°C and the triple point of water as 0.01°C, both points established at atmospheric pressure. A number of *primary* fixed points have been established above and below the two fundamental points. These are the boiling point of oxygen (-182.97°C), the boiling point of sulfur (444.6°C), the freezing point of silver (960.8°C), and the freezing point of gold (1,063°C). The numerical values of all these points are reproducible quantities at atmospheric pressure. The conversion between the Kelvin scale and the Celsius scale follows the relationship:

$$t(\degree\text{C}) = T(\text{K}) - T_0 \tag{3-4}$$

where $T_0 = 273.15$ degrees.

The primary *standard thermometer* is a platinum resistance thermometer of special construction so that the platinum wire is not subjected to strain. Interpolated values between the fundamental and primary fixed points on the scale are calculated by formulas based on the properties of the platinum resistance wire.

The primary *standard of luminous intensity* is a full radiator (black body or Planckian radiator), at the temperature of solidification of platinum (2,042 K approx.). The *candela* is then defined as one-sixtieth of the luminous intensity per cm^2 of the full radiator. Secondary standards of luminous intensity are special tungsten filament lamps, operated at a temperature whereby their spectral power distribution in the visible region matches that of the basic standard. These secondary standards are recalibrated against the basic standard at periodic intervals.

3-6 IEEE STANDARDS

A slightly different type of standard is published and maintained by the Institute of Electrical and Electronics Engineers, IEEE, an engineering society headquartered in New York City. These standards are not physical items that are available for comparison and checking of secondary standards but are standard procedures, nomenclature, definitions, etc. These standards have been kept updated, and some of the early standards were in use before World War II. Many of the IEEE standards have been adopted by other agencies and societies as standards for their organization, such as the American National Standards Institute.

A large group of the IEEE standards is the standard test methods for testing and evaluating various electronics systems and components. As an example, there is a standard method for testing and evaluating attenuators. Although any test method should result in the same value for the attenuation, when certain factors are introduced, such as high frequency or high attenuation, measurement errors are possible. Specifying a methodology for the measurement decreases the chances for disparity between measurements.

Another useful standard is the specifying of test equipment. The common laboratory oscilloscope becomes difficult to use when each manufacturer adopts a different arrangement of knobs and functions and, worst of all, different names for the same function. An IEEE standard addresses the laboratory oscilloscope and specifies the controls, functions, etc., so that an oscilloscope operator does not have to reeducate himself for each oscilloscope he uses.

There are various standards concerning the safety of wiring for power plants, ships, industrial buildings, etc. Not only is safety a factor, but standard voltages, current ratings, etc., are specified so that components may be interchanged without damage or danger.

Standard schematic and logic symbols are defined so that engineering drawings can be understood by all engineers.

Perhaps one of the most important standards is the IEEE 488 digital interface for programmable instrumentation for test and other equipment. Standardizing the interface between test equipment makes it possible to interface various pieces of laboratory test equipment, regardless of manufacture, to create sophisticated automatic test equipment systems. Applications of this standard will be discussed in Chapter 13.

REFERENCES

3-1. Kaye, G. W. C., and Laby, T. H., *Tables of Physical and Chemical Constants,* 13th ed. London: Longmans, Green & Co., Ltd., 1966.

3-2. Philco Technological Center, *Electronic Precision Measurement Techniques and Experiments.* Englewood Cliffs, N.J.: Prentice-Hall, Inc., 1964.

3-3. Stout, Melville B., *Basic Electrical Measurements,* 2nd ed. Englewood Cliffs, N.J.: Prentice-Hall, Inc., 1960.

3-4. *Time and Frequency User's Manual,* NBS Publication 559, November 1979.

PROBLEMS

3-1. What is the difference between a primary and secondary standard?

3-2. How is the standard meter defined?

3-3. What is atomic time? How does this differ from ephemeris time?

3-4. How can time and frequency standards be disseminated?

3-5. How is the Absolute Ampere determined?

3-6. A precision 1-Ω resistance standard has been calibrated at 25°C, and has an alpha factor of 0.6×10^{-6} and a beta factor of -4×10^{-7}. What would be the resistance of the standard be at 30°C?

3-7. A Josephson junction is irradiated with 10.25 GHz of microwave radiation. What would be the potential across the junction?

3-8. What are the disadvantages of transmitting time and frequency standards by high-frequency, 3–30-MHz, radio? What are some of the methods used to improve the dissemination of these standards?

3-9. What are IEEE standards? How do these standards differ from those maintained by national standards laboratories?

3-10. What is the normal emf of a Weston cell at 20°C, and how does this emf change when the cell is used at 0°C?

4

Electromechanical Indicating Instruments

4-1 SUSPENSION GALVANOMETER

Early measurements of direct current required a suspension galvanometer. This instrument was the forerunner of the moving-coil instrument, basic to most dc indicating movements currently used.

A *coil* of fine wire is suspended in a magnetic field produced by a *permanent magnet*. According to the fundamental law of electromagnetic force, the coil will rotate in the magnetic field when it carries an electric current. The fine filament suspension of the coil serves to carry current to and from it, and the elasticity of the filament sets up a moderate torque in opposition to the rotation of the coil. The coil will continue to deflect until its electromagnetic torque balances the mechanical countertorque of the suspension. The coil deflection therefore is a measure of the magnitude of the current carried by the coil. A *mirror* attached to the coil deflects a beam of light, cauisng a magnified light spot to move on a *scale* at some distance from the instrument. The optical effect is that of a pointer of great length but zero mass.

4-2 TORQUE AND DEFLECTION OF THE GALVANOMETER

4-2.1 Steady-State Deflection

Although the suspension galvanometer is neither a practical nor portable instrument, the principles governing its operation apply equally to its more modern version, the *permanent-magnet moving-coil mechanism* (PMMC). Figure 4-1 shows the construction of the PMMC mechanism. The different parts of the instrument are identified alongside the figure.

Here again we have a coil, suspended in the magnetic field of a permanent magnet, this time in the shape of a horseshoe. The coil is suspended so that it can rotate freely in the magnetic field. When current flows in the coil, the developed electromagnetic (EM) torque causes the coil to rotate. The EM torque is counterbalanced by the mechanical torque of control springs attached to the movable coil. The balance of torques, and therefore the angular position of the movable coil, is indicated by a pointer against a fixed reference, called a *scale*.

The equation for the developed torque, derived from the basic law for electromagnetic torque, is

$$T = B \times A \times I \times N \qquad (4\text{-}1)$$

where T = torque [newton-meter (N-m)]

B = flux density in the air gap [webers/square meter (tesla)]

A = effective coil area [square meters (m^2)]

I = current in the movable coil [amperes (A)]

N = turns of wire on the coil

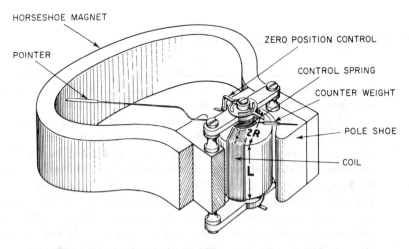

Figure 4-1 Construction details of the external magnet PMMC movement. (Courtesy of Weston Instruments, Inc.)

Equation (4-1) shows that the developed torque is directly proportional to the flux density of the field in which the coil rotates, the current in the coil, and the coil constants (area and turns). Since both flux density and coil area are *fixed* parameters for a given instrument, the developed torque is a direct indication of the current in the coil. This torque causes the pointer to deflect to a *steady-state* position where it is balanced by the opposing control-spring torque.

Equation (4-1) also shows that the designer may vary only the value of the control torque and the number of turns on the moving coil to measure a given full-scale current. The practical coil area generally ranges from approximately 0.5 to 2.5 cm². Flux densities for modern instruments usually range from 1,500 to 5,000 gauss (0.15 to 0.5 tesla). Thus a wide choice of mechanisms is available to the designer to meet many different measurement applications.

A typical panel PMMC instrument, with a $3\frac{1}{2}$-in. case, a 1-mA range, and full-scale deflection of 100 degrees of arc, would have the following characteristics:

$$A = 1.75 \text{ cm}^2$$

$$B = 2,000 \text{ G (0.2 tesla)}$$

$$N = 84 \text{ turns}$$

$$T = 2.92 \times 10^{-6} \text{ N-m}$$

$$\text{coil resistance} = 88 \ \Omega$$

$$\text{power dissipation} = 88 \ \mu\text{W}$$

4-2.2 Dynamic Behavior

In Sec. 4-2.1 we considered the galvanometer as a simple indicating instrument in which the deflection of the pointer is directly proportional to the magnitude of the current applied to the coil. This is perfectly satisfactory when we are dealing with a steady-state condition in which we are mainly interested in obtaining a reliable reading of a direct current. In some applications, however, the *dynamic* behavior of the galvanometer (such as speed of response, damping, overshoot) can be important. For example, when an alternating or varying current is applied to a recording galvanometer, the written record produced by the motion of the moving coil includes the response characteristics of the moving element itself and it is therefore important to consider its dynamic behavior.

The dynamic behavior of the galvanometer can be observed by suddenly interrupting the applied current, so that the coil swings back from its deflected position toward the zero position. It will be seen that as a result of inertia of the moving system the pointer swings past the zero mark in the opposite direction, and then oscillates back and forth around zero. These oscillations gradually die down as a result of the damping of the moving element, and the pointer will finally come to rest at zero.

The motion of a moving coil in a magnetic field is characterized by three quantities:

(a) The moment of inertia (J) of the moving coil about its axis of rotation
(b) The opposing torque (S) developed by the coil suspension
(c) The damping constant (D)

The differential equation that relates these three factors yields three possible solutions, each of which describes the dynamic behavior of the coil in terms of its deflection angle θ. The three types of behavior are shown in the curves of Fig. 4-2 and are known as *overdamped, underdamped,* and *critically damped.* Curve I of Fig. 4-2 shows the overdamped case in which the coil returns slowly to its rest position, without overshoot or oscillations. The pointer seems to approach the steady-state position in a sluggish manner. This case is of minor interest because we prefer to operate under the conditions of curve II or curve III for most applications. Curve II of Fig. 4-2 shows the underdamped case in which the motion of the coil is subject to damped sinusoidal oscillations. The rate at which these oscillations die away is determined by the damping constant (D), the moment of inertia (J), and the countertorque (S) produced by the coil suspension. Curve III of Fig. 4-2 shows the critically damped case in which the pointer returns promptly to its steady-state position, without oscillations.

Ideally, the galvanometer response should be such that the pointer travels to its final position without overshoot; hence, the movement should be critically damped. In practice, the galvanometer is usually slightly underdamped, causing the pointer to overshoot a little before coming to rest. This method is perhaps less direct than critical damping, but it assures the user that the movement has not been damaged because of rough handling, and it compensates for any additional friction that may develop in time because of dust or wear.

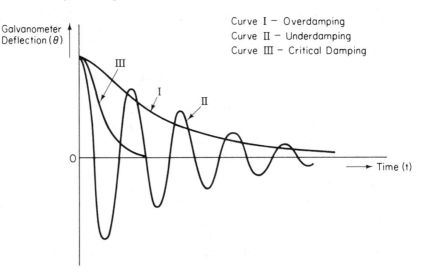

Figure 4-2 Dynamic behavior of a galvanometer.

4-2.3 Damping Mechanisms

Galvanometer damping is provided by two mechanisms: mechanical and electromagnetic. *Mechanical* damping is caused mainly by the motion of the coil through the air surrounding it; it is independent of any electrical current through the coil. Friction of the movement in its bearing and flexing of the suspension springs caused by the rotating coil also contribute to the mechanical damping effects. *Electromagnetic* damping is caused by induced effects in the moving coil as it rotates in the magnetic field, provided that the coil forms part of a closed electrical circuit.

PMMC instruments are generally constructed to produce as little viscous damping as possible and the required degree of damping is added. One of the simplest damping mechanisms is provided by an aluminum vane, attached to the shaft of the moving coil. As the coil rotates, the vane moves in an air chamber. The amount of clearance between the chamber walls and the air vane effectively controls the degree of damping.

Some instruments use the principle of electromagnetic damping (Lenz's law), where the movable coil is wound on a light aluminum frame. The rotation of the coil in the magnetic field sets up circulating currents in the conductive metal frame, causing a retarding torque that opposes the motion of the coil. Indeed, the same principle is often used to protect PMMC instruments during shipment by placing a metal shorting strap across the coil terminals to reduce deflection.

A galvanometer may also be damped by connecting a resistor across the coil. When the coil rotates in the magnetic field, a voltage is generated in the coil which circulates a current through the coil and the external resistor. This produces an opposing, or retarding, torque that damps the motion of the movement. For any galvanometer, a value for the external resistor can be found that produces critical damping. This resistance is called the *Critical Damping Resistance External* (CRDX); it is an important galvanometer constant. The dynamic damping torque produced by the CDRX depends on the total circuit resistance: the smaller the total circuit resistance, the larger the damping torque.

One way to determine the CDRX consists of observing the galvanometer swing when a current is applied or removed from the coil. Beginning with the oscillating condition, decreasing values of external resistances are tried until a value is found for which the overshoot just disappears. A determination like this is not very precise, but it is adequate for most practical purposes. The value of the CDRX may also be computed from known galvanometer constants.

4-3 PERMANENT-MAGNET MOVING-COIL MECHANISM

4-3.1 D'Arsonval Movement

The basic PMMC movement of Fig. 4-1 is often called the *d'Arsonval* movement, after its inventor. This design offers the largest magnet in a given space and is used when maximum flux in the air gap is required. It provides an instrument with very low power consumption and low current required for *full-scale deflection* (fsd).

Inspection of the diagram of Fig. 4-1 shows a *permanent magnet* of horse-shoe form, with soft iron *pole pieces* attached to it. Between the pole pieces is a *cylinder* of soft iron, which serves to provide a uniform magnetic field in the air gap between the pole pieces and the cylinder. The *coil* is wound on a light metal *frame* and is mounted so that it can rotate freely in the air gap. The *pointer*, attached to the coil, moves over a graduated *scale* and indicates the angular deflection of the coil and therefore the current through the coil.

Two phosphor-bronze *conductive springs*, normally equal in strength, provide the calibrated force opposing the moving-coil torque. Constancy of spring performance is essential to maintain instrument accuracy. The spring thickness is accurately controlled in manufacture to avoid permanent set of the springs. Current is conducted to and from the coil by the control springs.

The entire moving system is statically balanced for all deflection positions by three *balance weights*, as shown in Fig. 4-3. The pointer, springs, and pivots are assembled to the coil structure by means of pivot bases, and the entire movable-coil element is supported by jewel bearings. Different bearing systems are shown in Fig. 4-4.

The V-jewel, shown in Fig. 4-4(a), is almost universally used in instrument bearings. The pivot, bearing in the pit in the jewel, may have a radius at its tip from 0.01 mm to 0.02 mm, depending on the weight of the mechanism and the vibration the instrument will encounter. The radius of the pit in the jewel is

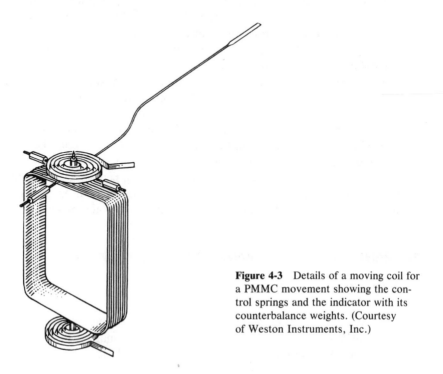

Figure 4-3 Details of a moving coil for a PMMC movement showing the control springs and the indicator with its counterbalance weights. (Courtesy of Weston Instruments, Inc.)

(a)

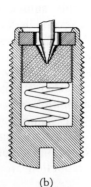

(b)

Figure 4-4 Details of instrument bearings: (a) V-jewel bearing; (b) spring-back jewel bearing. (Courtesy of Weston Instruments, Inc.)

slightly larger than the pivot radius, so that the contact area is circular, only a few microns across. The V-jewel design of Fig. 4-4(a) has the least friction of any practical type of instrument bearing. Although the moving elements of instruments are designed to have the smallest possible weight, the extremely minute area of contact between pivot and jewel results in stresses on the order of 10 kg/mm². If the weight of the moving element is further increased, the contact area does not increase in proportion so that the stress is even greater. Stresses set up by relatively moderate accelerations (like jarring or dropping an instrument) may consequently cause pivot damage. Specially protected (*ruggedized*) instruments use the spring-back (incabloc) jewel bearing, whose construction is shown in Fig. 4-4(b). It is located in its normal position by the spring and is free to move axially when the shock to the mechanism becomes severe.

The scale markings of the basic dc PMMC instrument are usually linearly spaced because the torque (and hence the pointer deflection) is directly proportional to the coil current. [See Eq. (4-1) for the developed torque.] The basic PMMC instrument is therefore a *linear-reading dc device*. The power requirements of the d'Arsonval movement are surprisingly small: typical values range from 25 μW to 200 μW. Accuracy of the instrument is generally on the order of 2 to 5 per cent of full-scale reading.

If low-frequency alternating current is applied to the movable coil, the deflection of the pointer would be up-scale for one half-cycle of the input waveform and down-scale (in the opposite direction) for the next half-cycle. At powerline frequencies (60 Hz) and above, the pointer could not follow the rapid variations in direction and would quiver slightly around the zero mark, seeking the *average* value of the alternating current (which equals zero). The PMMC instrument is therefore unsuitable for ac measurements, unless the current is rectified before application to the coil.

4-3.2 Core-Magnet Construction

In recent years, with the development of Alnico and other improved magnetic materials, it has become feasible to design a magnetic system in which the magnet itself serves as the core. These magnets have the obvious advantage of being relatively unaffected by external magnetic fields, eliminating the magnetic shunting effects in steel panel construction, where several meters operating side by side may affect each other's readings. The need for magnetic shielding, in the form of iron cases, is also eliminated by the *core-magnet* construction. Details of the core-magnet self-shielding movement are shown in Fig. 4-5.

Self-shielding makes the core-magnet mechanism particularly useful in aircraft and aerospace applications, where a multiplicity of instruments must be mounted in close proximity to each other. An example of this type of mounting may be found in the *cross-pointer indicator*, where as many as five mechanisms are mounted in one case to form a unified display. Obviously, the elimination of iron cases and the corresponding weight reduction are of great advantage in aircraft and aerospace instruments.

4-3.3 Taut-Band Suspension

The *suspension-type* galvanometer mechanism has been known for many years. Until recently the device was used only in the laboratory where high sensitivities were required and the torque was extremely low (because of small currents). It was desirable in such instruments to eliminate even the low friction of pivots and

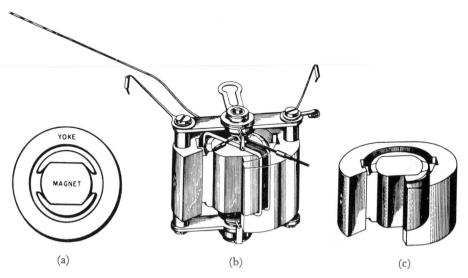

Figure 4-5 Construction of the core-magnet moving-coil mechanism: (a) magnet with its poleshoes is surrounded by the yoke, which acts as a magnetic shield; (b) assembled movement; (c) cutaway view of the yoke, the core, and the poleshoes. (Courtesy of Weston Instruments, Inc.)

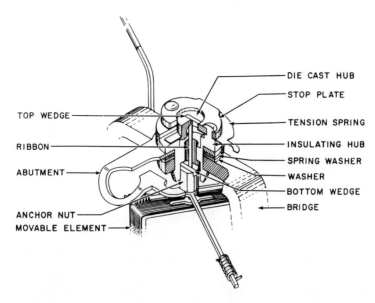

Figure 4-6 Taut-band suspension eliminates the friction of conventional pivot-and-jewel suspensions. This figure shows some construction details, in particular the torsion ribbon with its tension-spring mechanism. (Courtesy of Weston Instruments, Inc.)

jewels. The suspension galvanometer had to be used in the upright position, because sag in the low-torque ligaments caused the moving system to come in contact with stationary members of the mechanism in any other position. This increase in friction caused errors.

The *taut-band* instrument of Fig. 4-6 has the advantage of eliminating the friction of the jewel-pivot suspension. The movable coil is suspended by means of two *torsion ribbons*. The ribbons are placed under sufficient tension to eliminate any sag, as was the case in the suspension galvanometer. This tension is provided by a tension spring, so that the instrument can be used in any position. Generally speaking, taut-band suspension instruments can be made with higher sensitivities than those using pivots and jewels, and they can be used in almost every application served by pivoted instruments. Furthermore, taut-band instruments are relatively insensitive to shock and temperature and are capable of withstanding greater overloads then previous types described.

4-3.4 Temperature Compensation

The PMMC basic movement is not inherently insensitive to temperature, but it may be *temperature-compensated* by the appropriate use of series and shunt resistors of copper and manganin. Both the magnetic fieldstrength and spring tension decrease with an increase in temperature. The coil resistance increases with an increase in temperature. These changes tend to make the pointer read low for a given current with respect to magnetic fieldstrength and coil resistance. The

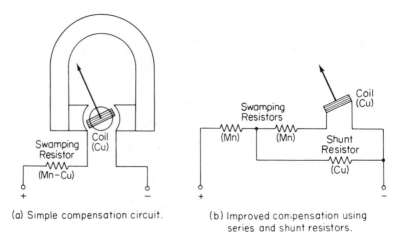

(a) Simple compensation circuit.

(b) Improved compensation using series and shunt resistors.

Figure 4-7 Placement of swamping resistors for temperature compensation of a meter movement.

spring change, conversely, tends to cause the pointer to read high with an increase in temperature. The effects are not identical, however; hence an *uncompensated meter tends to read low* by approximately 0.2 per cent per °C rise in temperature. For purposes of instrument specification, the movement is considered to be compensated when the change in accuracy, due to a 10°C-change in temperature, is not more than one-fourth of the total allowable error.*

Compensation may be accomplished by using *swamping resistors* in series with the movable coil, as shown in Fig. 4-7(a). The swamping resistor is made of manganin (which has a temperature coefficient of practically zero) combined with copper in the ratio of 20/1 to 30/1. The total resistance of coil and swamping resistor increases slightly with a rise in temperature, but only just enough to counteract the change of springs and magnet, so that the overall temperature effect is zero.

A more complete cancellation of temperature effects is obtained with the arrangement of Fig. 4-7(b). Here the *total* circuit resistance increases slightly with a rise in temperature, owing to the presence of the copper coil and the copper shunt resistor. For a fixed applied voltage, therefore, the total current decreases slightly with a rise in temperature. The resistance of the copper shunt resistor increases more than the series combination of coil and manganin resistor; hence a larger fraction of the total current passes through the coil circuit. By correct proportioning of the copper and manganin parts in the circuit, complete cancellation of temperature effects may be accomplished. One *disadvantage* of the use of swamping resistors is a reduction in the full-scale sensitivity of the movement, because a higher applied voltage is necessary to sustain the full-scale current.

* *PMMC Data Sheets*, Weston Instruments, Inc., Newark, N.J.

4-4 DC AMMETERS

4-4.1 Shunt Resistor

The basic movement of a dc ammeter is a PMMC galvanometer. Since the coil winding of a basic movement is small and light, it can carry only very small currents. When large currents are to be measured, it is necessary to bypass the major part of the current through a resistance, called a *shunt*, as shown in Fig. 4-8.

The resistance of the shunt can be calculated by applying conventional circuit analysis to Fig. 4-8, where

R_m = internal resistance of the movement (the coil)

R_s = resistance of the shunt

I_m = full-scale deflection current of the movement

I_s = shunt current

I = full-scale current of the ammeter including the shunt

Since the shunt resistance is in parallel with the meter movement, the voltage drops across the shunt and movement must be the same and we can write

$$V_{\text{shunt}} = V_{\text{movement}}$$

or

$$I_s R_s = I_m R_m \quad \text{and} \quad R_s = \frac{I_m R_m}{I_s} \tag{4-2}$$

Since $I_s = I - I_m$, we can write

$$R_s = \frac{I_m R_m}{I - I_m} \tag{4-3}$$

For each required value of full-scale meter current we can then solve for the value of the shunt resistance required.

EXAMPLE 4-1

A 1-mA meter movement with an internal resistance of 100 Ω is to be converted into a 0–100-mA ammeter. Calculate the value of the shunt resistance required.

SOLUTION

$$I_s = I - I_m = 100 - 1 = 99 \text{ mA}$$

$$R_s = \frac{I_m R_m}{I_s} = \frac{1 \text{ mA} \times 100 \ \Omega}{99 \text{ mA}} = 1.01 \ \Omega$$

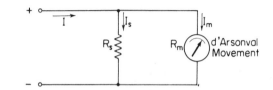

Figure 4-8   Basic dc ammeter circuit.

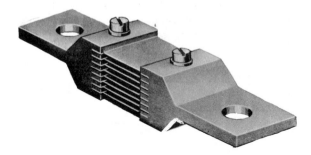

Figure 4-9   High-current shunt for a switchboard instrument. (Courtesy of Weston Instruments, Inc.)

The shunt resistance used with a basic movement may consist of a length of constant-temperature resistance wire within the case of the instrument or it may be an external (manganin or constantan) shunt having a very low resistance. Figure 4-9 shows an external shunt. It consists of evenly spaced sheets of resistive material welded into a large block of heavy copper on each end of the sheets. The resistance material has a very low temperature coefficient, and a low thermoelectric effect exists between the resistance material and the copper. External shunts of this type are normally used for measuring very large currents.

4-4.2 Ayrton Shunt

The current range of the dc ammeter may be further extended by a number of shunts, selected by a *range switch*. Such a meter is called a *multirange* ammeter. Figure 4-10 shows the schematic diagram of a multirange ammeter. The circuit has four shunts, R_a, R_b, R_c, and R_d, which can be placed in parallel with the movement to give four different current ranges. Switch S is a multiposition, *make-before-break* type switch, so that the movement will not be damaged, unprotected in the circuit, without a shunt as the range is changed.

The *universal*, or *Ayrton, shunt* of Fig. 4-11 eliminates the possibility of having the meter in the circuit without a shunt. This advantage is gained at the price of a slightly higher overall meter resistance. The Ayrton shunt provides an excellent opportunity to apply basic network theory to a practical circuit.

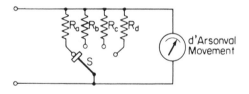

Figure 4-10   Schematic diagram of a simple multirange ammeter.

EXAMPLE 4-2

Design an Ayrton shunt to provide an ammeter with current ranges of 1 A, 5 A, and 10 A. A d'Arsonval movement with an interval resistance $R_m = 50 \ \Omega$ and full-scale deflection current of 1 mA is used in the configuration of Fig. 4-11.

SOLUTION *On the 1-A range:* $R_a + R_b + R_c$ are in parallel with the 50-Ω movement. Since the movement requires 1 mA for full-scale deflection, the shunt will be required to pass a current of 1 A − 1 mA = 999 mA. Using Eq. (4-2), we get

$$R_a + R_b + R_c = \frac{1 \times 50}{999} = 0.05005 \ \Omega \qquad \text{(I)}$$

On the 5-A range: $R_a + R_b$ are in parallel with $R_c + R_m$ (50 Ω). In this case there will be a 1-mA current through the movement and R_c in series, and 4,999 mA through $R_a + R_b$. Again using Eq. (4-2), we get

$$R_a + R_b = \frac{1 \times (R_c + 50 \ \Omega)}{4,999} \qquad \text{(II)}$$

On the 10-A range: R_a now serves as the shunt and $R_b + R_c$ are in series with the movement. The current through the movement again is 1 mA, and the shunt passes the remaining 9,999 mA. Using Eq. (4-2) again, we get

$$R_a = \frac{1 \times (R_b + R_c + 50 \ \Omega)}{9,999} \qquad \text{(III)}$$

Solving the three simultaneous equations (I), (II), and (III), we obtain

$$4,999 \times \text{(I)}: \quad 4,999R_a + 4,999R_b + 4,999R_c = 250.2$$

$$\text{(II)}: \quad 4,999R_a + 4,999R_b - \quad R_c = 50$$

Subtracting (II) from (I), we obtain

$$5,000R_c = 200.2$$

$$R_c = 0.04004 \ \Omega$$

Similarly,

$$9,999 \times \text{(I)}: \quad 9,999R_a + 9,999R_b + 9,999R_c = 500.45$$

$$\text{(III)}: \quad 9,999R_a - \quad R_b - \quad R_c = 50$$

Subtracting (III) from (I), we obtain

$$10,000R_b + 10,000R_c = 450.45$$

Substituting the previously calculated value for R_c into this expression yields

$$10,000R_b = 450.45 - 400.4$$

$$R_b = 0.005005 \ \Omega$$

$$R_a = 0.005005 \ \Omega$$

This calculation indicates that for larger currents the value of the shunt resistor may become very small.

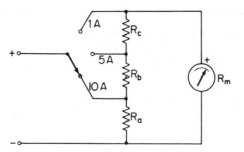

Figure 4-11 Universal or Ayrton shunt.

Direct-current ammeters are commercially available in a large number of ranges, from 20 μA to 50 A full-scale for a self-contained meter and to 500 A for a meter with external shunt. Laboratory-type precision ammeters are provided with a calibration chart, so that the user may correct his readings for any scale errors.

The following precautions should be observed when using an ammeter in measurement work:

(a) *Never* connect an ammeter *across* a source of emf. Because of its low resistance it would draw damaging high currents and destroy the delicate movement. *Always* connect an ammeter in series with a load capable of limiting the current.

(b) Observe the correct *polarity*. Reverse polarity causes the meter to deflect against the mechanical stop and this may damage the pointer.

(c) When using a multirange meter, first use the highest current range; then decrease the current range until substantial deflection is obtained. To increase accuracy of the observation (see Chapter 1), use the range that will give a reading as near to full-scale as possible.

4-5 DC VOLTMETERS

4-5.1 Multiplier Resistor

The addition of a series resistor, or *multiplier*, converts the basic d'Arsonval movement into a *dc voltmeter*, as shown in Fig. 4-12. The multiplier limits the current through the movement so as not to exceed the value of the full-scale deflection current (I_{fsd}). A dc voltmeter measures the potential difference be-

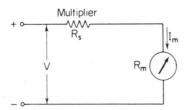

Figure 4-12 Basic dc voltmeter circuit.

tween two points in a dc circuit and is therefore connected *across* a source of emf or a circuit component. The meter terminals are generally marked "pos" and "neg," since polarity must be observed.

The value of a multiplier, required to extend the voltage range, is calculated from Fig. 4-12, where

$$I_m = \text{deflection current of the movement } (I_{\text{fsd}})$$

$$R_m = \text{internal resistance of the movement}$$

$$R_s = \text{multiplier resistance}$$

$$V = \text{full-range voltage of the instrument}$$

For the circuit of Fig. 4-12,

$$V = I_m(R_s + R_m)$$

Solving for R_s gives

$$R_s = \frac{V - I_m R_m}{I_m} = \frac{V}{I_m} - R_m \tag{4-4}$$

The multiplier is usually mounted inside the case of the voltmeter for moderate ranges up to 500 V. For higher voltages, the multiplier may be mounted separately outside the case on a pair of binding posts to avoid excessive heating inside the case.

4-5.2 Multirange Voltmeter

The addition of a number of multipliers, together with a *range switch*, provides the instrument with a workable number of voltage ranges. Figure 4-13 shows a *multirange* voltmeter using a four-position switch and four multipliers, R_1, R_2, R_3, and R_4, for the voltage ranges V_1, V_2, V_3, and V_4, respectively. The values of the multipliers can be calculated using the method shown earlier or, alternatively, by the *sensitivity method*. The sensitivity method is illustrated by Example 4-4 in Sec. 4-6, where sensitivity is discussed.

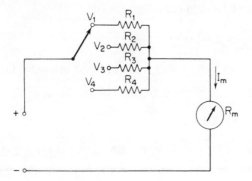

Figure 4-13 Multirange voltmeter.

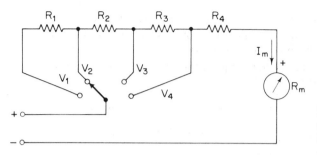

Figure 4-14 More practical arrangement of multiplier resistors in the multirange voltmeter.

A variation of the circuit of Fig. 4-13 is shown in Fig. 4-14, where the multipliers are connected in a series string and the range selector switches the appropriate amount of resistance in series with the movement. This system has the advantage that all multipliers except the first have standard resistance values and can be obtained commercially in precision tolerances. The low-range multiplier, R_4, is the only special resistor that must be manufactured to meet the specific circuit requirements.

EXAMPLE 4-3

A basic d'Arsonval movement with internal resistance, $R_m = 100 \; \Omega$, and full-scale current, $I_{\text{fsd}} = 1$ mA, is to be converted into a multirange dc voltmeter with voltage ranges of 0–10 V, 0–50 V, 0–250 V, and 0–500 V. The circuit arrangement of Fig. 4-16 is to be used for this voltmeter.

SOLUTION For the 10-V range (V_4 position of range switch), the total circuit resistance is

$$R_T = \frac{10 \text{ V}}{1 \text{ mA}} = 10 \text{ k}\Omega$$

$$R_4 = R_T - R_m = 10 \text{ k}\Omega - 100 \; \Omega = 9,900 \; \Omega$$

For the 50-V range (V_3 position of range switch),

$$R_T = \frac{50 \text{ V}}{1 \text{ mA}} = 50 \text{ k}\Omega$$

$$R_3 = R_T - (R_4 + R_m) = 50 \text{ k}\Omega - 10 \text{ k}\Omega = 40 \text{ k}\Omega$$

For the 250-V range (V_2 position of range switch),

$$R_T = \frac{250 \text{ V}}{1 \text{ mA}} = 250 \text{ k}\Omega$$

$$R_2 = R_T - (R_3 + R_4 + R_m) = 250 \text{ k}\Omega - 50 \text{ k}\Omega = 200 \text{ k}\Omega$$

For the 500-V range (V_1 position of range switch),

$$R_T = \frac{500 \text{ V}}{1 \text{ mA}} = 500 \text{ k}\Omega$$

$$R_1 = R_T - (R_2 + R_3 + R_4 + R_m) = 500 \text{ k}\Omega - 250 \text{ k}\Omega = 250 \text{ k}\Omega$$

Notice in Example 4-3 that only the low-range multiplier R_4 has a nonstandard value.

4-6 VOLTMETER SENSITIVITY

4-6.1 Ohms-per-Volt Rating

In Sec. 4-5 it was shown that the full-scale deflection current I_{fsd} was reached on all voltage ranges when the corresponding full-scale voltage was applied. As shown in Example 4-3, a current of 1 mA is obtained for voltages of 10 V, 50 V, 250 V, and 500 V across the meter terminals. For each voltage range, the quotient of the total circuit resistance R_T and the range voltage V is always $1,000 \ \Omega/V$. This figure is often referred to as the *sensitivity*, or the *ohms-per-volt rating*, of the voltmeter. Note that the sensitivity, S, is essentially the *reciprocal* of the full-scale deflection current of the basic movement, or

$$S = \frac{1}{I_{fsd}} \frac{\Omega}{V} \tag{4-5}$$

The sensitivity S of the voltmeter may be used to advantage in the *sensitivity method* of calculating the resistance of the multiplier in a dc voltmeter. Consider the circuit of Fig. 4-14, where

S = sensitivity of the voltmeter (Ω/V)

V = the voltage range, as set by the range switch

R_m = internal resistance of the movement (plus the previous series resistors)

R_s = resistance of the multiplier

For the circuit of Fig. 4-14,

$$R_T = S \times V \tag{4-6}$$

and

$$R_s = (S \times V) - R_m \tag{4-7}$$

Use of the sensitivity method is illustrated in Example 4-4.

EXAMPLE 4-4

Repeat Example 4-3, now using the sensitivity method for calculating the multiplier resistances.

SOLUTION

$$S = \frac{1}{I_{fsd}} = \frac{1}{0.001 \ A} = 1,000 \ \frac{\Omega}{V}$$

$$R_4 = (S \times V) - R_m = \frac{1,000 \ \Omega}{V} \times 10 \ V - 100 \ \Omega = 9,900 \ \Omega$$

$$R_3 = (S \times V) - R_m = \frac{1{,}000\ \Omega}{V} \times 50\ V - 10{,}000\ \Omega = 40\ k\Omega$$

$$R_2 = (S \times V) - R_m = \frac{1{,}000\ \Omega}{V} \times 250\ V - 50\ k\Omega = 200\ k\Omega$$

$$R_1 = (S \times V) - R_m \frac{1{,}000\ \Omega}{V} \times 500\ V - 250\ k\Omega = 250\ k\Omega$$

4-6.2 Loading Effect

The sensitivity of a dc voltmeter is an important factor when selecting a meter for a certain voltage measurement. A low-sensitivity meter may give correct readings when measuring voltages in low-resistance circuits, but it is certain to produce very unreliable readings in high-resistance circuits. A voltmeter, when connected across two points in a highly resistive circuit, acts as a shunt for that portion of the circuit and thus reduces the equivalent resistance in that portion of the circuit. The meter will then give a lower indication of the voltage drop than actually existed before the meter was connected. This effect is called the *loading effect* of an instrument; it is caused principally by *low-sensitivity* instruments. The loading effect of a voltmeter is illustrated in Example 4-5.

EXAMPLE 4-5

It is desired to measure the voltage across the 50-kΩ resistor in the circuit of Fig. 4-15. Two voltmeters are available for this measurement: voltmeter 1 with a sensitivity of 1,000 Ω/V and voltmeter 2 with a sensitivity of 20,000 Ω/V. Both meters are used on their 50-V range. Calculate (a) the reading of each meter; (b) the error in each reading, expressed as a percentage of the true value.

SOLUTION Inspection of the circuit indicates that the voltage across the 50-kΩ resistor is

$$\frac{50\ k\Omega}{150\ k\Omega} \times 150\ V = 50\ V$$

This is the *true* value of voltage across the 50-kΩ resistor.

(a) *Voltmeter 1 (S = 1,000 Ω/V)* has a resistance of 50 V $\times$ 1,000 Ω/V = 50 kΩ on its 50-V range. Connecting the meter across the 50-kΩ resistor causes the equivalent parallel resistance to be decreased to 25 kΩ and the total circuit resistance to 125 kΩ. The potential difference across the combination of meter and 50-kΩ resistor is

$$V_1 = \frac{25\ k\Omega}{125\ k\Omega} \times 150\ V = 30\ V$$

Hence the voltmeter indicates a voltage of 30 V. *Voltmeter 2 (S = 20 kΩ/V)* has a resistance of 50 V $\times$ 20 kΩ/V = 1 megohm on its 50-V range. When this meter is connected across the 50-kΩ resistor, the equivalent parallel resistance equals 47.6 kΩ. This combination produces a voltage of

$$V_2 = \frac{47.6 \text{ k}\Omega}{147.6 \text{ k}\Omega} \times 150 \text{ V} = 48.36 \text{ V}$$

which is indicated on the voltmeter.

(b) The error in the reading of voltmeter 1 is

$$\% \text{ error} = \frac{\text{true voltage} - \text{apparent voltage}}{\text{true voltage}} \times 100\%$$

$$= \frac{50 \text{ V} - 30 \text{ V}}{50 \text{ V}} \times 100\% = 40\%$$

The error in the reading of voltmeter 2 is

$$\% \text{ error} = \frac{50 \text{ V} - 48.36 \text{ V}}{50 \text{ V}} \times 100\% = 3.28\%$$

The calculation of Example 4-5 indicates that the meter with the higher sensitivity or ohms-per-volt rating gives the most reliable result. It is important to realize the factor of sensitivity, particularly when voltage measurements are made in high-resistance circuits.

The matter of reliability and accuracy of the test result raises an interesting point. When an *insensitive, yet highly accurate,* dc voltmeter is placed across the terminals of a high resistance, the meter accurately reflects the voltage condition produced by loading. The error is a human or gross error (Sec. 1-4), because the proper instrument was not selected. The meter "disturbs" the circuit, and the ideal of instrumentation, at all times, is to measure a condition without affecting it in any way. The human investigator has the responsibility to select an instrument which is precise, reliable, and sufficiently sensitive not to disturb that which is being measured. The fault lies not with the highly accurate instrument but with the investigator, who is using it incorrectly. In fact, the sophisticated instrument user could calculate the true voltage by using an insensitive yet accurate meter. Therefore *accuracy* is always required in instruments; *sensitivity* is needed only in special applications where loading disturbs that which is being measured. Example 4-6 illustrates how an insensitive yet accurate instrument is used to perform an entirely valid measurement.

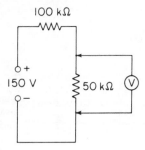

Figure 4-15 Voltmeter loading effect.

EXAMPLE 4-6

The only voltmeter available in a laboratory has a sensitivity of 100 Ω/V and three scales, 50 V, 150 V, and 300 V. When connected in the circuit of Fig. 4-16, the meter reads 4.65 V on its lowest (50-V) scale. Calculate the value of R_x, where R_V is the voltmeter resistance.

SOLUTION The equivalent resistance of the voltmeter on its 50-V scale is

$$R_V = 100 \frac{\Omega}{V} \times 10 \text{ V} = 5 \text{ k}\Omega$$

Let R_p = the parallel resistance of R_x and R_V.

$$R_p = \frac{V_p}{V_s} \times R_s = \frac{4.65}{95.35} \times 100 \text{ }\Omega = 4.878 \text{ k}\Omega$$

Then

$$R_x = \frac{R_p \times R_V}{R_V - R_p} = \frac{4.878 \text{ k}\Omega \times 5 \text{ k}\Omega}{0.122 \text{ k}\Omega} = 200 \text{ k}\Omega$$

Example 4-6 shows that when the instrument user is aware of the limitations of his instrument, he can still make allowances provided that the voltmeter is accurate.

The following general precautions should be observed when using a voltmeter:

(a) Observe the correct polarity. Wrong polarity causes the meter to deflect against the mechanical stop and this may damage the pointer.

(b) Place the voltmeter *across* the circuit or component whose voltage is to be measured.

(c) When using a multirange voltmeter, always use the highest voltage range and then decrease the range until a good up-scale reading is obtained.

(d) Always be aware of the loading effect. The effect can be minimized by using as high a voltage range (and highest sensitivity) as possible. The precision of measurement decreases if the indication is at the low end of the scale (Sec. 1-4).

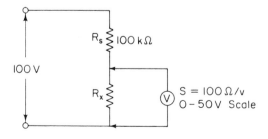

Figure 4-16 Use of an accurate but insensitive voltmeter to determine resistance of R_x.

4-7 SERIES-TYPE OHMMETER

The series-type ohmmeter essentially consists of a d'Arsonval movement connected in series with a resistance and a battery to a pair of terminals to which the unknown is connected. The current through the movement then depends on the magnitude of the unknown resistor, and the meter indication is proportional to the value of the unknown, provided that calibration problems are taken into account. Figure 4-17 shows the elements of a simple single-range series ohmmeter. In Fig. 4-17,

$$R_1 = \text{current-limiting resistor}$$

$$R_2 = \text{zero adjust resistor}$$

$$E = \text{internal battery}$$

$$R_m = \text{internal resistance of the d'Arsonval movement}$$

$$R_x = \text{unknown resistor}$$

When the unknown resistor $R_x = 0$ (terminals A and B shorted), maximum current flows in the circuit. Under this condition, shunt resistor R_2 is adjusted until the movement indicates full-scale current (I_{fsd}). The full-scale current position of the pointer is marked "0 Ω" on the scale. Similarly, when $R_x = \infty$ (terminals A and B open), the current in the circuit drops to zero and the movement indicates zero current, which is then marked "∞" on the scale. Intermediate markings may be placed on the scale by connecting different known values of R_x to the instrument. The accuracy of these scale markings depends on the repeating accuracy of the movement and the tolerances of the calibrating resistors.

Although the series-type ohmmeter is a popular design and is used extensively in portable instruments for general-service work, it has certain disadvantages. Important among these is the internal battery whose voltage decreases gradually with time, so that the full-scale current drops and the meter does not read "0" when A and B are shorted. The variable shunt resistor R_2 in Fig. 4-17 provides an adjustment to counteract the effect of battery change. Without R_2, it would be possible to bring the pointer back to full scale by adjusting R_1, but this would change the calibration all along the scale. Adjustment by R_2 is a superior solution, since the parallel resistance of R_2 and the coil R_m is always low compared to R_1 and therefore the change in R_2 needed for adjustment does not change the

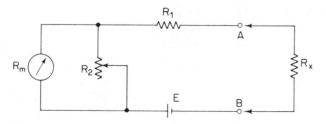

Figure 4-17   Series-type ohmmeter.

calibration very much. The circuit of Fig. 4-17 does not compensate completely for aging of the battery, but it does a reasonably good job within the expected limits of accuracy of the instrument.

A convenient quantity to use in the design of a series-type ohmmeter is the value of R_x which causes half-scale deflection of the meter. At this position, the resistance across terminals A and B is defined as the half-scale position resistance R_h. Given the full-scale current I_{fsd} and the internal resistance of the movement R_m, the battery voltage E, and the desired value of the half-scale resistance R_h, the circuit can be analyzed; i.e., values can be found for R_1 and R_2.

The design can be approached by recognizing that, if introducing R_h reduces the meter current to $\frac{1}{2}I_{fsd}$, the unknown resistance must be equal to the total internal resistance of the ohmmeter. Therefore

$$R_h = R_1 + \frac{R_2 R_m}{R_2 + R_m} \tag{4-8}$$

The total resistance presented to the battery then equals $2R_h$, and the battery current needed to supply the half-scale defection is

$$I_h = \frac{E}{2R_h} \tag{4-9}$$

To produce full-scale deflection, the battery current must be doubled, and therefore

$$I_t = 2I_h = \frac{E}{R_h} \tag{4-10}$$

The shunt current through R_2 is

$$I_2 = I_t - I_{fsd} \tag{4-11}$$

The voltage across the shunt (E_{sh}) is equal to the voltage across the movement and

$$E_{sh} = E_m \quad \text{or} \quad I_2 R_2 = I_{fsd} R_m$$

and

$$R_2 = \frac{I_{fsd} R_m}{I_2} \tag{4-12}$$

Substituting Eq. (4-11) into Eq. (4-12), we obtain

$$R_2 = \frac{I_{fsd} R_m}{I_t - I_{fsd}} = \frac{I_{fsd} R_m R_h}{E - I_{fsd} R_h} \tag{4-13}$$

Solving Eq. (4-8) for R_1 gives

$$R_1 = R_h - \frac{R_2 R_m}{R_2 + R_m} \tag{4-14}$$

Substituting Eq. (4-13) into Eq. (4-14) and solving for R_1 yields

$$R_1 = R_h - \frac{I_{\text{fsd}} R_m R_h}{E} \qquad (4\text{-}15)$$

A typical calculation for the series-type ohmmeter is given in Example 4-7.

EXAMPLE 4-7

The ohmmeter of Fig. 4-17 uses a 50-Ω basic movement requiring a full-scale current of 1 mA. The internal battery voltage is 3 V. The desired scale marking for half-scale deflection is 2,000 Ω. Calculate (a) the values of R_1 and R_2; (b) the maximum value of R_2 to compensate for a 10% drop in battery voltage; (c) the scale error at the half-scale mark (2,000 Ω) when R_2 is set as in (b).

SOLUTION   (a) The total battery current at full-scale deflection is

$$I_t = \frac{E}{R_h} = \frac{3 \text{ V}}{2,000 \ \Omega} = 1.5 \text{ mA} \qquad (4\text{-}16)$$

The current through the zero-adjust resistor R_2 then is

$$I_2 = I_t - I_{\text{fsd}} = 1.5 \text{ mA} - 1 \text{ mA} = 0.5 \text{ mA} \qquad (4\text{-}17)$$

The value of the zero-adjust resistor R_2 is

$$R_2 = \frac{I_{\text{fsd}} R_m}{I_2} = \frac{1 \text{ mA} \times 50 \ \Omega}{0.5 \text{ mA}} = 100 \ \Omega \qquad (4\text{-}18)$$

The parallel resistance of the movement and the shunt (R_p) is

$$R_p = \frac{R_2 R_m}{R_2 + R_m} = \frac{50 \times 100}{150} = 33.3 \ \Omega$$

The value of the current-limiting resistor R_1 is

$$R_1 = R_h - R_p = 2,000 - 33.3 = 1,966.7 \ \Omega$$

(b) At a 10% drop in battery voltage,

$$E = 3 \text{ V} - 0.3 \text{ V} = 2.7 \text{ V}$$

The total battery current I_t then becomes

$$I_t = \frac{E}{R_h} = \frac{2.7 \text{ V}}{2,000 \ \Omega} = 1.35 \text{ mA}$$

The shunt current I_2 is

$$I_2 = I_t - I_{\text{fsd}} = 1.35 \text{ mA} - 1 \text{ mA} = 0.35 \text{ mA}$$

and the zero-adjust resistor R_2 equals

$$R_2 = \frac{I_{\text{fsd}} R_m}{I_2} = \frac{1 \text{ mA} \times 50 \ \Omega}{0.35 \text{ mA}} = 143 \ \Omega$$

(c) The parallel resistance of the meter movement and the new value of R_2 becomes

$$R_p = \frac{R_2 R_m}{R_2 + R_m} = \frac{50 \times 143}{193} = 37 \ \Omega$$

Since the half-scale resistance R_h is equal to the total internal circuit resistance, R_h will increase to

$$R_h = R_1 + R_p = 1,966.7 \ \Omega + 37 \ \Omega = 2,003.7 \ \Omega$$

Therefore the true value of the half-scale mark on the meter is 2,003.7 Ω whereas the actual scale mark is 2,000 Ω. The percentage error is then

$$\% \ \text{error} = \frac{2,000 - 2,003.7}{2,003.7} \times 100\% = -0.185\%$$

The negative sign indicates that the meter reading is low.

The ohmmeter of Example 4-7 could be designed for other values of R_h, within limits. If $R_h = 3,000 \ \Omega$, the battery current would be 1 mA, which is required for the full-scale deflection current. If the battery voltage would decrease owing to aging, the total battery current would fall below 1 mA and there would then be no provision for adjustment.

4-8 SHUNT-TYPE OHMMETER

The circuit diagram of a *shunt-type ohmmeter* is shown in Fig. 4-18. It consists of a battery in series with an adjustable resistor R_1 and a d'Arsonval movement. The unknown resistance is connected across terminals A and B, in *parallel* with the meter. In this circuit it is necessary to have an *off-on switch* to disconnect the battery from the circuit when the instrument is not used. When the unknown resistor $R_x = 0 \ \Omega$ (A and B shorted), the meter current is zero. If the unknown resistor $R_x = \infty$ (A and B open), the current finds a path only through the meter, and by appropriate selection of the value of R_1, the pointer can be made to read full scale. The ohmmeter therefore has the "zero" mark at the left-hand side of the scale (no current) and the "infinite" mark at the right-hand side of the scale (full-scale deflection current).

The shunt-type ohmmeter is particularly suited to the measurement of *low-value resistors*. It is not a commonly used test instrument, but it is found in laboratories or for special low-resistance applications.

The analysis of the shunt-type ohmmeter is similar to that of the series-type ohmmeter (Sec. 4-7). In Fig. 4-18, when $R_x = \infty$, the full-scale meter current will

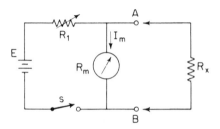

Figure 4-18 Shunt-type ohmmeter.

be

$$I_{fsd} = \frac{E}{R_1 + R_m} \qquad (4\text{-}19)$$

where E = internal battery voltage

R_1 = current-limiting resistor

R_m = internal resistance of the movement

Solving for R_1, we find

$$R_1 = \frac{E}{I_{fsd}} - R_m \qquad (4\text{-}20)$$

For any value of R_x connected across the meter terminals, the meter current decreases and is given by

$$I_m = \frac{E}{R_1 + [R_m R_x/(R_m + R_x)]} \times \frac{R_x}{R_m + R_x}$$

or

$$I_m = \frac{E R_x}{R_1 R_m + R_x(R_1 + R_m)} \qquad (4\text{-}21)$$

The meter current for any value of R_x, expressed as a fraction of the full-scale current, is

$$s = \frac{I_m}{I_{fsd}} = \frac{R_x(R_1 + R_m)}{R_1(R_m + R_x) + R_m R_x}$$

or

$$s = \frac{R_x(R_1 + R_m)}{R_x(R_1 + R_m) + R_1 R_m} \qquad (4\text{-}22)$$

Defining

$$\frac{R_1 R_m}{R_1 + R_m} = R_p \qquad (4\text{-}23)$$

and substituting Eq. (4-23) into Eq. (4-22), we obtain

$$s = \frac{R_x}{R_x + R_p} \qquad (4\text{-}24)$$

If Eq. (4-24) is used, the meter can be calibrated by calculating s in terms of R_x and R_p.

At half-scale reading of the meter ($I_m = 0.5\,I_{fsd}$), Eq. (4-21) reduces to

$$0.5 I_{fsd} = \frac{E R_h}{R_1 R_m + R_h(R_1 + R_m)} \qquad (4\text{-}25)$$

where R_h = external resistance causing half-scale deflection. To determine the relative scale values for a given value of R_1, the half-scale reading may be found by dividing Eq. (4-19) by Eq. (4-25) and solving for R_h:

$$R_h = \frac{R_1 R_m}{R_1 + R_m} \tag{4-26}$$

The analysis shows that the half-scale resistance is determined by limiting resistor R_1 and the internal resistance of the movement, R_m. The limiting resistance, R_1, is in turn determined by the meter resistance R_m, and the full-scale deflection current, I_{fsd}.

To illustrate that the shunt-type ohmmeter is particularly suited to the measurement of very low resistances, consider Example 4-8.

EXAMPLE 4-8

The circuit of Fig. 4-18 uses a 10-mA basic d'Arsonval movement with an internal resistance of 5 Ω. The battery voltage $E = 3$ V. It is desired to modify the circuit by adding an appropriate resistor R_{sh} across the movement, so that the instrument will indicate 0.5 Ω at the midpoint on its scale. Calculate (a) the value of the shunt resistor, R_{sh}; (b) the value of the current-limiting resistor, R_1.

SOLUTION   (a) For half-scale deflection of the movement,

$$I_m = 0.5 I_{fsd} = 5 \text{ mA}$$

The voltage across the movement is

$$E_m = 5 \text{ mA} \times 5 \text{ Ω} = 25 \text{ mA}$$

Since this voltage also appears across the unknown resistor, R_x, the current through R_x is

$$I_x = \frac{25 \text{ mV}}{0.5 \text{ Ω}} = 50 \text{ mA}$$

The current through the movement (I_m) plus the current through the shunt (I_{sh}) must be equal to the current through the unknown (I_x). Therefore

$$I_{sh} = I_x - I_m = 50 \text{ mA} - 5 \text{ mA} = 45 \text{ mA}$$

The shunt resistance then is

$$R_{sh} = \frac{E_m}{I_{sh}} = \frac{25 \text{ mV}}{45 \text{ mA}} = \frac{5}{9} \text{ Ω}$$

(b) The total battery current is

$$I_t = I_m + I_{sh} + I_x = 5 \text{ mA} + 45 \text{ mA} + 50 \text{ mA} = 100 \text{ mA}$$

The voltage drop across limiting resistor R_1 equals 3 V − 25 mV = 2.975 V. Therefore

$$R_1 = \frac{2.975 \text{ V}}{100 \text{ mA}} = 29.75 \text{ Ω}$$

Electromechanical Indicating Instruments   Chap. 4

4-9 MULTIMETER OR VOM

The ammeter, the voltmeter, and the ohmmeter all use a d'Arsonval movement. The difference between these instruments is the circuit in which the basic movement is used. It is therefore obvious that a single instrument can be designed to perform the three measurement functions. This instrument, which contains a *function switch* to connect the appropriate circuits to the d'Arsonval movement, is often called a *multimeter* or *volt-ohm-milliammeter* (VOM).

A representative example of a commercial multimeter is shown in Fig. 4-19. The circuit diagram of this meter is given in Fig. 4-20. The meter is a combination of a dc milliammeter, a dc voltmeter, an ac voltmeter, a multirange ohmmeter, and an output meter. (The circuits of the ac voltmeter and the output meter are discussed in Sec. 4-11.2.)

Figure 4-21 shows the circuit for the dc voltmeter section, where the common input terminals are used for voltage ranges of 0–1.5 to 0–1,000 V. An external voltage jack, marked "DC 5,000 V," is used for dc voltage measurements to 5,000 V. The operation of this circuit is similar to the circuit of Fig. 4-12, which was discussed in Sec. 4-5.

The basic movement of the multimeter of Fig. 4-19 has a full-scale current of 50 μA and an internal resistance of 2,000 Ω. The values of the multipliers are given in Fig. 4-21. Notice that on the 5,000-V range, the range switch should be set to the 1,000-V position, but the test lead should be connected to the external

Figure 4-19 General-purpose multimeter. This instrument has been a familiar sight in electronics laboratories for many years. (Courtesy of Simpson Electric Company.)

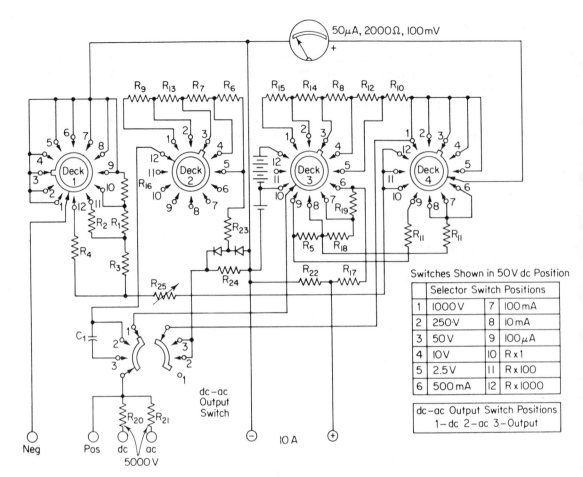

Figure 4-20 Schematic diagram of the Simpson Model 260 multimeter. (Courtesy of Simpson Electric Company.)

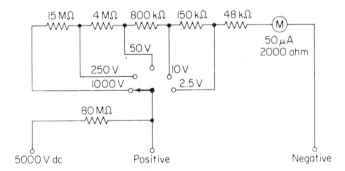

Figure 4-21 Dc voltmeter section of the Simpson Model 260 multimeter. (Courtesy of Simpson Electric Company.)

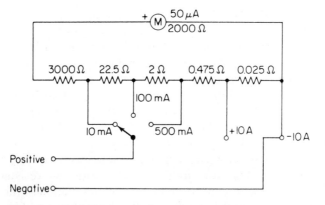

Figure 4-22 Dc ammeter section of the Simpson Model 260 multimeter. (Courtesy of Simpson Electric Company.)

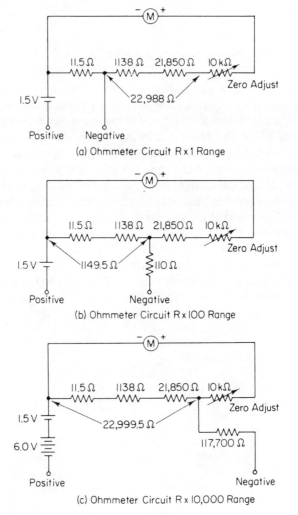

(a) Ohmmeter Circuit R x 1 Range

(b) Ohmmeter Circuit R x 100 Range

(c) Ohmmeter Circuit R x 10,000 Range

Figure 4-23 Ohmmeter section of the Simpson Model 260 multimeter. (Courtesy of Simpson Electric Company.)

jack marked "DC 5,000 V." The normal precautions for measuring voltage should be taken. Because of its fairly high sensitivity (20 kΩ/V), the instrument is suited to general-service work in the electronics field.

The circuit for measuring dc milliamperes and amperes is given in Fig. 4-22 and again the circuit is self-explanatory. The positive (+) and "negative" (−) terminals are used for current measurements up to 500 mA and the jacks marked "+10 A" and "−10 A" are used for the 0–10-A range.

Details of the ohmmeter section of the VOM are shown in Fig. 4-23. The circuit in Fig. 4-23 gives the ohmmeter circuit for a scale multiplication of 1. Before any measurement is made, the instrument is short-circuited and the "zero-adjust" control is varied until the meter reads zero resistance (full-scale current). Notice that the circuit takes the form of a variation of the shunt-type ohmmeter. Scale multiplications of 100 and 10,000 are shown in Fig. 4-23(b) and (c).

The ac voltmeter section of the meter is selected by setting the "ac-dc" switch to the "ac" position. The operation of this circuit is discussed in Sec. 4-11.2.

4-10 CALIBRATION OF DC INSTRUMENTS

Although detailed calibration techniques are beyond the scope of this chapter, some general procedures for the calibration of basic dc instruments are given.

Calibration of a *dc ammeter* can most easily be carried out by the arrangement of Fig. 4-24. The value of the current through the ammeter to be calibrated is determined by measuring the potential difference across a standard resistor by the *voltmeter method* and then calculating the current by Ohm's law. The result of this calculation is compared to the actual reading of the ammeter under calibration and inserted in the circuit. A good source of constant current is required and is usually provided by storage cells or a precision power supply. A rheostat is placed in the circuit to control the current to any desired value, so that different points on the meter scale can be calibrated.

A simple method of calibrating a *dc voltmeter* is shown in Fig. 4-25, where the voltage across dropping resistor *R* is accurately measured with a potentiometer. The meter to be calibrated is connected across the same two points as the

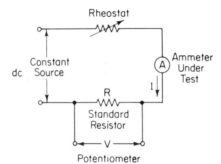

Figure 4-24 Potentiometer method of calibrating a dc ammeter.

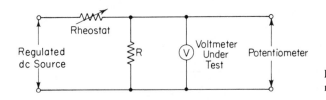

Figure 4-25 Potentiometer method of calibrating a dc voltmeter.

potentiometer and should therefore indicate the same voltage. A rheostat is placed in the circuit to control the amount of current and therefore the drop across the resistor, *R,* so that several points on the voltmeter scale can be calibrated. Voltmeters tested with the method of Fig. 4-25 can be calibrated with an accuracy of ±0.01 per cent, which is well beyond the usual accuracy of a d'Arsonval movement.

The *ohmmeter* is generally considered to be an instrument of moderate accuracy and low precision. A rough calibration may be done by measuring a standard resistance and noting the reading of the ohmmeter. Doing this for several points on the ohmmeter scale and on several ranges allows one to obtain an indication of the correct operation of the instrument.

4-11 ALTERNATING-CURRENT INDICATING INSTRUMENTS

The d'Arsonval movement responds to the *average* or *dc* value of the current through the moving coil. If the movement carries an alternating current with positive and negative half-cycles, the driving torque would be in one direction for the positive alternation and in the other direction for the negative alternation. If the frequency of the ac is very low, the pointer would swing back and forth around the zero point on the meter scale. At higher frequencies, the inertia of the coil is so great that the pointer cannot follow the rapid reversals of the driving torque and hovers around the zero mark, vibrating slightly.

To measure ac on a d'Arsonval movement, some means must be devised to obtain a *unidirectional* torque that does not reverse each half-cycle. One method involves rectification of the ac, so that the rectified current deflects the coil. Other methods use the heating effect of the alternating current to produce an indication of its magnitude. Some of these methods are discussed in this chapter.

4-11.1 Electrodynamometer

One of the most important ac movements is the *electrodynamometer*. It is often used in accurate ac voltmeters and ammeters, not only at the powerline frequency but also in the lower audiofrequency range. With some slight modifications, the electrodynamometer can be used as a wattmeter, a VARmeter, a power-factor meter, or a frequency meter. The electrodynamometer movement may also serve as a *transfer* instrument, because it can be calibrated on dc and then used directly on ac, establishing a direct means of equating ac and dc measurements of voltage and current.

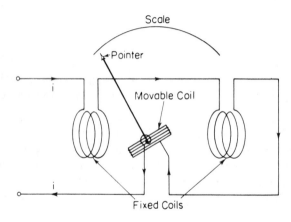

Scale

Pointer

Movable Coil

Fixed Coils

Figure 4-26 Schematic diagram of an electrodynamometer movement.

Where the d'Arsonval movement uses a permanent magnet to provide the magnetic field in which the movable coil rotates, the electrodynamometer uses the current under measurement to produce the necessary field flux. Figure 4-26 shows a schematic arrangement of the parts of this movement. A fixed coil, split into two equal halves, provides the magnetic field in which the movable coil rotates. The two coil halves are connected in series with the moving coil and are fed by current under measurement. The fixed coils are spaced far enough apart to allow passage of the shaft of the movable coil. The movable coil carries a pointer, which is balanced by counterweights. Its rotation is controlled by springs, similar to the d'Arsonval movement construction. The complete assembly is surrounded by a laminated shield to protect the instrument from stray magnetic fields which may affect its operation. Damping is provided by aluminum air vanes, moving in sector-shaped chambers. The entire movement is very solid and rigidly constructed in order to keep its mechanical dimensions stable and its calibration intact. A cutaway view of the electrodynamometer is shown in Fig. 4-27.

The operation of the instrument may be understood by returning to the expression for the torque developed by a coil suspended in a magnetic field. We previously stated [Eq. (4-1)] that

$$T = B \times A \times I \times N$$

indicating that the torque, which deflects the movable coil, is directly proportional to the coil constants (A and N), the strength of the magnetic field in which the coil moves (B), and the current through the coil (I). In the electrodynamometer the flux density (B) depends on the current through the fixed coil and is therefore directly proportional to the deflection current (I). Since the coil dimensions and the number of turns on the coil frame are fixed quantities for any given meter, the developed torque becomes a function of the current squared (I^2).

If the electrodynamometer is exclusively designed for dc use, its square-law scale is easily noticed, with crowded scale markings at the very low current values, progressively spreading out at the higher current values. For ac use, the developed torque at any instant is proportional to the *instantaneous* current squared (i^2). The instantaneous value of i^2 is always positive and torque pulsa-

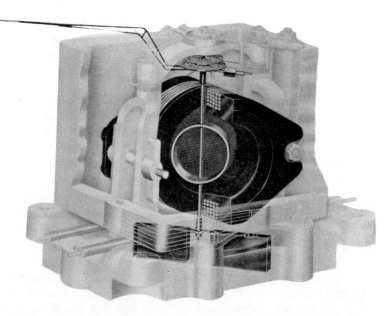

Figure 4-27 Phantom photograph of an electrodynamometer, showing the arrangement of fixed and movable coils. The rigidly constructed mechanism is surrounded by a laminated shield to minimize the effect of external magnetic fields on the meter indication. (Courtesy of Weston Instruments, Inc.)

tions are therefore produced. The movement, however, cannot follow the rapid variations of the torque and takes up a position in which the *average* torque is balanced by the torque of the control springs. The meter deflection is therefore a function of the *mean* of the *squared current*. The scale of the electrodynamometer is usually calibrated in terms of the square root of the average current squared, and the meter therefore reads the *rms* or *effective value* of the ac.

The transfer properties of the electrodynamometer become apparent when we compare the effective value of alternating current and direct current in terms of their heating effect or transfer of power. An alternating current that produces heat in a given resistance at the same average rate as a direct current (I) has, by definition, a value of I amperes. The average rate of producing heat by a dc of I amperes in a resistance R is I^2R watts. The average rate of producing heat by an ac of i amperes during one cycle in the same resistance R is $\frac{1}{T}\int_0^T i^2R\ dt$. By definition, therefore,

$$I^2R = \frac{1}{T}\int_0^T i^2R\ dt$$

and

$$I = \sqrt{\frac{1}{T}\int_0^T i^2\ dt} = \sqrt{\text{average } i^2}$$

This current, I, is then called the root-mean-square (rms) or effective value of the alternating current and is often referred to as the *equivalent* dc value.

If the electrodynamometer is calibrated with a direct current of 1 A and a mark is placed on the scale to indicate this 1-A dc value, then that alternating current which causes the pointer to deflect to the same mark on the scale must have an rms value of 1 A. We can therefore "transfer" a reading made with dc to its corresponding ac value and have thereby established a direct connection between ac and dc. The electrodynamometer then becomes very useful as a *calibration* instrument and is often used for this purpose because of its inherent accuracy.

The electrodynamometer, however, has certain disadvantages. One of these is its high power consumption, a direct result of its construction. The current under measurement must not only pass through the movable coil, but it must also provide the field flux. To get a sufficiently strong magnetic field, a high mmf is required and the source must supply a high current and power. In spite of this high power consumption, the magnetic field is very much weaker than that of a comparable d'Arsonval movement because there is no iron in the circuit, i.e., the entire flux path consists of air. Some instruments have been designed using special laminated steel for part of the flux path, but the presence of metal introduces calibration problems caused by frequency and waveform effects. Typical values of electrodynamometer flux density are in the range of approximately 60 gauss. This compares very unfavorably with the high flux densities (1,000–4,000 gauss) of a good d'Arsonval movement. The low flux density of the electrodynamometer immediately affects the developed torque and therefore the sensitivity of the instrument is typically very low.

The addition of a series resistor converts the electrodynamometer into a voltmeter, which again can be used to measure dc and ac voltages. For reasons previously mentioned, the sensitivity of the electrodynamometer voltmeter is low, approximately 10 to 30 Ω/V (compare this to the 20 kΩ/V of a d'Arsonval meter). The reactance and resistance of the coils also increase with increasing frequency, limiting the application of the electrodynamometer voltmeter to the lower frequency ranges. It is, however, very accurate at the powerline frequencies and is therefore often used as a *secondary* standard.

The electrodynamometer movement (even unshunted) may be regarded as an ammeter, but it becomes rather difficult to design a moving coil which can carry more than approximately 100 mA. Larger current would have to be carried to the moving coil through heavy lead-in wires, which would lose their flexibility. A shunt, when used, is usually placed across the movable coil only. The fixed coils are then made of heavy wire which can carry the large total current and it is feasible to build ammeters for currents up to 20 A. Larger values of ac currents are usually measured by using a current transformer and a standard 5-A ac ammeter (Sec. 4-16).

4-11.2 Rectifier-Type Instruments

One obvious answer to the question of ac measurement is found by using a rectifier to convert ac into a unidirectional dc and then to use a dc movement to indicate the value of the rectified ac. This method is very attractive, because a dc movement generally has a higher sensitivity than either the electrodynamometer or the moving-iron instrument.

Rectifier-type instruments generally use a PMMC movement in combination with some rectifier arrangement. The rectifier element usually consists of a germanium or a silicon diode. Copper oxide and selenium rectifiers have become obsolete, because they have small inverse voltage ratings and can handle only limited amounts of current. Germanium diodes have a peak inverse voltage (PIV) on the order of 300 V and a current rating of approximately 100 mA. Low-current silicon diode rectifiers have a PIV of up to 1,000 V and a current rating on the order of 500 mA.

Rectifiers for instrument work sometimes consist of four diodes in a bridge configuration, providing full-wave rectification. Figure 4-28 shows an ac voltmeter circuit consisting of a multiplier, a bridge rectifier, and a PMMC movement.

The bridge rectifier produces a pulsating unidirectional current through the meter movement over the complete cycle of the input voltage. Because of the inertia of the moving coil, the meter will indicate a steady deflection proportional to the *average* value of the current. Since alternating currents and voltages are usually expressed in *rms* values, the meter scale is *calibrated* in terms of the rms value of a sinusoidal waveform.

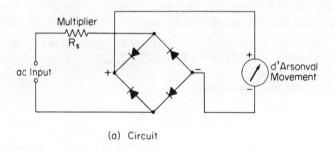

(a) Circuit

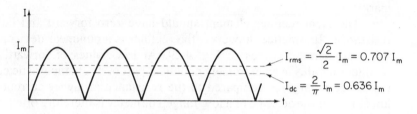

$$I_{rms} = \frac{\sqrt{2}}{2} I_m = 0.707 \, I_m$$

$$I_{dc} = \frac{2}{\pi} I_m = 0.636 \, I_m$$

(b) Rectified Current Through Meter Movement

Figure 4-28 Full-wave rectifier ac voltmeter.

EXAMPLE 4-9

An experimental ac voltmeter uses the circuit of Fig. 4-28(a), where the PMMC movement has an internal resistance of 50 Ω and requires a dc current of 1 mA for full-scale deflection. Assuming ideal diodes (zero forward resistance and infinite reverse resistance), calculate the value of the multiplier R_s to obtain full-scale meter deflection with 10 V ac (rms) applied to the input terminals.

SOLUTION   For full-wave rectification,

$$E_{dc} = \frac{2}{\pi} E_m = \frac{2\sqrt{2}}{\pi} E_{rms} = 0.9 E_{rms}$$

and

$$E_{dc} = 0.9 \times 10 \text{ V} = 9 \text{ V}$$

The total circuit resistance, neglecting the forward diode resistance, is

$$R_t = R_s + R_m = \frac{9 \text{ V}}{1 \text{ mA}} = 9 \text{ k}\Omega$$

$$R_s = 9{,}000 \text{ }\Omega - 50 \text{ }\Omega = 8{,}950 \text{ }\Omega$$

A nonsinusoidal waveform has an average value that may differ considerably from the average value of a pure sine wave (for which the meter is calibrated) and the indicated reading may be very erroneous. The *form factor* relates the average value and the rms value of time-varying voltages and currents:

$$\text{form factor} = \frac{\text{effective value of the ac wave}}{\text{average value of the ac wave}}$$

For a sinusoidal waveform:

$$\text{form factor} = \frac{E_{rms}}{E_{av}} = \frac{(\sqrt{2}/2)E_m}{(2/\pi)E_m} = 1.11 \qquad (4\text{-}27)$$

Note that the voltmeter of Example 4-9 has a scale suitable only for sinusoidal ac measurements. The form factor of Eq. (4-27) is therefore also the factor by which the actual (average) dc current is multiplied to obtain the equivalent rms scale markings.

The ideal rectifier element should have zero forward and infinite reverse resistance. In practice, however, the rectifier is a nonlinear device, indicated by the characteristic curves of Fig. 4-29. At low values of forward current, the rectifier operates in an extremely nonlinear part of its characteristic curve, and the resistance is large as compared to the resistance at higher current values. The lower part of the ac scale of a low-range voltmeter is therefore often crowded, and most manufacturers provide a separate low-voltage scale, calibrated especially for this purpose. The high resistance in the early part of the rectifier characteristics also sets a limit on the sensitivity which can be obtained in microammeters and voltmeters.

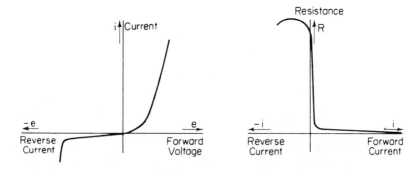

Figure 4-29 Characteristic curves of a solid-state rectifier.

The resistance of the rectifying element changes with varying temperature, one of the major drawbacks of rectifier-type ac instruments. The meter accuracy is usually satisfactory under normal operating conditions at room temperature and is generally on the order of ±5 per cent of full-scale reading for sinusoidal waveforms. At very much higher or lower temperatures, the resistance of the rectifier changes the total resistance of the measuring circuit sufficiently to cause the meter to be gravely in error. If large temperature variations are expected, the meter should be enclosed in a temperature-controlled box.

Frequency also affects the operation of the rectifier elements. The rectifier exhibits capacitive properties and tends to bypass the higher frequencies. Meter readings may be in error by as much as 0.5 per cent decrease for every 1-kHz rise in frequency.

4-11.3 Typical Multimeter Circuits

General rectifier-type ac voltmeters often use the arrangement shown in Fig. 4-30. Two diodes are used in this circuit, forming a full-wave rectifier with the movement so connected that it receives only half of the rectified current. Diode D_1 conducts during the positive half-cycle of the input waveform and causes the meter to deflect according to the average value of this half-cycle. The meter movement is shunted by a resistance R_{sh}, in order to draw more current through the diode D_1 and move its operating point into the linear portion of the characteristic curve. In the absence of diode D_2, the negative half-cycle of the input voltage would apply a reverse voltage across diode D_1, causing a small leakage current in

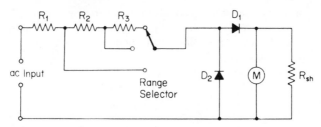

Figure 4-30 Typical ac voltmeter section of a commercial multimeter.

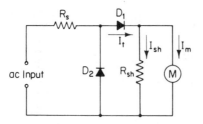

Figure 4-31 Computation of the multiplier resistor and the ac voltmeter sensitivity.

the reverse direction. The average value of the complete cycle would therefore be lower than it should be for half-wave rectification. Diode D_2 deals with this problem. On the negative half-cycle, D_2 conducts heavily, and the current through the measuring circuit, which is now in the opposite direction, bypasses the meter movement.

The commercial multimeter often uses the same scale markings for both its dc and ac voltage ranges. Since the dc component of a sine wave for half-wave rectification equals 0.45 times the rms value, a problem arises immediately. In order to obtain the same deflection on corresponding dc and ac voltage ranges, the multiplier for the ac range must be lowered proportionately. The circuit of Fig. 4-31 illustrates a solution to the problem and is discussed in some detail in Example 4-10.

EXAMPLE 4-10

A meter movement has an internal resistance of 100 Ω and requires 1 mA dc for full-scale deflection. Shunting resistor R_{sh}, placed across the movement, has a value of 100 Ω. Diodes D_1 and D_2 have an average forward resistance of 400 Ω each and are assumed to have infinite resistance in the reverse direction. For a 10-V ac range, calculate (a) the value of multiplier R_s; (b) the voltmeter sensitivity on the ac range.

SOLUTION (a) Since R_m and R_{sh} are both 100 Ω, the total current the source must supply for full-scale deflection is $I_t = 2$ mA. For half-wave rectification the equivalent dc value of the rectified ac voltage will be

$$E_{dc} = 0.45E_{rms} = 0.45 \times 10 \text{ V} = 4.5 \text{ V}$$

The total resistance of the instrument circuit then is

$$R_t = \frac{E_{dc}}{I_t} = \frac{4.5 \text{ V}}{2 \text{ mA}} = 2,250 \ \Omega$$

This total resistance is made up of several parts. Since we are interested only in the resistance of the circuit during the half-cycle that the movement receives current, we can eliminate the infinite resistance of reverse-biased diode D_2 from the circuit. Therefore

$$R_t = R_s + R_{D_1} + \frac{R_m R_{sh}}{R_m + R_{sh}}$$

and

$$R_t = R_s + 400 + \frac{100 \times 100}{200} = R_s + 450 \ \Omega$$

The value of the multiplier therefore is

$$R_s = 2,250 - 450 = 1,800 \ \Omega$$

(b) The sensitivity of the voltmeter on this 10-V ac range is

$$S = \frac{2,250 \ \Omega}{10 \ \text{V}} = 225 \ \Omega/\text{V}$$

The same movement, used in a dc voltmeter, would have given a sensitivity figure of $1,000 \ \Omega/\text{V}$.

Section 4-10 dealt with the dc circuitry of a typical multimeter, using the simplified circuit diagram of Fig. 4-20. The circuit for measuring ac volts (subtracted from Fig. 4-20) is reproduced in Fig. 4-32. Resistances R_9, R_{13}, R_7, and R_6 form a chain of multipliers for the voltage ranges of 1,000 V, 250 V, 50 V, and 10 V, respectively, and their values are indicated in the diagram of Fig. 4-32. On the 2.5-V ac range, resistor R_{23} acts as the multiplier and corresponds to the multiplier R_s of Example 4-10 shown in Fig. 4-31. Resistor R_{24} is the meter shunt and again acts to improve the rectifier operation. Both values are unspecified in the diagram and are factory selected. A little thought, however, will convince us that the shunt resistance could be 2,000 Ω, equal to the meter resistance. If the average forward resistance of the rectifier elements is 500 Ω (a reasonable assumption), then resistance R_{23} must have a value of 1,000 Ω. This follows because the meter sensitivity on the ac ranges is given as 1,000 Ω/V; on the 2.5-V ac range, the circuit must therefore have a total resistance of 2,500 Ω. This value is made up of the sum of R_{23}, the diode forward resistance, and the combination of movement and shunt resistance, as shown in Example 4-10.

4-12 THERMOINSTRUMENTS

Figure 4-33 shows a combination of a thermocouple and a PMMC movement that can be used to measure both ac and dc. This combination is called a *thermocouple instrument,* since its operation is based on the action of the thermocouple ele-

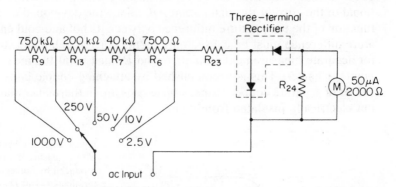

Figure 4-32 Multirange ac voltmeter circuit of the Simpson Model 260 multimeter. (Courtesy of Simpson Electric Company.)

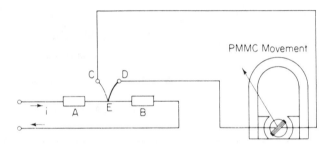

Figure 4-33 Schematic representation of a basic thermocouple instrument using thermocouple CDE and a PMMC movement.

ment. When two dissimilar metals are mutually in contact, a voltage is generated at the junction of the two dissimilar metals. This voltage rises in proportion to the temperature of the junction. In Fig. 4-33, *CE* and *DE* represent the two dissimilar metals, joined at point *E,* and are drawn as a light and a heavy line, to indicate dissimilarity. The potential difference between *C* and *D* depends on the temperature of the so-called *cold junction,* *E*. A rise in temperature causes an increase in the voltage and this is used to advantage in the thermocouple. Heating element *AB,* which is in mechanical contact with the junction of the two metals at point *E,* forms part of the circuit in which the current is to be measured. *AEB* is called the *hot junction.* Heat energy generated by the current in the heating element raises the temperature of the cold junction and causes an increase in the voltage generated across terminals *C* and *D*. This potential difference causes a dc current through the PMMC-indicating instrument. The heat generated by the current is directly proportional to the current squared (I^2R), and the temperature rise (and hence the generated dc voltage) is proportional to the square of the rms current. The deflection of the indicating instrument will therefore follow a square-law relationship, causing crowding at the lower end of the scale and spreading at the high end. The arrangement of Fig. 4-33 does not provide compensation for ambient temperature changes.

The *compensated thermoelement,* shown schematically in Fig. 4-34, produces a thermoelectric voltage in thermocouple *CED,* which is directly proportional to the current through circuit *AB*. Since the developed couple voltage is a function of the temperature difference between its hot and cold ends, this temperature difference must be caused only by the current being measured. Therefore, for accurate measurements, points *C* and *D* must be at the mean temperature of points *A* and *B*. This is accomplished by attaching couple ends *C* and *D* to the center of separate copper strips, whose ends are in thermal contact with *A* and *B*, but electrically insulated from them.

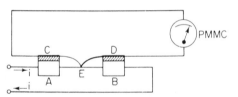

Figure 4-34 Compensated thermocouple to measure the thermo voltage produced by current *i* alone. Couple terminals *C* and *D* are in thermal contact with heater terminals *A* and *B*, but are electrically insulated from them.

Electromechanical Indicating Instruments   Chap. 4

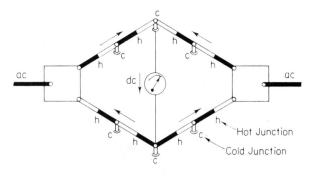

Figure 4-35 Bridge-type thermocouple instrument.

Self-contained thermoelectric instruments of the compensated type are available in the 0.5–20-A range. Higher current ranges are available, but in this case the heating element is external to the indicator. Thermoelements used for current ranges over 60 A are generally provided with air cooling fins.

Current measurements in the lower ranges, from approximately 0.1–0.75 A, use a *bridge-type thermoelement,* shown schematically in Fig. 4-35. This arrangement does not use a separate heater: the current to be measured passes directly through the thermoelements and raises their temperature in proportion to I^2R. The cold junctions (marked c) are at the pins which are embedded in the insulating frame, and the hot junctions (marked h) are at splices midway between the pins. The couples are arranged as shown in Fig. 4-35, and the resultant thermal voltage generates a dc potential difference across the indicating instrument. Since the bridge arms have equal resistances, the ac voltage across the meter is 0 V, and no ac passes through the meter. The use of several thermocouples in series provides a greater output voltage and deflection than is possible with a single element, resulting in an instrument with increased sensitivity.

Thermoinstruments may be converted into voltmeters using low-current thermocouples and suitable series resistors. Thermocouple voltmeters are available in ranges of up to 500 V and sensitivities of approximately 100 to 500 Ω/V.

A major advantage of a thermocouple instrument is that its accuracy can be as high as 1 per cent, up to frequencies of approximately 50 MHz. For this reason, it is classified as an *RF instrument.* Above 50 MHz, the skin effect tends to force the current to the outer surface of the conductor, increasing the effective resistance of the heating wire and reducing instrument accuracy. For small currents (up to 3 A), the heating wire is solid and very thin. Above 3 A the heating element is made from tubing to reduce the errors due to skin effect.

4-13 ELECTRODYNAMOMETERS IN POWER MEASUREMENTS

The electrodynamometer movement is used extensively in measuring power. It may be used to indicate *both* dc and ac power for any waveform of voltage and current and it is not restricted to sinusoidal waveforms. As described in

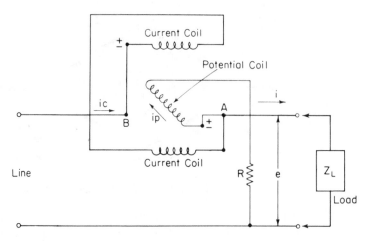

Figure 4-36 Diagram of an electrodynamometer wattmeter connected to measure the power of a single-phase load.

Sec. 4-11.1, the electrodynamometer used as a voltmeter or an ammeter has the fixed coils and the movable coil connected in series, thereby reacting to the effect of the current squared. When used as a *single-phase power meter,* the coils are connected in a different arrangement (see Fig. 4-36).

The fixed coils, or *field coils,* shown here as two separate elements, are connected in series and carry the total line current (i_c). The movable coil, located in the magnetic field of the fixed coils, is connected in series with a current-limiting resistor across the power line and carries a small current (i_p). The instantaneous value of the current in the movable coil is $i_p = e/R_p$, where e is the instantaneous voltage across the power line, and R_p is the total resistance of the movable coil and its series resistor. The deflection of the movable coil is proportional to the product of these two currents, i_c and i_p, and we can write for the average deflection over one period:

$$\theta_{av} = K \frac{1}{T} \int_0^T i_c i_p \, dt \qquad (4\text{-}28)$$

where θ_{av} = average angular deflection of the coil

K = instrument constant

i_c = instantaneous current in the field coils

i_p = instantaneous current in the potential coil

Assuming for the moment that i_c is equal to the load current, i (actually, $i_c = i_p + i$), and using the value for $i_p = e/R_p$, we see that Eq. (4-28) reduces to

$$\theta_{av} = K \frac{1}{T} \int_0^T i \frac{e}{R_p} \, dt = K_2 \frac{1}{T} \int_0^T ei \, dt \qquad (4\text{-}29)$$

By definition, the average power in a circuit is

$$P_{av} = \frac{1}{T} \int_0^T ei \, dt \qquad (4\text{-}30)$$

which indicates that the electrodynamometer movement, connected in the configuration of Fig. 4-36, has a deflection proportional to the average power. If e and i are sinusoidally varying quantities of the form $e = E_m \sin \omega t$ and $i = I_m \sin (\omega t \pm \theta)$, Eq. (4-29) reduces to

$$\theta_{av} = K_3 EI \cos \theta \qquad (4\text{-}31)$$

where E and I represent the rms values of the voltage and the current, and θ represents the phase angle between voltage and current. Equations (4-29) and (4-30) show that the electrodynamometer indicates the average power delivered to the load.

Wattmeters have one voltage terminal and one current terminal marked "$\pm$." When the marked current terminal is connected to the incoming line, and the marked voltage terminal is connected to the line side in which the current coil is connected, the meter will always read up-scale when power is connected to the load. If for any reason (as in the two-wattmeter method of measuring three-phase power), the meter should read backward, the *current* connections (not the voltage connections) should be reversed.

The electrodynamometer wattmeter consumes some power for maintenance of its magnetic field, but this is usually so small, compared to the load power, that it may be neglected. If a correct reading of the load power is required, the current coil should carry exactly the load current, and the potential coil should be connected across the load terminals. With the potential coil connected to point A, as in Fig. 4-36, the load voltage is properly metered, but the current through the field coils is greater by the amount i_p. The wattmeter therefore reads high by the amount of additional power loss in the potential circuit. If, however, the potential coil is connected to point B in Fig. 4-38, the field coils meter the correct load current, but the voltage across the potential coil is higher by the amount of the drop across the field coils. The wattmeter will again read high, but now by the amount of the I^2R losses in the field windings. Choice of the correct connection depends on the situation. Generally, connection of the potential coil at point A is preferred for high-current, low-voltage loads; connection at B is preferred for low-current, high-voltage loads.

The difficulty in placing the connection of the potential coil is overcome in the *compensated* wattmeter, shown schematically in Fig. 4-37. The current coil consists of two windings, each winding having the same number of turns. One winding uses heavy wire that carries the load current plus the current for the potential coil. The other winding uses thin wire and carries only the current to the voltage coil. This current, however, is in a direction opposite to the current in the heavy winding, causing a flux that opposes the main flux. The effect of i_p is therefore canceled out, and the wattmeter indicates the correct power.

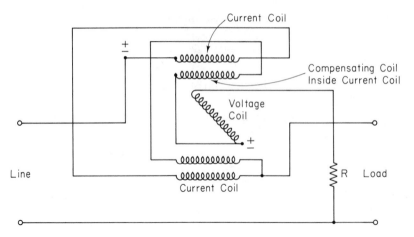

Figure 4-37 Diagram of a compensated wattmeter in which the effect of the current in the potential coil is canceled by the current in the compensating winding.

4-14 WATTHOUR METER

The watthour meter is not often found in a laboratory situation but it is widely used for the commercial measurement of electrical energy. In fact, it is evident wherever a power company supplies the industrial or domestic consumer with electrical energy. Figure 4-38 shows the elements of a single-phase watthour meter in schematic form.

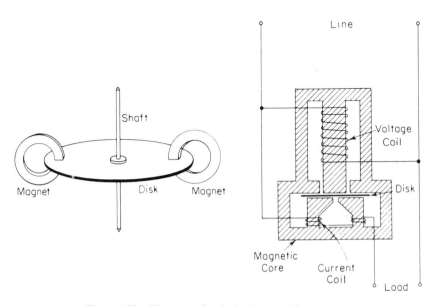

Figure 4-38 Elements of a single-phase watthour meter.

Electromechanical Indicating Instruments   Chap. 4

The current coil is connected in series with the line, and the voltage coil is connected across the line. Both coils are wound on a metal frame of special design, providing two magnetic circuits. A light aluminum disk is suspended in the air gap of the current-coil field, which causes eddy currents to flow in the disk. The reaction of the eddy currents and the field of the voltage coil creates a torque (motor action) on the disk, causing it to rotate. The developed torque is proportional to the fieldstrength of the voltage coil and the eddy currents in the disk which are in turn a function of the fieldstrength of the current coil. The number of rotations of the disk is therefore proportional to the energy consumed by the load in a certain time interval, and is measured in terms of kilowatthours (kWh). The shaft that supports the aluminum disk is connected by a gear arrangement to the clock mechanism on the front of the meter, providing a decimally calibrated readout of the number of kWh.

Damping of the disk is provided by two small permanent magnets located opposite each other at the rim of the disk. Whenever the disk rotates, the permanent magnets induce eddy currents in it. These eddy currents react with the magnetic fields of the small permanent magnets, damping the motion of the disk. A typical single-phase watthour meter is shown in Fig. 4-39.

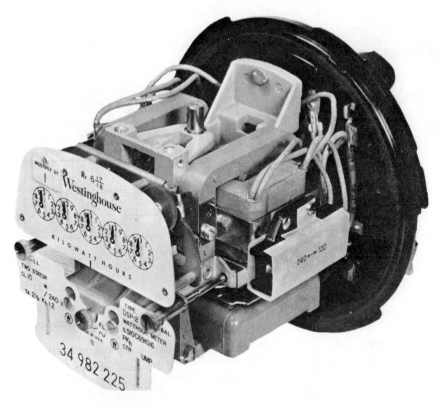

Figure 4-39 Watthour meter for industrial or domestic application. (Courtesy of Westinghouse Electric Corporation.)

Calibration of the watthour meter is performed under conditions of full rated load and 10 per cent of rated load. At full load, the calibration consists of adjustment of the position of the small permanent magnets until the meter reads correctly. At very light loads, the voltage component of the field produces a torque that is not directly proportional to the load. Compensation for the error is provided by inserting a shading coil or plate over a portion of the voltage coil, with the meter operating at 10 per cent of rated load. Calibration of the meter at these two positions usually provides satisfactory readings at all other loads.

The *floating-shaft* watthour meter uses a unique design to suspend the disk. The rotating shaft has a small magnet at each end. The upper magnet of the shaft is attracted to a magnet in the upper bearing, and the lower magnet of the shaft is attracted to a magnet in the lower bearing. The movement thus floats without touching either bearing surface, and the only contact with the movement is that of the gear connecting the shaft with the gear train.

Measurements of energy in three-phase systems are performed with polyphase watthour meters. Each phase of the watthour meter has its own magnetic circuit and its own disk, but all the disks are mounted on a common shaft. The developed torque on each disk is mechanically summed and the total number of revolutions per minute of the shaft is proportional to the total three-phase energy consumed.

4-15 POWER-FACTOR METERS

The power factor, by definition, is the cosine of the phase angle between voltage and current, and power-factor measurements usually involve the determination of this phase angle. This is demonstrated in the operation of the *crossed-coil power-factor meter*. The instrument is basically an electrodynamometer movement, where the moving element consists of two coils, mounted on the same shaft but at right angles to each other. The moving coils rotate in the magnetic field provided by the field coil that carries the line current.

The connections for this meter in a single-phase circuit are shown in the circuit diagram of Fig. 4-40. The field coil is connected as usual in series with the line and carries the line current. One coil of the movable element is connected in series with a *resistor* across the lines and receives its current from the applied potential difference. The second coil of the movable element is connected in series with an *inductor* across the lines. Since no control springs are used, the balance position of the movable element depends on the resulting torque developed by the two crossed coils. When the movable element is in a balanced position, the contribution to the total torque by each element must be equal but of opposite sign. The developed torque in each coil is a function of the current through the coil and therefore depends on the impedance of that coil circuit. The torque is also proportional to the mutual inductance between each part of the crossed coil and the stationary field coil. This mutual inductance depends on the angular position of the crossed-coil elements with respect to the position of the stationary field coil. When the movable element is at balance, it can be shown that

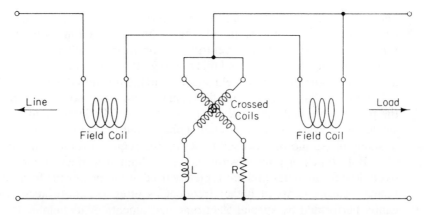

Figure 4-40 Connections for a single-phase crossed-coil power-factor meter.

its angular displacement is a function of the phase angle between line current (field coil) and line voltage (crossed coils). The indication of the pointer, which is connected to the movable element, is calibrated directly in terms of the phase angle or power factor.

The *polarized-vane power-factor meter* is shown in the construction sketch of Fig. 4-41. This instrument is used primarily in three-phase power systems, because its operating principle depends on the application of three-phase voltage. The outside coil is the potential coil, which is connected to the three phase

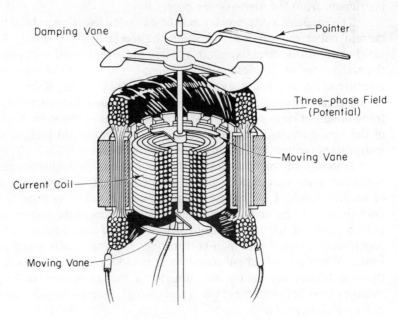

Figure 4-41 Polarized-vane power-factor meter.

lines of the system. The application of three-phase voltage to the potential coil causes it to act like the stator of a three-phase induction motor in setting up a *rotating magnetic flux*. The central coil, or current coil, is connected in series with one of the phase lines, and this *polarizes* the iron vanes. The polarized vanes move in a rotating magnetic field and take up the position that the rotating field has at the instant that the polarizing flux is maximum. This position is an indication of the phase angle and therefore the power factor. The instrument may be used in single-phase systems, provided that a phase-splitting network (similar to that used in single-phase motors) is used to set up the required rotating magnetic field.

Both types of power-factor meter are limited to measurement at comparatively low frequencies and are typically used at the powerline frequency (60 Hz). Phase measurements at higher frequencies often are more accurately and elegantly performed by special electronic instruments or techniques.

4-16 INSTRUMENT TRANSFORMERS

Instrument transformers are used to measure ac at generating stations, transformer stations, and at transmission lines, in conjunction with ac measuring instruments (voltmeters, ammeters, wattmeters, VARmeters, etc.). Instrument transformers are classified according to their use and are referred to as *current transformers* (CT) and *potential transformers* (PT).

Instrument transformers perform two important functions: They serve to *extend the range* of the ac measuring instrument, much as the shunt or the multiplier extends the range of a dc meter; they also serve to *isolate* the measuring instrument from the high-voltage power line.

The *range* of a dc ammeter may be extended by using a shunt that divides the current under measurement between the meter and the shunt. This method is satisfactory for dc circuits, but in ac circuits current division depends not only on the resistances of the meter and the shunt but also on their reactances. Since ac measurements are made over a wide frequency range, it becomes difficult to obtain great accuracy. A CT provides the required range extension through its transformation ratio and in addition produces almost the same reading regardless of the meter constants (reactance and resistance) or, in fact, of the number of instruments (within limits) connected in the circuit.

Isolation of the measuring instrument from the high-voltage power line is important when we consider that ac power systems frequently operate at voltages of several hundred kilovolts. It would be impractical to bring the high-voltage lines directly to an instrument panel in order to measure voltage or current, not only because of the safety hazards involved but also because of the insulation problems connected with high-voltage lines running closely together in a confined space. When an instrument transformer is used, only the low-voltage wires from the transformer secondary are brought to the instrument panel and only low voltages exist between these wires and ground, thereby minimizing safety hazards and insulation problems.

Many textbooks develop in detail the theory underlying the operation of

Figure 4-42 High-voltage potential transformer.
(Courtesy of Westinghouse Electric Corporation.)

transformers. Here these instrument transformers are merely described and their use in measurement situations is shown.*

Figure 4-42 shows a *potential transformer*; Fig. 4-43 shows a *current transformer*. The *potential transformer* (PT) is used to transform the high voltage of a power line to a lower value suitable for direct connection to an ac voltmeter or the potential coil of an ac wattmeter. The usual secondary transformer voltage is 120 V. Primary voltages are standardized to accommodate the usual transmission line voltages which include 2,400 V, 4,160 V, 7,200 V, 13.8 kV, 44 kV, 66 kV, and 220 kV. The PT is rated to deliver a certain power to the secondary load or *burden*. Different load capacities are available to suit individual applications; a general capacity is 200 VA at a frequency of 60 Hz.

The PT must satisfy certain design requirements that include accuracy of the turns ratio, small leakage reactance, small magnetizing current, and minimal voltage drop. Furthermore, since we may be working with very high primary voltages, the insulation between the primary and secondary windings must be able to

* For fuller treatment of ac machines and circuits, consult textbooks like the following: Michael Liwshitz-Garik and Clyde C. Whipple, *AC Machines,* 2nd ed. (Princeton, N.J.: D. Van Nostrand Company, Inc., 1961), chaps. 2–5; Russell M. Kerchner and George F. Corcoran, *Alternating Current Circuits,* 4th ed. (New York: John Wiley & Sons, Inc., 1961), pp. 291–317.

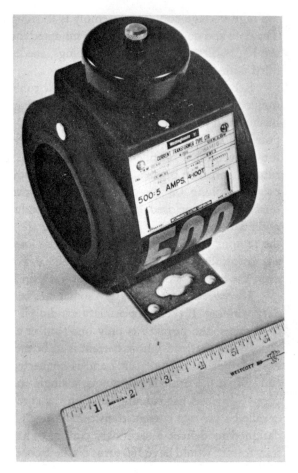

Figure 4-43 Current transformer. (Courtesy of Westinghouse Electric Corporation.)

withstand large potential differences, and the dielectric requirements are very high. In the usual case, the high-voltage coil is of a circular pancake construction, shielded to avoid localized dielectric stresses. The low-voltage coil or coils are wound on a paper form and assembled inside the high-voltage coil. The assembly is thoroughly dried and oil impregnated. The core and coil assembly is then mounted inside a steel case, which supports the high-voltage terminals or porcelain bushings. The case is then filled with an insulating oil.

Developments in the synthetic rubber industry have introduced the molded rubber potential transformer, replacing the insulating oil and porcelain bushings in some applications. Figure 4-42 shows a rubber-molded 25-kV potential transformer suitable for outdoor use. This unit is less expensive than the conventional oil-filled PT, and since the bushings are made of molded rubber, porcelain breakage is eliminated. A white polarity dot is placed on the proper bushing on the front of the transformer. Two stud-type secondary terminals are enclosed in a removable conduit box. The power rating of a potential transformer is based on considerations other than load capacity, for the reasons previously outlined. A typical load rating is 200 VA at 60 Hz for a transformer having a ratio of 2,400/120 V. For most metering purposes, however, the burden will be significantly less than 200 VA.

The *current transformer* (CT) sometimes has a primary and always has a secondary winding. If there is a primary winding, it has a small number of turns. In most cases, the primary is only one turn or a single conductor connected in series with the load whose current is to be measured. The secondary winding has a larger number of turns and is connected to a current meter or a relay coil. Often the primary winding is a single conductor in the form of a heavy copper or brass bar running through the core of the transformer. Such a CT is called a *bar-type* current transformer. The CT secondary winding is usually designed to deliver a secondary current of 5 A. An 800/5-A bar-type current transformer would have 160 turns on the secondary coil.

The primary winding of the current transformer is connected directly in the load circuit. When the secondary winding is open-circuited, the voltage developed across the open terminals may be very high (because of the step-up ratio) and could easily break down the insulation between the secondary windings. The secondary winding of a current transformer should therefore always be short-circuited, or connected to a meter or relay coil. A current transformer should *never* have its secondary open while the primary is carrying current; it should *always* be closed through a current meter, relay coil, wattmeter current coil, or simply a short. Failure to observe this precaution may cause serious damage to either equipment or operating personnel.

The current transformer shown in Fig. 4-43 consists of a core with the secondary winding encased in molded-rubber insulation. The window in the core allows for the insertion of one or more turns of the current-carrying high-voltage conductor. A single conductor constitutes a one-turn primary winding. The nominal ratio of the transformer is given on the nameplate; this is not the turns ratio (since more than one turn can be used as the primary) but only indicates that a primary current of 500 A will cause a secondary current of 5 A when the secondary coil is connected to a 5-A ammeter. Within practical limits, the current in the secondary winding is determined by the primary excitation current and not by the secondary circuit impedance. Since the primary current is determined by the load in the ac system, the secondary current is related to the primary current by approximately the inverse of the turns-ratio. This is true within rather wide limits of the nature of the secondary burden.

Figure 4-44 indicates the use of instrument transformers in a typical measurement application. This diagram illustrates the connection of instrument transformers in a three-wire three-phase circuit, including two wattmeters, two voltmeters, and two ammeters. The potential transformers are connected across phase lines *A* and *B,* and phase lines *C* and *B*; the current transformers are in phase lines *A* and *D*. The secondary windings of the potential transformers are connected to the voltmeter coils and the potential coils of the wattmeters; the current transformer secondaries feed the ammeters and the current coils of the wattmeters.

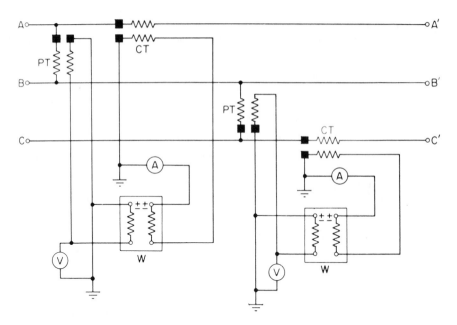

Figure 4-44 Instrument transformers in a three-phase measurement application. Polarity markings of the potential and current transformers are indicated by black squares.

The polarity markings on the transformers, indicated by a dot at the transformer leads, aid in making the correct polarity connections to the measuring instruments. At any given instant of the ac cycle, the dot-marked terminals have the *same* polarity and the marked wattmeter terminals must be connected to these transformer leads as shown.

REFERENCES

4-1. Bartholomew, Davis, *Electrical Measurements and Instrumentation,* chap. 5. Boston: Allyn and Bacon, Inc., 1963.

4-2. Geczy, Steven, *Basic Electrical Measurements.* Englewood Cliffs, N.J.: Prentice-Hall, Inc., 1984.

4-3. Jackson, Herbert W., *Introduction to Electric Circuits,* 5th ed., chap. 19. Englewood Cliffs, N.J.: Prentice-Hall, Inc., 1981.

4-4. Prensky, Sol D., and Castellucis, Richard L., *Electronic Instrumentation,* 3rd ed., chaps. 2 and 3. Englewood Cliffs, N.J.: Prentice-Hall, Inc., 1982.

4-5. Stout, Melville B., *Basic Electrical Measurements,* 2nd ed., chap. 17. Englewood Cliffs, N.J.: Prentice-Hall, Inc., 1960.

PROBLEMS

4-1. Determine the resistor value required to use a 0–1-mA meter with an internal resistance of 125 Ω for a 0–1-V meter.

4-2. What value of shunt resistance is required for using a 50-μA meter movement, with an internal resistance of 250 Ω, for measuring 0–500 mA?

4-3. What series resistance must be used to extend the 0–200-V range of a 20,000-Ω/V meter to 0–2000 V? What power rating must this resistor have?

4-4. What will a 5,000-Ω/V meter read on a 0–5-V scale when connected to the circuit of Fig. P4-4?

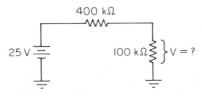

Figure P4-4

4-5. Draw the schematic, including values, for an Ayrton shunt for a meter movement having a full-scale deflection of 1 mA and an internal resistance of 500 Ω to cover the current ranges of 10, 50, 100, and 500 mA.

4-6. Many electronic voltage measuring instruments have a fixed input resistance of 1 MΩ. Which settings of the range switch of the multimeter shown in Figs. 4-21 and 4-22 would have a higher input resistance than the typical electronic instrument for dc measurements?

4-7. The resistance of a 50-kΩ resistor is measured using the multimeter shown in Figs. 4-21, 4-22, and 4-25. **(a)** How much power is dissipated in the resistor if the $R \times 10,000$ range is used? **(b)** How much power is dissipated in the resistor if the $R \times 100$ range is used? Assume that the zero control is set to its midpoint.

4-8. A series-type ohmmeter, designed to operate with a 6-V battery, has a circuit diagram as shown in Fig. 4-19. The meter movement has an internal resistance of 2,000 Ω and requires a current of 100 μA for full-scale deflection. The value of R_1 is 49 kΩ. **(a)** Assuming the battery voltage has fallen to 5.9 V, calculate the value of R_2 required to zero the meter. **(b)** Under the conditions mentioned in part (a), an unknown resistor is connected to the meter causing a 60 per cent meter deflection. Calculate the value of the unknown resistance.

4-9. How low must the battery voltage of the 1.5-V cell in the multimeter ohmmeter section shown in Fig. 4-25(a) fall before it is impossible to zero the meter?

4-10. What is a transfer instrument? Why is an electrodynamometer a transfer instrument?

4-11. Why is sensitivity (ohms per volt) of the ac scales of a multimeter less than the dc section?

4-12. What is meant by a waveform error? Which ac meters are most likely to be affected by this form of error?

4-13. What are the advantages of a thermocouple meter?

4-14. What is the midscale point of a 10-A full-scale thermocouple meter?

4-15. The circuit diagram of Fig. 4-30 shows a full-wave rectifier ac voltmeter. The meter movement has an internal resistance of 250 Ω and requires 1 mA for full deflection. The diodes each have a forward resistance of 50 Ω and infinite reverse resistance. Calculate **(a)** the series resistance required for full-scale meter deflection when 25 V rms is applied to the meter terminals; **(b)** the ohms-per-volt rating of this ac voltmeter.

4-16. Calculate the indication of the meter in Problem 4-15 when a triangular waveform with a peak value of 20 V is applied to the meter terminals.

4-17. If an electrodynamometer is used to measure power with a full-scale reading of 100 W, what is the one-quarter scale reading?

5

Bridge Measurements

5-1 INTRODUCTION

Precision measurments of component values have been made for many years using various forms of bridges. The simplest form of bridge is for the purpose of measuring resistance and is called the Wheatstone bridge. There are variations of the Wheatstone bridge for measuring very high and very low resistances. There is an entire group of ac bridges for measuring inductance, capacitance, admittance, conductance, and any of the impedance parameters.

General-purpose bridges are hardly used any more. Some specialized measurements, such as impedance at high frequencies are still made with a bridge.

The bridge circuit still forms the backbone of some measurements and for the interfacing of transducers. As an example, there are fully automatic bridges that electronically null a bridge to make precision component measurements. For this reason, a chapter is devoted to bridge measurements. Also, in this chapter, the concept of guarded measurements and three-terminal resistance measurement is covered.

5-2 WHEATSTONE BRIDGE

5-2.1 Basic Operation

Figure 5-1 shows the schematic of a Wheatstone bridge. The bridge has four resistive arms, together with a source of emf (a battery) and a null detector, usually a galvanometer or other sensitive current meter. The current through the galvanometer depends on the potential difference between points c and d. The bridge is said to be *balanced* when the potential difference across the galvanometer is 0 V so that there is no current through the galvanometer. This condition occurs when the voltage from point c to point a equals the voltage from point d to point a; or by referring to the other battery terminal, when the voltage from point c to point b equals the voltage from point d to point b. Hence the bridge is balanced when

$$I_1 R_1 = I_2 R_2 \tag{5-1}$$

If the galvanometer current is zero, the following conditions also exist:

$$I_1 = I_3 = \frac{E}{R_1 + R_3} \tag{5-2}$$

and

$$I_2 = I_4 = \frac{E}{R_2 + R_4} \tag{5-3}$$

Combining Eqs. (5-1), (5-2), and (5-3) and simplifying, we obtain

$$\frac{R_1}{R_1 + R_3} = \frac{R_2}{R_2 + R_4} \tag{5-4}$$

from which

$$R_1 R_4 = R_2 R_3 \tag{5-5}$$

Equation (5-5) is the well-known expression for balance of the Wheatstone bridge. If three of the resistances have known values, the fourth may be determined from Eq. (5-5). Hence, if R_4 is the unknown resistor, its resistance R_x can

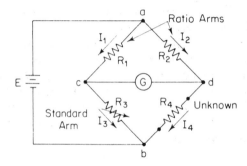

Figure 5-1 Wheatstone bridge used for the precision measurement of resistances ranging from fractions of an ohm to several megohms. The ratio control switches the ratio arms in decade steps. The remaining four step switches set the resistance of the standard arm.

Bridge Measurements   Chap. 5

be expressed in terms of the remaining resistors as follows:

$$R_x = R_3 \frac{R_2}{R_1} \qquad (5\text{-}6)$$

Resistor R_3 is called the *standard arm* of the bridge, and resistors R_2 and R_1 are called the *ratio arms*.

The measurement of the unknown resistance R_x is independent of the characteristics or the calibration of the null-detecting galvanometer, provided that the null detector has sufficient sensitivity to indicate the balance position of the bridge with the required degree of precision.

5-2.2 Measurement Errors

The Wheatstone bridge is widely used for precision measurement of resistance from approximately 1 Ω to the low megohm range. The main source of measurement error is found in the limiting errors of the three known resistors. Other errors may include the following:

(a) Insufficient sensitivity of the null detector. This problem is discussed more fully in Sec. 5-2.3.

(b) Changes in resistance of the bridge arms due to the heating effect of the current through the resistors. Heating effect (I^2R) of the bridge arm currents may change the resistance of the resistor in question. The rise in temperature not only affects the resistance during the actual measurement, but *excessive* currents may cause a permanent change in resistance values. This may not be discovered in time and subsequent measurements could well be erroneous. The power dissipation in the bridge arms must therefore be computed in advance, particularly when low-resistance values are to be measured, and the current must be limited to a safe value.

(c) Thermal emfs in the bridge circuit or the galvanometer circuit can also cause problems when low-value resistors are being measured. To prevent thermal emfs, the more sensitive galvanometers sometimes have copper coils and copper suspension systems to avoid having dissimilar metals in contact with one another and generating thermal emfs.

(d) Errors due to the resistance of leads and contacts exterior to the actual bridge circuit play a role in the measurement of very low-resistance values. These errors may be reduced by using a Kelvin bridge (see Sec. 5-3).

5-2.3 Thévenin Equivalent Circuit

To determine whether or not the galvanometer has the required *sensitivity* to detect an unbalance condition, it is necessary to calculate the galvanometer current. Different galvanometers not only may require different currents per unit deflection (current sensitivity), but they also may have a different internal resistance. It is impossible to say, without prior computation, which galvanometer will make the bridge circuit more sensitive to an unbalance condition. This sensitivity can be calculated by "solving" the bridge circuit for a *small* unbalance. The solution is approached by converting the Wheatstone bridge of Fig. 5-1 to its Thévenin equivalent.

Since we are interested in the current through the galvanometer, the Thévenin equivalent circuit is determined by looking into galvanometer terminals *c* and *d* in Fig. 5.1 Two steps must be taken to find the Thévenin equivalent: The first step involves finding the *equivalent voltage* appearing at terminals *c* and *d* when the galvanometer is removed from the circuit. The second step involves finding the *equivalent resistance* looking into terminals *c* and *d*, with the battery replaced by its internal resistance. For convenience, the circuit of Fig. 5-1(b) is redrawn in Fig. 5-2(a).

The Thévenin, or open-circuit, voltage is found by referring to Fig. 5-2(a), and we can write

$$E_{cd} = E_{ac} - E_{ad} = I_1R_1 - I_2R_2$$

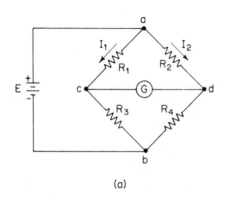

(a)

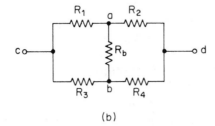

(b)

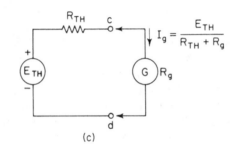

(c)

Figure 5-2 Application of Thévenin's theorem to the Wheatstone bridge. (a) Wheatstone bridge configuration. (b) Thévenin resistance looking into terminals *c* and *d*. (c) Complete Thévenin circuit, with the galvanometer connected to terminals *c* and *d*.

where

$$I_1 = \frac{E}{R_1 + R_3} \quad \text{and} \quad I_2 = \frac{E}{R_2 + R_4}$$

Therefore

$$E_{cd} = E \left(\frac{R_1}{R_1 + R_3} - \frac{R_2}{R_2 + R_4} \right) \tag{5-7}$$

This is the voltage of the Thévenin generator.

The resistance of the Thévenin equivalent circuit is found by looking back into terminals c and d and replacing the battery by its internal resistance. The circuit of Fig. 5-2(b) represents the Thévenin resistance. Notice that the internal resistance, R_b, of the battery has been included in Fig. 5-2(b). Converting this circuit into a more convenient form requires use of the *delta-wye transformation* theorem. Readers interested in this approach should consult texts on circuit analysis where this theorem is derived and applied.* In most cases, however, the extremely low internal resistance of the battery can be neglected and this simplifies the reduction of Fig. 5-2(a) to its Thévenin equivalent considerably.

Referring to Fig. 5-2(b), we see that a short circuit exists between points a and b when the internal resistance of the battery is assumed to be 0 Ω. The Thévenin resistance, looking into terminals c and d, then becomes

$$R_{\text{TH}} = \frac{R_1 R_3}{R_1 + R_3} + \frac{R_2 R_4}{R_2 + R_4} \tag{5-8}$$

The Thévenin equivalent of the Wheatstone bridge circuit therefore reduces to a Thévenin generator with an emf described by Eq. (5-7) and an internal resistance given by Eq. (5-8). This is shown in the circuit of Fig. 5-2(c).

When the null detector is now connected to the output terminals of the Thévenin equivalent circuit, the galvanometer current is found to be

$$I_g = \frac{E_{\text{TH}}}{R_{\text{TH}} + R_g} \tag{5-9}$$

where I_g is the galvanometer current and R_g its resistance.

EXAMPLE 5-1

Figure 5-3(a) shows the schematic diagram of a Wheatstone bridge with values of the bridge elements as shown. The battery voltage is 5 V and its internal resistance negligible. The galvanometer has a current sensitivity of 10 mm/μA and an internal resistance of 100 Ω. Calculate the deflection of the galvanometer caused by the 5-Ω unbalance in arm BC.

SOLUTION Bridge balance occurs if arm BC has a resistance of 2,000 Ω. The diagram shows arm BC as a resistance of 2,005 Ω, representing a small unbalance ($\ll$2,000 Ω). The first step in the solution consists of converting the

---

* Herbert W. Jackson, *Introduction to Electric Circuits*, 5th ed. (Englewood Cliffs, N.J.: Prentice-Hall, Inc., 1981), pp. 448ff.

bridge circuit into its Thévenin equivalent circuit. Since we are interested in finding the current in the galvanometer, the Thévenin equivalent is determined with respect to galvanometer terminals B and D. The potential difference from B to D, with the galvanometer removed from the circuit, is the Thévenin voltage. Using Eq. (5-7), we obtain

$$E_{TH} = E_{AD} - E_{AB} = 5 \text{ V} \times \left(\frac{100}{100 + 200} - \frac{1,000}{1,000 + 2,005} \right)$$

$$\cong 2.77 \text{ mV}$$

The second step of the solution involves finding the equivalent Thévenin resistance, looking into terminals B and D, and replacing the battery with its internal resistance. Since the battery resistance is 0 Ω, the circuit is represented by the configuration of Fig. 5-3(b) from which we find

$$R_{TH} = \frac{100 \times 200}{300} + \frac{1,000 \times 2,005}{3,005} = 734 \ \Omega$$

The Thévenin equivalent circuit is given in Fig. 5-2(c). When the galvanometer is now connected to the output terminals of the equivalent circuit, the current through the galvanometer is

$$I_g = \frac{E_{TH}}{R_{TH} + R_g} = \frac{2.77 \text{ mV}}{734 \ \Omega + 100 \ \Omega} = 3.32 \ \mu A$$

The galvanometer deflection is

$$d = 3.34 \ \mu A \times \frac{10 \text{ mm}}{\mu A} = 33.2 \text{ mm}$$

At this point the merit of the Thévenin equivalent circuit for the solution of an unbalanced bridge becomes evident. If a different galvanometer is used (with a different current sensitivity and internal resistance), the computation of its deflection is very simple, as is clear from Fig. 5-3(c). Conversely, if the galvanometer sensitivity is given, we can solve for the unbalance voltage needed to give a unit deflection (say 1 mm). This value is of interest when we want to determine the sensitivity of the bridge to unbalance, or in response to the question: "Is the galvanometer selected capable of detecting a certain small unbalance?" The Thévenin method is used to find the galvanometer response, which in most cases is of prime interest.

EXAMPLE 5-2

The galvanometer of Example 5-1 is replaced by one with an internal resistance of 500 Ω and a current sensitivity of 1 mm/μA. Assuming that a deflection of 1 mm can be observed on the galvanometer scale, determine if this new galvanometer is capable of detecting the 5-Ω unbalance in arm BC of Fig. 5-3(a).

SOLUTION Since the bridge constants have not been changed, the equivalent circuit is again represented by a Thévenin generator of 2.77 mV and a

Thévenin resistance of 734 Ω. The new galvanometer is now connected to the output terminals, resulting in a galvanometer current

$$I_g = \frac{E_{TH}}{R_{TH} + R_g} = \frac{2.77 \text{ mA}}{734 \ \Omega + 500 \ \Omega} = 2.24 \ \mu A$$

The galvanometer deflection therefore equals 2.24 $\mu A \times 1$ mm/μA = 2.24 mm, indicating that this galvanometer produces a deflection that can be easily observed.

The Wheatstone bridge is limited to the measurement of resistances ranging from a few ohms to several megohms. The *upper* limit is set by the reduction in sensitivity to unbalance, caused by high resistance values, because in this case the equivalent Thévenin resistance of Fig. 5-3(c) becomes high, thus reducing the galvanometer current. The *lower* limit is set by the resistance of the connecting

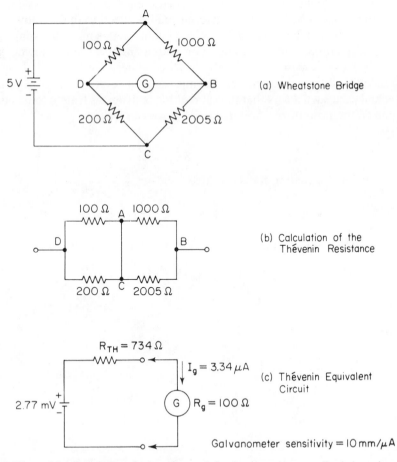

(a) Wheatstone Bridge

(b) Calculation of the Thévenin Resistance

(c) Thévenin Equivalent Circuit

Galvanometer sensitivity = 10 mm/μA

Figure 5-3 Calculation of galvanometer deflection caused by a small unbalance in arm BC, using the simplified Thévenin approach.

leads and the contact resistance at the binding posts. The resistance of the leads could be calculated or measured, and the final result modified, but contact resistance is very hard to compute or measure. For low-resistance measurements, therefore, the Kelvin bridge is generally the preferred instrument.

5-3 KELVIN BRIDGE

5-3.1 Effects of Connecting Leads

The Kelvin bridge is a modification of the Wheatstone bridge and provides greatly increased accuracy in the measurement of *low-value resistances*, generally below 1Ω. Consider the bridge circuit shown in Fig. 5-4, where R_y represents the resistance of the connecting lead from R_3 to R_x. Two galvanometer connections are possible, to point m or to point n. When the galvanometer is connected to point m, the resistance R_y of the connecting lead is added to the unknown R_x, resulting in too high an indication for R_x. When connection is made to point n, R_y is added to bridge arm R_3 and the resulting measurement of R_x will be lower than it should be, because now the actual value of R_3 is higher than its nominal value by resistance R_y. If the galvanometer is connected to a point p, in between the two points m and n, in such a way that the ratio of the resistances from n to p and from m to p equals the ratio of resistors R_1 and R_2, we can write

$$\frac{R_{np}}{R_{mp}} = \frac{R_1}{R_2} \tag{5-10}$$

The balance equation for the bridge yields

$$R_x + R_{np} = \frac{R_1}{R_2}(R_3 + R_{mp}) \tag{5-11}$$

Substituting Eq. (5-10) into Eq. (5-11), we obtain

$$R_x + \left(\frac{R_1}{R_1 + R_2}\right) R_y = \frac{R_1}{R_2}\left[R_3 + \left(\frac{R_2}{R_1 + R_2}\right) R_y\right] \tag{5-12}$$

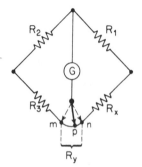

Figure 5-4 Wheatstone bridge circuit, showing resistance R_y of the lead from point m to point n.

which reduces to

$$R_x = \frac{R_1}{R_2} R_3 \qquad (5\text{-}13)$$

Equation (5-13) is the usual balance equation developed for the Wheatstone bridge and it indicates that the effect of the resistance of the connecting lead from point m to point n has been eliminated by connecting the galvanometer to the intermediate position p.

This development forms the basis for construction of the Kelvin double bridge, commonly known as the *Kelvin bridge*.

5-3.2 Kelvin Double Bridge

The term *double* bridge is used because the circuit contains a second set of ratio arms, as shown in the schematic diagram of Fig. 5-5. This second set of arms, labeled a and b in the diagram, connects the galvanometer to a point p at the appropriate potential between m and n, and it eliminates the effect of the yoke resistance R_y. An initially established condition is that the resistance ratio of a and b is the same as the ratio of R_1 and R_2.

The galvanometer indication will be zero when the potential at k equals the potential at p, or when $E_{kl} = E_{lmp}$, where

$$E_{kl} = \frac{R_2}{R_1 + R_2} E = \frac{R_2}{R_1 + R_2} I\left[R_3 + R_x + \frac{(a+b)R_y}{a+b+R_y}\right] \qquad (5\text{-}14)$$

and

$$E_{lmp} = I\left\{R_3 + \frac{b}{a+b}\left[\frac{(a+b)R_y}{a+b+R_y}\right]\right\} \qquad (5\text{-}15)$$

We can solve for R_x by equating E_{kl} and E_{lmp} in the following manner:

$$\frac{R_2}{R_1 + R_2} I\left[R_3 + R_x + \frac{(a+b)R_y}{a+b+R_y}\right] = I\left[R_3 + \frac{b}{a+b} \cdot \frac{(a+b)R_y}{a+b+R_y}\right]$$

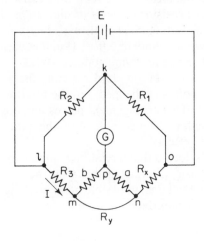

Figure 5-5 Basic Kelvin double bridge circuit.

Or simplifying, we get

$$R_3 + R_x + \frac{(a+b)R_y}{a+b+R_y} = \frac{R_1+R_2}{R_2}\left[R_3 + \frac{bR_y}{a+b+R_y}\right]$$

and expanding the right-hand member yields

$$R_3 + R_x + \frac{(a+b)R_y}{a+b+R_y} = \frac{R_1 R_3}{R_2} + R_3 + \frac{R_1+R_2}{R_2} \cdot \frac{bR_y}{a+b+R_y}$$

Solving for R_x yields

$$R_x = \frac{R_1 R_3}{R_2} + \frac{R_1}{R_2} \cdot \frac{bR_y}{a+b+R_y} + \frac{bR_y}{a+b+R_y} - \frac{(a+b)R_y}{a+b+R_y}$$

so that

$$R_x = \frac{R_1 R_3}{R_2} + \frac{bR_y}{a+b+R_y}\left(\frac{R_1}{R_2} - \frac{a}{b}\right) \qquad (5\text{-}16)$$

Using the initially established condition that $a/b = R_1/R_2$, we see that Eq. (5-16) reduces to the well-known relationship

$$R_x = R_3 \frac{R_1}{R_2} \qquad (5\text{-}17)$$

Equation (5-17) is the usual working equation for the Kelvin bridge. It indicates that the resistance of the yoke has no effect on the measurement, provided that the two sets of ratio arms have equal resistance ratios.

The Kelvin bridge is used for measuring very low resistances, from approximately 1 Ω to as low as 0.00001 Ω. Figure 5-6 shows the simplified circuit diagram of a commercial Kelvin bridge capable of measuring resistances from 10 Ω to 0.00001 Ω. In this bridge, resistance R_3 of Eq. (5-17) is represented by the variable *standard* resistor in Fig. 5-6. The ratio arms (R_1 and R_2) can usually be switched in a number of decade steps.

Contact potential drops in the measuring circuit may cause large errors and to reduce this effect the standard resistor consists of nine steps of 0.001 Ω each plus a calibrated manganin bar of 0.0011 Ω with a sliding contact. The total resistance of the R_3 arm therefore amounts to 0.0101 Ω and is variable in steps of 0.001 Ω plus fractions of 0.0011 Ω by the sliding contact. When both contacts are switched to select the suitable value of standard resistor, the voltage drop between the ratio-arm connection points is changed, but the total resistance around the battery circuit is unchanged. This arrangement places any contact resistance in series with the relatively high-resistance values of the ratio arms, and the contact resistance has negligible effect.

The ratio R_1/R_2 should be selected that a relatively large part of the standard resistance is used in the measuring circuit. In this way the value of unknown resistance R_x is determined with the largest possible number of significant figures, and the measurement accuracy is improved.

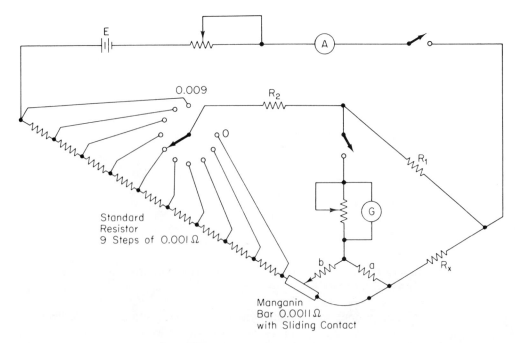

Figure 5-6 Simplified circuit of a Kelvin double bridge used for the measurement of very low resistances.

5-4 GUARDED WHEATSTONE BRIDGE

5-4.1 Guard Circuits

The measurement of extremely high resistances, such as the insulation resistance of a cable or the leakage resistance of a capacitor (often on the order of several thousands of megohms), is beyond the capability of the ordinary dc Wheatstone bridge. One of the major problems in high-resistance measurements is the leakage that occurs over and around the component or specimen being measured, or over the binding posts by which the component is attached to the instrument, or within the instrument itself. These leakage currents are undesired because they can enter the measuring circuit and affect the measurement accuracy to a considerable extent. Leakage currents, whether inside the instrument itself or associated with the test specimen and its mounting, are particularly noticeable in high-resistance measurements where high voltages are often necessary to obtain sufficient deflection sensitivity. Also, leakage effects are generally variable from day to day, depending on the humidity of the atmosphere.

The effects of leakage paths on the measurement are usually removed by some form of guard circuit. The principle of a simple guard circuit in the R_x arm of a Wheatstone bridge is explained with the aid of Fig. 5-7. Without a guard circuit, leakage current I_l along the insulated surface of the binding post adds to current I_x

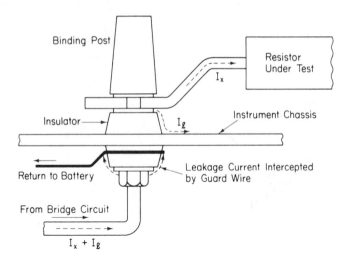

Figure 5-7 Simple guard wire on the R_x terminal of a guarded Wheatstone bridge eliminates surface leakage.

through the component under measurement to produce a total circuit current that can be considerably larger than the actual device current. A guard wire, completely surrounding the surface of the insulated post, intercepts this leakage current and returns it to the battery. The guard must be carefully placed so that the leakage current always meets some portion of the guard wire and is prevented from entering the bridge circuit.

In the schematic diagram of Fig. 5-8 the guard around the R_x binding post, indicated by a small circle around the terminal, does not touch any part of the bridge circuitry and is connected directly to the battery terminal. The principle of the guard wire on the binding post can be applied to any internal part of the bridge circuit where leakage affects the measurement; we then speak of a guarded Wheatstone bridge.

5-4.2 Three-Terminal Resistance

To avoid the effects of leakage currents external to the bridge circuitry, the junction of ratio arms R_A and R_B is usually brought out as a separate guard

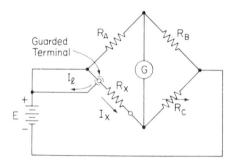

Figure 5-8 Guarded terminal returns leakage current to the battery.

Bridge Measurements   Chap. 5

terminal on the front panel of the instrument. This guard terminal can be used to connect a so-called *three-terminal resistance*, as shown in Fig. 5-9. The high resistance is mounted on two insulating posts that are fastened to a metal plate. The two main terminals of the resistor are connected to the R_x terminals of the bridge in the usual manner. The third terminal of the resistor is the common point of resistances R_1 and R_2, which represent the leakage paths from the main terminals along the insulating posts to the metal plate, or guard. The guard is connected to the guard terminal on the front panel of the bridge, as indicated in the schematic of Fig. 5-9. This connection puts R_1 in parallel with ratio arm R_A, but since R_1 is very much larger than R_A, its shunting effect is negligible. Similarly, leakage resistance R_2 is in parallel with the galvanometer, but the resistance of R_2 is so much higher than that of the galvanometer that the only effect is a slight reduction in galvanometer sensitivity. The effects of external leakage paths are therefore removed by using the guard circuit on the three-terminal resistance.

If the guard circuit were not used, leakage resistance R_1 and R_2 would be directly across R_x and the measured value of R_x would be considerably in error.

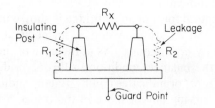

(a)   Three-terminal resistance

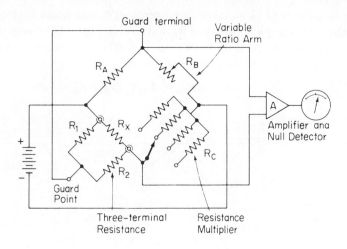

(b) Guarded bridge circuit

Figure 5-9   Three-terminal resistance, connected to a guarded high-voltage megohm bridge.

Assuming, for example, that the unknown is 100 MΩ and that the leakage resistance from each terminal to the guard is also 100 MΩ, resistance R_x would be measured as 67 MΩ, an error of approximately 33 per cent.

5-5 AC BRIDGES AND THEIR APPLICATION

5-5.1 Conditions for Bridge Balance

The ac bridge is a natural outgrowth of the dc bridge and in its basic form consists of four bridge arms, a source of excitation, and a null detector. The power source supplies an ac voltage to the bridge at the desired frequency. For measurements at low frequencies, the power line may serve as the source of excitation; at higher frequencies, an oscillator generally supplies the excitation voltage. The null detector must respond to ac unbalance currents and in its cheapest (but very effective) form consists of a pair of headphones. In other applications, the null detector may consist of an ac amplifier with an output meter, or an electron ray tube (tuning eye) indicator.

The general form of an ac bridge is shown in Fig. 5-10. The four bridge arms Z_1, Z_2, Z_3, and Z_4 are indicated as unspecified impedances and the detector is represented by headphones. As in the case of the Wheatstone bridge for dc measurements, the balance condition in this ac bridge is reached when the detector response is zero, or indicates a null. Balance adjustment to obtain a null response is made by varying one or more of the bridge arms.

The general equation for bridge balance is obtained by using *complex notation* for the impedances of the bridge circuit. (Boldface type is used to indicate quantities in complex notation.) These quantities may be impedances or admittances as well as voltages or currents. The condition for bridge balance requires that the potential difference from A to C in Fig. 5-10 be zero. This will be the case when the voltage drop from B to A equals the voltage drop from B to C, in both *magnitude* and *phase*. In complex notation we can write

$$\mathbf{E}_{BA} = \mathbf{E}_{BC} \quad \text{or} \quad \mathbf{I}_1\mathbf{Z}_1 = \mathbf{I}_2\mathbf{Z}_2 \qquad (5\text{-}18)$$

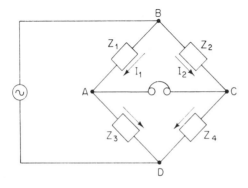

Figure 5-10   General form of the ac bridge.

For zero detector current (the balance condition), the currents are

$$I_1 = \frac{E}{Z_1 + Z_3} \tag{5-19}$$

and

$$I_2 = \frac{E}{Z_2 + Z_4} \tag{5-20}$$

Substitution of Eqs. (5-19) and (5-20) into Eq. (5-18) yields

$$Z_1 Z_4 + Z_2 Z_3 \tag{5-21}$$

or when using admittances instead of impedances.

$$Y_1 Y_4 = Y_2 Y_3 \tag{5-22}$$

Equation (5-21) is the most convenient form in most cases and is the *general equation for balance* of the ac bridge. Equation (5-22) can be used to advantage when dealing with parallel components in bridge arms.

Equation (5-21) states that the product of impedances of one pair of opposite arms must equal the product of impedances of the other pair of opposite arms, with the impedances expressed in complex notation. If the impedance is written in the form $\mathbf{Z} = Z \underline{/\theta}$, where Z represents the magnitude and θ the phase angle of the complex impedance, Eq. (5-21) can be written in the form

$$(Z_1 \underline{/\theta_1})(Z_4 \underline{/\theta_4}) = (Z_2 \underline{/\theta_2})(Z_3 \underline{/\theta_3}) \tag{5-23}$$

Since in multiplication of complex numbers the magnitudes are *multiplied* and the phase angles *added*, Eq. (5-23) can also be written as

$$Z_1 Z_4 \underline{/(\theta_1 + \theta_4)} = Z_2 Z_3 \underline{/(\theta_2 + \theta_3)} \tag{5-24}$$

Equation (5-24) shows that *two conditions* must be met *simultaneously* when balancing an ac bridge. The first condition is that the magnitudes of the impedances satisfy the relationship

$$Z_1 Z_4 = Z_2 Z_3 \tag{5-25}$$

or, in words:

The products of the magnitudes of the opposite arms must be equal.

The second condition requires that the phase angles of the impedances satisfy the relationship

$$\underline{/\theta_1} + \underline{/\theta_4} = \underline{/\theta_2} + \underline{/\theta_3} \tag{5-26}$$

Again, in words:

The sum of the phase angles of the opposite arms must be equal.

5-5.2 Application of the Balance Equations

The two balance conditions expressed in Eqs. (5-25) and (5-26) can be applied when the impedances of the bridge arms are given in polar form, with both magnitude and phase angle. In the usual case, however, the component values of the bridge arms are given, and the problem is solved by writing the balance equation in complex notation. The following examples illustrate the procedure.

EXAMPLE 5-3 —————

The impedances of the basic ac bridge of Fig. 5-10 are given as follows:

$$\mathbf{Z}_1 = 100\ \Omega\ \underline{/80°}\ \text{(inductive impedance)}$$

$$\mathbf{Z}_2 = 250\ \Omega\ \text{(pure resistance)}$$

$$\mathbf{Z}_3 = 400\ \Omega\ \underline{/30°}\ \text{(inductive impedance)}$$

$$\mathbf{Z}_4 = \text{unknown}$$

Determine the constants of the unknown arm.

SOLUTION   The first condition for bridge balance requires that

$$Z_1 Z_4 = Z_2 Z_3 \tag{5-25}$$

Substituting the magnitudes of the known components and solving for Z_4, we obtain

$$Z_4 = \frac{Z_2 Z_3}{Z_1} = \frac{250 \times 400}{100} = 1{,}000\ \Omega$$

The second condition for bridge balance requires that the sums of the phase angles of opposite arms be equal, or

$$\theta_1 + \theta_4 = \theta_2 + \theta_3 \tag{5-26}$$

Substituting the known phase angles and solving for θ_4, we obtain

$$\theta_4 = \theta_2 + \theta_3 - \theta_1 = 0 + 30 - 80 = -50°$$

Hence the unknown impedance Z_4 can be written in polar form as

$$\mathbf{Z}_4 = 1{,}000\ \Omega\ \underline{/-50°}$$

indicating that we are dealing with a capacitive element, possibly consisting of a series combination of a resistor and a capacitor.

The problem becomes slightly more complex when the component values of the bridge arms are specified and the impedances are to be expressed in complex notation. In this case, the inductive or capacitive reactances can only be calculated when the frequency of the excitation voltage is known, as Example 5-4 shows.

EXAMPLE 5-4

The ac bridge of Fig. 5-10 is in balance with the following constants: arm AB, $R = 450\ \Omega$; arm BC, $R = 300\ \Omega$ in series with $C = 0.265\ \mu\text{F}$; arm CD, unknown; arm DA, $R = 200\ \Omega$ in series with $L = 15.9$ mH. The oscillator frequency is 1 kHz. Find the constants of arm CD.

SOLUTION The general equation for bridge balance states that

$$\mathbf{Z}_1\mathbf{Z}_4 = \mathbf{Z}_2\mathbf{Z}_3$$

$$\mathbf{Z}_1 = R = 450\ \Omega$$

$$\mathbf{Z}_2 = R - j/\omega C = (300 - j600)\ \Omega \qquad (5\text{-}21)$$

$$\mathbf{Z}_3 = R + j\omega L = (200 + j100)\ \Omega$$

$$\mathbf{Z}_4 = \text{unknown}$$

Substituting the known values in Eq. (5-21) and solving for the unknown yields

$$\mathbf{Z}_4 = \frac{450 \times (200 + j100)}{300 - j600} = +j150\ \Omega$$

This result indicates that $\mathbf{Z}_4$ is a pure inductance with an inductive reactance of 150 Ω at a frequency of 1 kHz. Since the inductive reactance $X_L = 2\pi fL$, we solve for L and obtain $L = 23.9$ mH.

5-6 MAXWELL BRIDGE

The Maxwell bridge, whose schematic diagram is shown in Fig. 5-11, measures an unknown *inductance* in terms of a known *capacitance*. One of the ratio arms has a resistance and a capacitance in *parallel*, and it may now prove somewhat easier to write the balance equations using the *admittance* of arm 1 instead of its impedance.

Rearranging the general equation for bridge balance, as expressed in Eq. (5-21), we obtain

$$\mathbf{Z}_x = \mathbf{Z}_2\mathbf{Z}_3\mathbf{Y}_1 \qquad (5\text{-}27)$$

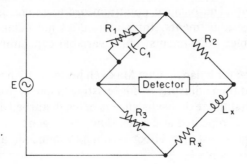

Figure 5-11 Maxwell bridge for inductance measurements.

where $\mathbf{Y}_1$ is the admittance of arm 1. Reference to Fig. 5-11 shows that

$$\mathbf{Z}_2 = R_2; \qquad \mathbf{Z}_3 = R_3; \qquad \text{and} \qquad \mathbf{Y}_1 = \frac{1}{R_1} + j\omega C_1$$

Substitution of these values in Eq. (5-27) gives

$$\mathbf{Z}_x = R_x + j\omega L_x = R_2 R_3 \left(\frac{1}{R} + j\omega C_1 \right) \tag{5-28}$$

Separation of the real and imaginary terms yields

$$R_x = \frac{R_2 R_3}{R_1} \tag{5-29}$$

and

$$L_x = R_2 R_3 C_1 \tag{5-30}$$

where the resistances are expressed in ohms, inductance in henrys, and capacitance in farads.

The Maxwell bridge is limited to the measurement of *medium-Q coils* ($1 < Q < 10$). This can be shown by considering the second balance condition which states that the sum of the phase angles of one pair of opposite arms must be equal to the sum of the phase angles of the other pair. Since the phase angles of the resistive elements in arm 2 and arm 3 add up to $0°$, the sum of the angles of arm 1 and arm 4 must also add up to $0°$. The phase angle of a *high-Q* coil will be very nearly $90°$ (positive), which requires that the phase angle of the capacitive arm must also be very nearly $90°$ (negative). This in turn means that the resistance of R_1 must be very large indeed, which can be very impractical. High-Q coils are therefore generally measured on the Hay bridge, presented in Sec. 5-7.

The Maxwell bridge is also unsuited for the measurement of coils with a very low Q-value ($Q < 1$) because of balance convergence problems. Very low Q-values occur in inductive resistors, for example, or in an RF coil if measured at low frequency. As can be seen from the equations for R_x and L_x, adjustment for inductive balance by R_3 upsets the resistive balance by R_1 and gives the effect known as *sliding balance*. Sliding balance describes the interaction between controls, so that when we balance with R_1 and then with R_3, then go back to R_1, we find a new balance point. The balance point appears to move or *slide* toward its final point after many adjustments. Interaction does not occur when R_1 and C_1 are used for the balance adjustments, but a variable capacitor is not always suitable.

The usual procedure for balancing the Maxwell bridge is by first adjusting R_3 for inductive balance and then adjusting R_1 for resistive balance. Returning to the R_3 adjustment, we find that the resistive balance is being disturbed and moves to a new value. This process is repeated and gives *slow* convergence to final balance. For medium-Q coils, the resistance effect is not pronounced, and balance is reached after a few adjustments.

5-7 HAY BRIDGE

The Hay bridge of Fig. 5-12 differs from the Maxwell bridge by having resistor R_1 in series with standard capacitor C_1 instead of in parallel. It is immediately apparent that for large phase angles, R_1 should have a very low value. The Hay circuit is therefore more convenient for measuring high-Q coils.

The balance equations are again derived by substituting the values of the impedances of the bridge arms into the general equation for bridge balance. For the circuit of Fig. 5-12, we find that

$$\mathbf{Z}_1 = R_1 - \frac{j}{\omega C_1}; \quad \mathbf{Z}_2 = R_2; \quad \mathbf{Z}_3 = R_3; \quad \mathbf{Z}_x = R_x + j\omega L_x$$

Substituting these values in Eq. (5-21), we get

$$\left(R_1 - \frac{j}{\omega C_1}\right)(R_x + j\omega L_x) = R_2 R_3 \tag{5-31}$$

which expands to

$$R_1 R_x + \frac{L_x}{C_1} - \frac{jR_x}{\omega C_1} + j\omega L_x R_1 = R_2 R_3$$

Separating the real and imaginary terms, we obtain

$$R_1 R_x + \frac{L_x}{C_1} = R_2 R_3 \tag{5-32}$$

and

$$\frac{R_x}{\omega C_1} = \omega L_x R_1 \tag{5-33}$$

Both Eq. (5-32) and Eq. (5-33) contain L_x and R_x, and we must solve these equations simultaneously. This yields

$$R_x = \frac{\omega^2 C_1^2 R_1 R_2 R_3}{1 + \omega^2 C_1^2 R_1^2} \tag{5-34}$$

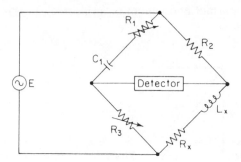

Figure 5-12 Hay bridge for inductance measurements.

$$L_x = \frac{R_2 R_3 C_1}{1 + \omega^2 C_1^2 R_1^2} \tag{5-35}$$

These expressions for the unknown inductance and resistance both contain the angular velocity ω and it therefore appears that the frequency of the voltage source must be known accurately. That this is not true when a high-Q coil is being measured follows from the following considerations: Remembering that the sum of the opposite sets of phase angles must be equal, we find that the inductive phase angle must be equal to the capacitive phase angle, since the resistive angles are zero. Figure 5-13 shows that the tangent of the inductive phase angle equals

$$\tan \theta_L = \frac{X_L}{R} = \frac{\omega L_x}{R_x} = Q \tag{5-36}$$

and that of the capacitive phase angle is

$$\tan \theta_C = \frac{X_C}{R} = \frac{1}{\omega C_1 R_1} \tag{5-37}$$

When the two phase angles are equal, their tangents are also equal and we can write

$$\tan \theta_L = \tan \theta_C \quad \text{or} \quad Q = \frac{1}{\omega C_1 R_1} \tag{5-38}$$

Returning now to the term $(1 + \omega^2 C_2^1 R_1^2)$ which appears in Eqs. (5-34) and (5-35), we find that, after submitting Eq. (5-38) in the expression for L_x, Eq. (5-35) reduces to

$$L_x = \frac{R_2 R_3 C_1}{1 + (1/Q)^2} \tag{5-39}$$

For a value of Q greater than ten, the term $(1/Q)^2$ will be smaller than $1/100$ and can be neglected. Equation (5-35) therefore reduces to the expression derived for the Maxwell bridge,

$$L_x = R_2 R_3 C_1$$

The Hay bridge is suited for the measurement of high-Q inductors, especially for those inductors having a Q greater than 10. For Q-values smaller than 10, the term $(1/Q)^2$ becomes important and cannot be neglected. In this case, the Maxwell bridge is more suitable.

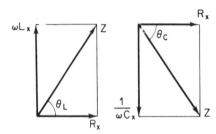

Figure 5-13 Impedance triangles illustrate inductive and capacitive phase angles.

Bridge Measurements   Chap. 5

5-8 SCHERING BRIDGE

The Schering bridge, one of the most important ac bridges, is used extensively for the measurement of *capacitors*. Although the Schering bridge is used for capacitance measurements in a general sense, it is particularly useful for measuring insulating properties, i.e., for phase angles very nearly 90°.

The basic circuit arrangement is shown in Fig. 5-14, and inspection of the circuit shows a strong resemblance to the comparison bridge. Notice that arm 1 now contains a parallel combination of a resistor and a capacitor, and the standard arm contains only a capacitor. The standard capacitor is usually a high-quality mica capacitor for general measurement work or an air capacitor for insulation measurements. A good-quality mica capacitor has very low losses (no resistance) and therefore a phase angle of approximately 90°. An air capacitor, when designed carefully, has a very stable value and a very small electric field; the insulating material to be tested can easily be kept out of any strong fields.

The balance conditions require that the sum of the phase angles of arms 1 and 4 equals the sum of the phase angles of arms 2 and 3. Since the standard capacitor is in arm 3, the sum of the phase angles of arm 2 and arm 3 will be 0° + 90° = 90°. In order to obtain the 90°-phase angle needed for balance, the sum of the angles of arm 1 and arm 4 must equal 90°. Since in general measurement work the unknown will have a phase angle smaller than 90°, it is necessary to give arm 1 a small capacitive angle by connecting capacitor C_1 in parallel with resistor R_1. A small capacitive angle is very easy to obtain, requiring a small capacitor across resistor R_1.

The balance equations are derived in the usual manner, and by substituting the corresponding impedance and admittance values in the general equation, we obtain

$$\mathbf{Z}_x = \mathbf{Z}_2\mathbf{Z}_3\mathbf{Y}_1$$

or

$$R_x - \frac{j}{\omega C_x} = R_2 \left(\frac{-j}{\omega C_3}\right) \left(\frac{1}{R_1} + j\omega C_1\right)$$

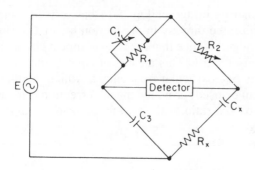

Figure 5-14 Schering bridge for the measurement of capacitance.

and expanding

$$R_x - \frac{j}{\omega C_x} = \frac{R_2 C_1}{C_3} - \frac{j R_2}{\omega C_3 R_1} \tag{5-40}$$

Equating the real terms and the imaginary terms, we find that

$$R_x = R_2 \frac{C_1}{C_3} \tag{5-41}$$

$$C_x = C_3 \frac{R_1}{R_2} \tag{5-42}$$

As can be seen from the circuit diagram of Fig. 5-13, the two variables chosen for the balance adjustment are capacitor C_1 and resistor R_2. There seems to be nothing unusual about the balance equations or the choice of variable components, but consider for a moment how the quality of a capacitor is defined.

The *power factor* (PF) of a series RC combination is defined as the cosine of the phase angle of the circuit. Therefore the PF of the unknown equals PF = $R_x/\mathbf{Z}_x$. For phase angles very close to 90°, the reactance is almost equal to the impedance and we can approximate the power factor to

$$PF \simeq \frac{R_x}{X_x} = \omega C_x R_x \tag{5-43}$$

The *dissipation factor* of a series RC circuit is defined as the cotangent of the phase angle and therefore, by definition, the dissipation factor

$$D = \frac{R_x}{X_x} = \omega C_x R_x \tag{5-44}$$

Incidentally, since the quality of a coil is defined by $Q = X_L/R_L$, we find that the dissipation factor, D, is the reciprocal of the quality factor, Q, and therefore $D = 1/Q$. The dissipation factor tells us something about the quality of a capacitor; i.e., how close the phase angle of the capacitor is to the ideal value of 90°. By substituting the value of C_x in Eq. (5-42) and of R_x in Eq. (5-41) into the expression for the dissipation factor, we obtain

$$D = \omega R_1 C_1 \tag{5-45}$$

If resistor R_1 in the Schering bridge of Fig. 5-14 has a fixed value, the dial of capacitor C_1 may be calibrated directly in dissipation factor D. This is the usual practice in a Schering bridge. Notice that the term ω appears in the expression for the dissipation factor [Eq. (5-45)]. This means, of course, that the calibration of the C_1 dial holds for only one particular frequency at which the dial is calibrated. A different frequency can be used, provided that a correction is made by multiplying the C_1 dial reading by the ratio of the two frequencies. Figure 5-15 shows a modern automatic bridge.

Figure 5-15 Modern bridges measure a variety of parameters and are automatic, such as this unit. (Courtesy of Electro Scientific Industries, Inc.)

5-9 UNBALANCE CONDITIONS

It sometimes happens that an ac bridge cannot be balanced at all simply because one of the stated balance conditions (Sec. 5-5) cannot be met. Consider for example, the circuit of Fig. 5-16, where Z_1 and Z_4 are inductive elements (positive phase angles), Z_2 is a pure capacitance ($-90°$ phase angle), and Z_3 is a variable resistance (zero phase angle). The resistance of R_3 needed to obtain bridge balance can be determined by applying the first balance condition (magnitudes) and

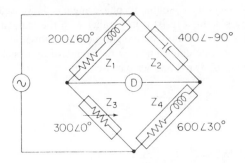

Figure 5-16 Ac bridge that cannot be balanced.

we find that

$$R_3 = \frac{Z_1 Z_4}{Z_2} = \frac{200 \times 600}{400} = 300\ \Omega$$

Hence adjusting R_3 to a value of 300 Ω will satisfy the first condition.

Considering the second balance condition (phase angles) yields the following situation:

$$\theta_1 + \theta_4 = +60° + 30° = +90°$$

$$\theta_2 + \theta_3 = -90° + 0° = -90°$$

Obviously, $\theta_1 + \theta_4 \neq \theta_2 + \theta_3$, and the second condition is not satisified. In this case, bridge balance cannot be obtained.

An interesting illustration of a bridge balancing problem is given in Example 5-5, where minor adjustments to one or more of the bridge arms result in a situation where balance can be obtained.

EXAMPLE 5-5

Consider the circuit of Fig. 5-17(a) and determine whether or not the bridge is in complete balance. If not, show two ways in which it can be made to balance and specify numerical values for any additional components. Assume that bridge arm 4 is the unknown that cannot be modified.

SOLUTION   Inspection of the circuit shows that the first balance condition (magnitudes) can easily be met by slightly increasing the resistance of R_3. The second balance condition requires that $\theta_1 + \theta_4 = \theta_2 + \theta_3$ where

$$\theta_1 = -90° \text{ (pure capacitance)}$$

$$\theta_2 = \theta_3 = 0° \text{ (pure resistance)}$$

$$\theta_4 < +90° \text{ (inductive impedance)}$$

Obviously, balance is not possible with the configuration of Fig. 5-17(a) because the sum of θ_1 and θ_4 will be slightly negative while $\theta_2 + \theta_3$ will be exactly 0°. Balance can be restored by modifying the circuit in such a way that the phase angle condition is satisfied. There are basically two methods to accomplish this: The first option is to modify Z_1 so that its phase angle is decreased to less than 90° (equal to θ_4) by placing a resistor in parallel with the capacitor. This modification results in a Maxwell bridge configuration, as shown in Fig. 5-17(b). The resistance of R_1 can be determined by the standard approach of Sec. 5-6, using the admittance of arm 1, and we can write

$$Y_1 = \frac{Z_4}{Z_2 Z_3}$$

where

$$Y_1 = \frac{1}{R_1} + \frac{j}{1,000}$$

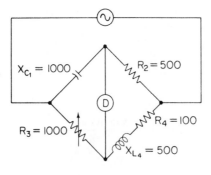

(a) Unbalanced condition

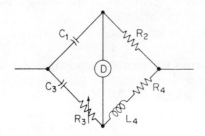

(b) Bridge balance is restored by adding a resistor to arm 1. (Maxwell configuration).

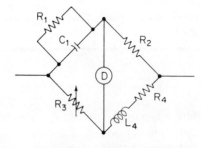

(c) Alternative method of restoring bridge balance, by adding a capacitor to arm 3.

Figure 5-17 Bridge balancing problem.

Substituting the known values and solving for R_1, we obtain

$$\frac{1}{R_1} + \frac{j}{1,000} = \frac{100 + j500}{500 \times 1,000}$$

and

$$R_1 = 5,000 \ \Omega$$

It should be noted that the addition of R_1 upsets the first balance condition of the circuit (the magnitude of $\mathbf{Z}_1$ has changed) and variable resistor R_3 should be adjusted to compensate for this effect.

The second option is to modify the phase angle of arm 2 or arm 3 by adding a series capacitor, as shown in Fig. 5-17(c). Again writing the general balance equation, using impedances this time, we obtain

$$\mathbf{Z}_3 = \frac{\mathbf{Z}_1 \mathbf{Z}_4}{\mathbf{Z}_2}$$

Substituting the component values and solving for X_C yields

$$1,000 - jX_C = \frac{-j1,000(100 + j500)}{500}$$

or

$$X_C = 200 \ \Omega$$

In this case, also, the magnitude of $\mathbf{Z}_3$ has increased so that the first balance condition has changed. A small readjustment of R_3 is necessary to restore balance.

5-10 WIEN BRIDGE

The Wien bridge is presented here not only for its use as an ac bridge to measure *frequency,* but also for its application in various other useful circuits. We find, for example, a Wien bridge in the harmonic distortion analyzer, where it is used as a *notch filter,* discriminating against one specific frequency. The Wien bridge also finds application in audio- and HF oscillators as the *frequency-determining* element. In this chapter, however, the Wien bridge is discussed in its basic form, designed to measure frequency; in other chapters it is shown as an element of different types of instrument.

The Wien bridge has a series RC combination in one arm and a parallel RC combination in the adjoining arm (see Fig. 5-18). The impedance of arm 1 is $\mathbf{Z}_1 = R_1 - j/\omega C_1$. The admittance of arm 3 is $\mathbf{Y}_3 = 1/R_3 + j\omega C_3$. Using the basic equation for bridge balance and substituting the appropriate values, we obtain

$$R_2 = \left(R_1 - \frac{j}{\omega C_1}\right) R_4 \left(\frac{1}{R_3} + j\omega C_3\right) \tag{5-46}$$

Expanding this expression, we get

$$R_2 = \frac{R_1 R_4}{R_3} + j\omega C_3 R_1 R_4 - \frac{jR_4}{\omega C_1 R_3} + \frac{R_4 C_3}{C_1} \tag{5-47}$$

Equating the *real* terms, we obtain

$$R_2 = \frac{R_1 R_4}{R_3} + \frac{R_4 C_3}{C_1} \tag{5-48}$$

which reduces to

$$\frac{R_2}{R_4} = \frac{R_1}{R_3} + \frac{C_3}{C_1} \tag{5-49}$$

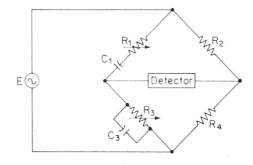

Figure 5-18 Frequency measurement with the Wien bridge.

Equating the *imaginary* terms, we obtain

$$\omega C_3 R_1 R_4 = \frac{R_4}{\omega C_1 R_3} \tag{5-50}$$

where $\omega = 2\pi f$, and solving for f, we get

$$f = \frac{1}{2\pi \sqrt{C_1 C_3 R_1 R_3}} \tag{5-51}$$

Notice that the two conditions for bridge balance now result in an expression determining the required resistance ratio, R_2/R_4, and another expression determining the frequency of the applied voltage. In other words, if we satisfy Eq. (5-49), and also excite the bridge with a frequency described by Eq. (5-51), the bridge will be in balance.

In most Wien bridge circuits, the components are chosen such that $R_1 = R_3$ and $C_1 = C_3$. This reduces Eq. (5-49) to $R_2/R_4 = 2$ and Eq. (5-51) to

$$f = \frac{1}{2\pi RC} \tag{5-52}$$

which is the general expression for the frequency of the Wien bridge. In a practical bridge, capacitors C_1 and C_3 are fixed capacitors, and resistors R_1 and R_3 are variable resistors controlled by a common shaft. Provided now that $R_2 = 2R_4$, the bridge may be used as a frequency-determining device balanced by a single control. This control may be calibrated directly in terms of frequency.

Because of its frequency sensitivity, the Wien bridge may be difficult to balance (unless the waveform of the applied voltage is purely sinusoidal). Since the bridge is *not* balanced for any harmonics present in the applied voltage, these harmonics will sometimes produce an output voltage masking the true balance point.

5-11 WAGNER GROUND CONNECTION

The discussion so far has assumed that the four bridge arms consist of simple lumped impedances which do not interact in any way. In practice, however, *stray capacitances* exist between the various bridge elements and ground, and also

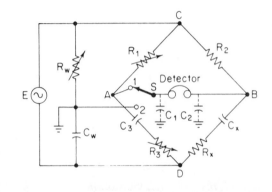

(a)

(b)

Figure 5-19 (a) The Wagner ground connection eliminates the effect of stray capacitances across the detector; (b) completely automatic capacitance bridge that can be computer interfaced. (Photograph courtesy of Boonton Electronics Corporation.)

between the bridge arms themselves. These stray capacitances shunt the bridge arms and cause measurement errors, particularly at the higher frequencies or when small capacitors or large inductors are measured. One way to control stray capacitances is by shielding the arms and connecting the shields to ground. This does not eliminate the capacitances but at least makes them constant in value, and they can therefore be compensated.

One of the most widely used methods for eliminating some of the effects of stray capacitance in a bridge circuit is the *Wagner ground connection*. This circuit eliminates the troublesome capacitance which exists between the detector terminals and ground. Figure 5-19(a) shows the circuit of a capacitance bridge, where C_1 and C_2 represent these stray capacitances. The oscillator is removed from its usual ground connection and bridged by a series combination of resistor R_w and capacitor C_w. The junction of R_w and C_w is grounded and is called the Wagner ground connection. The procedure for initial adjustment of the bridge is as follows: The detector is connected to point 1, and R_1 is adjusted for null or minimum sound in the headphones. The switch is then thrown to position 2, which connects the detector to the Wagner ground point. Resistor R_w is now adjusted for minimum sound. When the switch is thrown to position 1 again, some unbalance will probably be shown. Resistors R_1 and R_3 are then adjusted for minimum detector response, and the switch is again thrown to position 2. A few adjustments of R_w and R_1 (and R_3) may be necessary before a null is reached on both switch positions. When null is finally obtained, points 1 and 2 are at the same potential, and this is ground potential. Stray capacitances C_1 and C_2 are then effectively shorted out and have no effect on normal bridge balance. There are also capacitances from points C and D to ground, but the addition of the Wagner ground point eliminates them from the detector circuit, since current through these capacitances will enter through the Wagner ground connection.

The capacitances across the bridge arms are not eliminated by the Wagner ground connection and they will still affect the accuracy of the measurement. The idea of the Wagner ground can also be applied to other bridges, as long as care is taken that the grounding arms duplicate the impedance of one pair of bridge arms across which they are connected. Since the addition of the Wagner ground connection does not affect the balance conditions, the procedure for measurement remains unaltered.

REFERENCES

5-1. ITT Staff, *Reference Data for Radio Engineers,* 7th ed., chap. 12. Indianapolis, Ind.: Howard W. Sams & Company, Inc., 1985.

5-2. Maloney, Timothy, *Electrical Circuits: Principles and Applications,* chap. 6. Englewood Cliffs, N.J., Prentice-Hall, Inc., 1984.

5-3. Prensky, Sol D., and Castellucis, Richard L., *Electronic Instrumentation,* 3rd ed., chaps. 4 and 5. Englewood Cliffs, N.J.: Prentice-Hall, Inc., 1982.

PROBLEMS

5-1. The standard resistor arm of the bridge shown in Fig. P5-1 has a range from 0 to 100 Ω with a resolution of 0.001 Ω. The galvanometer has an internal resistance of 100 Ω and can be read to 0.5 μA. When the unknown resistance is 50 Ω, what is the resolution of the bridge in both ohms and per cent of the unknown?

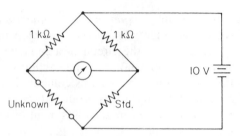

Figure P5-1

5-2. The ratio arms of the Kelvin bridge of Fig. 5-5 are 100 Ω each. The galvanometer has an internal resistance of 500 Ω and a current sensitivity of 200 mm/μA. The unknown resistance $R_x = 0.1002$ Ω and the standard resistance is set at 0.1000 Ω. A dc current of 10 A is passed through the standard and the unknown from a 2.2-V battery in series with a rheostat. The resistance of the yoke may be neglected. Calculate **(a)** the deflection of the galvanometer, and **(b)** the resistance unbalance required to produce a galvanometer deflection of 1 mm.

5-3. The ratio arms of a Kelvin bridge are 1,000 Ω each. The galvanometer has an internal resistance of 100 Ω and a current sensitivity of 500 mm/μA. A dc current of 10 A is passed through the standard arm and the unknown from a 2.2-V battery in series with a rheostat. The standard resistance is set at 0.1000 Ω and the galvanometer deflection is 30 mm. Neglecting the resistance of the yoke, determine the value of the unknown.

5-4. A balanced ac bridge has the following constants: arm AB, $R = 2,000$ Ω in parallel with $C = 0.047$ μF; arm BC, $R = 1,000$ Ω in series with $C = 0.47$ μF; arm CD, unknown; arm DA, $C = 0.5$ μF. The frequency of the oscillator is 1,000 Hz. Find the constants of arm CD.

5-5. A bridge is balanced at 1,000 Hz and has the following constants: AB, 0.2 μF pure capacitance; BC, 500 Ω pure resistance; CD, unknown; DA, $R = 300$ Ω in parallel with $C = 0.1$ μF. Find the R and C or L constants of arm CD, considered as a series circuit.

5-6. A 1,000-Hz bridge has the following constants: arm AB, $R = 1,000$ Ω in parallel with $C = 0.5$ μF; BC, $R = 1,000$ Ω in series with $C = 0.5$ μF; CD, $L = 30$ mH in series with $R = 200$ Ω. Find the constants of arm DA to balance the bridge. Express the result as a pure R in series with a pure C or L and also as a pure R in parallel with a pure C or L.

5-7. An ac bridge has in arm AB a pure capacitance of 0.2 μF; in arm BC, a pure resistance of 500 Ω; in arm CD, a series combination of $R = 50$ Ω and $L = 0.1$ H. Arm DA consists of a capacitor $C = 0.4$ μF in series with a variable resistor R_s. $w = 5,000$ rad/s. **(a)** Find the value of R_s to obtain bridge balance. **(b)** Can complete balance be attained by the adjustment of R_s? If not, specify the position and value of an adjustable resistance to complete the balance.

5-8. An ac bridge has the following constants: arm AB, $R = 1,000$ Ω in parallel with $C = 0.159$ μF; BC, $R = 1,000$ Ω; CD, $R = 500$ Ω; DA, $C = 0.636$ μF in series with an unknown resistance. Find the frequency for which this bridge is in balance and determine the value of the resistance in arm DA to produce this balance.

6

Electronic Instruments
for Measuring
Basic Parameters

6-1 INTRODUCTION

The measuring instruments discussed in the previous chapters used the movement of an electromagnetic meter to measure voltage, current, resistance, power, etc. Although the bridges and multimeters used electrical components for these measurements, the instruments described used no amplifiers to increase the sensitivity of the measurements. The heart of these instruments was the d'Arsonval meter, which typically cannot be constructed with a full-scale sensitivity of less than about 50 μA. Any measurement system using the d'Arsonval meter, without amplifiers, must obtain at least 50 μA from the circuit under test for a full-scale deflection. For the measurement of currents of less than 50 μA full scale, an amplifier must be employed. The resistance of a (very) sensitive meter, such as a 50-μA meter for a volt-ohm-milliammeter, is on the order of a few hundred ohms and represents a small but finite amount of power. As an example, 50 μA through a 200-Ω meter represents $\frac{1}{2}$ microwatt (μW). This represents the power required for the meter for full-scale deflection and does not represent the power dissipated in the series resistor, and thus the total power required by the example meter would be greater than $\frac{1}{2}$ μW and would depend on the voltage range. This doesn't sound like much power, but many electronic circuits cannot tolerate this much power being drained from them. Consider, also, the voltage across a 200-Ω, 50-

μA meter at full scale, which, by Ohm's law, is 10 mV. The most sensitive voltmeter that could be constructed from the 50-μA meter, without an amplifier, would be 10 mV full scale. The schematic of this sensitive meter would have no external resistor but only the internal resistance of the meter itself.

As shown above, an amplifier is required to increase the current sensitivity below 50 μA, the voltage below 10 mV, and the power required below $\frac{1}{2}$ μW. For the case of ac measurements, the amplifier is even more necessary for sensitive measurements.

In addition to instruments for making measurements of small currents and voltages, included in this chapter are electronic instruments for measuring other parameters, such as resistance, inductance, and capacitance.

6-2 AMPLIFIED DC METER

A simple amplified voltmeter is shown in Fig. 6-1. This meter decreases the amount of power drawn from a circuit under test by increasing the input impedance using an amplifier with unity gain. A source follower drives an emitter follower. This combination is capable of a thousand-fold or more increase in impedance, while maintaining a voltage gain of very nearly one. The input impedance of this meter is 10 MΩ, which would require 0.025 μW of power for a 0.5-V deflection, as compared to 25 μW for an unamplified meter, an increase in sensitivity of 100 times.

Because the emitter follower must have some bias current present, the emitter voltage does not go to zero volts with zero input voltage. Thus the meter must be returned not to ground, but to a voltage that can be set to be equal to the quiescent point of the emitter-follower output. This tends to vary somewhat with temperature, and in many practical meters this is made adjustable from the front

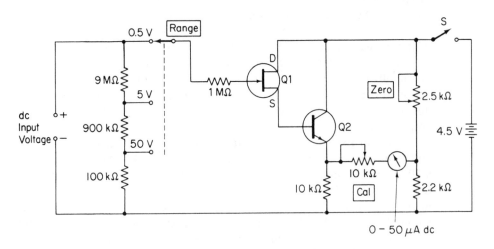

Figure 6-1 Basic dc voltmeter circuit with FET input.

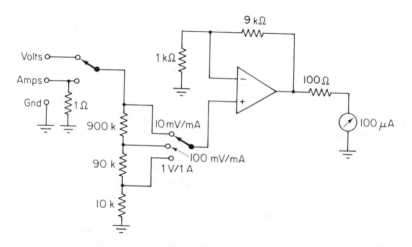

Figure 6-2 Amplified voltage and current meter.

panel of the meter. Because the setting of the Zero control affects the total resistance in series with the meter, a Cal (calibrate) control is also supplied. This control is not necessary for amplified meters using a differential amplifier because there is no interaction between the zero adjustment and the calibration of the meter.

A block diagram of a meter capable of measuring small voltages and currents is shown in Fig. 6-2. The input voltage is amplified and applied to a meter. If the amplifier has a gain of 10, the sensitivity of the measurement is increased by a like amount. A dc-coupled amplifier, that is, an amplifier with no coupling capacitors and having a well-controlled dc gain, is used to provide the necessary amplification. An amplifier capable of a fixed dc gain of 10 is not difficult to construct and to keep stable. A simple op-amp plus the required feedback components will do a suitable job for this application.

Dc gains of much more than 10 are required to use a standard d'Arsonval meter movement to measure very small currents and voltages such as microvolts and nanoamperes. To amplify nanoamperes to drive a milliampere meter requires a gain of 10^6. In theory, this requires an op-amp and two resistors, and a simple circuit. However, when gains this large are desired, all the defects of an operational amplifier become significant. Offset current, offset voltage, and bias currents become so troublesome that it is practically impossible to achieve acceptable performance with a standard op-amp. Many of these defects can be reduced or eliminated by the use of trim adjustments accessible from the front panel in a similar fashion as the Cal and Zero functions discussed above. However, temperature- and time-induced drifts would soon render the amplifier unusable, and the adjustment would have to be repeated. Direct-coupled amplifiers that have been optimized for low-temperature drift and low offset and bias currents are called *instrumentation amplifiers* and are manufactured by semiconductor suppliers.

6-2.1 Chopper-Stabilized Amplifier

One technique for amplifying direct currents and relatively low-frequency alternating currents is the chopper-stabilized amplifier. This circuit eliminates the effects of dc offset currents and the drift of other dc parameters by using an ac-coupled amplifier for the necessary gain. The technique, as shown in Fig. 6-3, is to convert the input signal to an ac signal and, after high-gain amplification, reconstruct the dc from the amplified ac sigmal.

The input signal is converted to an ac signal by *chopping*, which simply involves switching the input of an amplifier between the input and ground with an electronic switch or an electromechanical chopper, which is similar to a relay. The output of the chopper is an ac signal with a peak value equal to the input dc voltage. Because the chopped input has a negative peak of zero and a positive peak of the input voltage, the resulting ac waveform has a dc component of approximately one-half of the input dc voltage. The actual dc component of the chopped waveform is not important, as this will be fed to an ac-coupled amplifier where the dc component will be lost.

The amplified signal is chopped in a similar fashion as the input and in synchronism with the input chopper. The synchronized chopping restores the dc value of the input signal amplified by the ac gain of the amplifier. Because the amplifier did not provide any dc gain, the effects of dc offset voltages and currents are eliminated.

Enormous gains can be achieved in this fashion, and the chopper-stabilized amplifier can provide gains of more than 10^6 with excellent dc stability. All this does not come without problems. First, when dealing with very small currents and voltages, unusual problems can occur. One significant problem is with the chopper. This device must be specially made to avoid generating voltages from

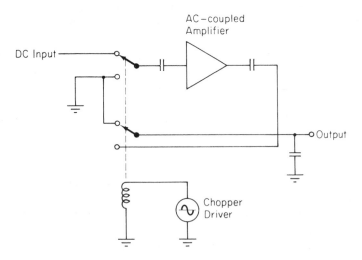

Figure 6-3 Ac-coupled amplifier can be used to amplify dc signals if the input and output are chopped using the circuit shown.

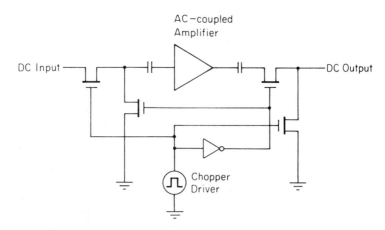

Figure 6-4 All-electric chopper circuit using field effect transistors.

thermocouple effects. When two dissimilar metals are joined, depending on the temperature, small voltages can be generated. The chopper is specially made to reduce these thermally generated voltages.

The electromechanical chopper, being a mechanical device, has a relatively short life span when compared to other electronic devices. Various types of all-electronic choppers have been devised to replace the venerable mechanical chopper. The most important characteristic of the chopper is that it must not inject any current into the circuit being chopped, especially for the input chopper. Bipolar transistors, light-activated devices, and field effect transistors have been used for choppers with the MOS field effect transistor being the most successful. Because the MOS transistor has no junction as a source of leakage current, very little current is transmitted from the chopping signal to the input. Figure 6-4 shows a series-shunt chopper using two MOS field effect transistors. The chopping signal is fed to the inverter, which drives the two chopper FETs, one on each half of the chopping cycle.

The input impedance of the chopper-stabilized amplifier is very high for direct current. Looking into the chopper-stabilized amplifier, the series chopper switches the input to the ac-coupled amplifier every half-cycle; however, because the amplifier is ac coupled, it appears as an infinite resistance to direct current. The series chopper switch is always open before the shunt switch is closed, and thus there is no path to ground.

6-3 AC VOLTMETER USING RECTIFIERS

Electronic ac voltmeters are basically identical to dc voltmeters except that the ac input voltage must be *rectified* before it can be applied to the dc meter circuit. In some instances, rectification takes place *before* amplification, in which case a simple diode rectifier circuit precedes the amplifier and meter, as in Fig. 6-5(a).

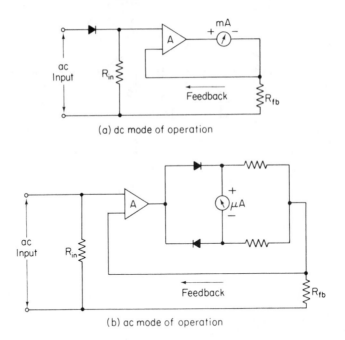

(a) dc mode of operation

(b) ac mode of operation

Figure 6-5 Basic ac voltmeter circuits: (a) the ac input signal is first rectified and then applied to a dc amplifier and meter movement; (b) the ac input signal is first amplified and then applied to a full-wave rectifier in the meter circuit.

This approach ideally requires a dc amplifier with zero drift characteristics and unity voltage gain, and a dc meter movement with adequate sensitivity.

In another approach the ac signal is rectified *after* amplification, as in Fig. 6-5(b) where full-wave rectification takes place in the meter circuit connected to the output terminals of the ac amplifier. This approach generally requires an ac amplifier with high open-loop gain and large amounts of negative feedback to overcome the nonlinearity of the rectifier diodes.

Ac voltmeters are usually of the *average-responding* type, with the meter scale *calibrated* in terms of the rms value of a sine wave. Since so many waveforms in electronics are sinusoidal, this is an entirely satisfactory solution and certainly much less expensive than a true rms-responding voltmeter. Nonsinusoidal waveforms, however, will cause this type of meter to read high or low, depending on the form factor of the waveform.

A few basic rectifier circuits are shown in Fig. 6-6. The series-connected diode of Fig. 6-6(a) provides half-wave rectification, and the average value of the half-wave voltage is developed across the resistor and applied to the input terminals of the dc amplifier. Full-wave rectification can be obtained by the bridge circuit of Fig. 6-6(b), where the average value of the sine wave is applied to the amplifier and meter circuit. In some cases, there may be a requirement to measure the peak value of a waveform instead of the average value; the circuit of Fig. 6-6(c) may then be used. In this circuit the rectifier diode charges the small

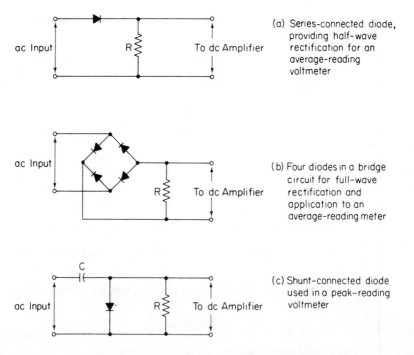

(a) Series-connected diode, providing half-wave rectification for an average-reading voltmeter

(b) Four diodes in a bridge circuit for full-wave rectification and application to an average-reading meter

(c) Shunt-connected diode used in a peak-reading voltmeter

Figure 6-6 Rectifier circuits used in ac voltmeters.

capacitor to the peak of the applied input voltage and the meter will therefore indicate the peak voltage. In most cases, the meter scale is calibrated in terms of both the rms and peak values of the sinusoidal input waveform.

The rms value of a voltage wave that has equal positive and negative excursions is related to the average value by the *form factor*. The form factor, as the ratio of the rms value of the average value of this waveform, for a sinusoid can be expressed as

$$k = \frac{\sqrt{(1/T) \int_0^T e^2 \, dt}}{(2/T) \int_0^{T/2} e \, dt} = \frac{\sqrt{(\omega/2\pi) \int_0^{2\pi/\omega} E_m \sin^2 \omega t \, dt}}{(\omega/\pi) \int_0^{\pi/\omega} E_m \sin \omega t \, dt}$$

$$= \frac{\sqrt{(E_m^2/4\pi) \, [\omega t - \sin \omega t \cos \omega t]_0^{2\pi/\omega}}}{(E_m/\pi) \, [-\cos \omega t]_0^{\pi/\omega}} = \frac{E_m \, 0.707}{E_m \, 0.636} = 1.11 \qquad (6\text{-}1)$$

Therefore when an average-responding voltmeter has scale markings corresponding to the rms value of the applied sinusoidal input waveform, those markings are actually corrected by a factor of 1.11 from the true (average) value of applied voltage.

Nonsinusoidal waveforms, when applied to this voltmeter, will cause the meter to read either high or low, depending on the form factor of the waveform. An illustration of the effect of nonsinusoidal waveforms on ac voltmeter is given in Examples 6-1 and 6-2.

EXAMPLE 6-1

The symmetrical square-wave voltage of Fig. 6-7(a) is applied to an average-responding ac voltmeter with a scale calibrated in terms of the rms value of a sine wave. Calculate (a) the form factor of the square-wave voltage; (b) the error in the meter indication.

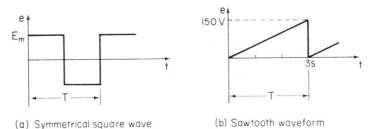

(a) Symmetrical square wave   (b) Sawtooth waveform

Figure 6-7   Waveforms used in Examples 6-1 and 6-2.

SOLUTION   (a) The rms value of the square-wave voltage is

$$E_{rms} = \sqrt{\frac{1}{T} \int_0^T e^2 \, dt} = E_m$$

and the average value is

$$E_{av} = \frac{2}{T} \int_0^{T/2} e \, dt = E_m$$

so that the form factor equals, by definition,

$$k = \frac{E_{rms}}{E_{av}} = 1$$

(b) The meter scale is calibrated in terms of the rms value of a sine-wave voltage, where $E_{rms} = k \times E_{av} = 1.11E_{av}$. For the square-wave voltage, $E_{rms} = E_{av}$, since $k = 1$. Therefore the meter indication for the square-wave voltage is *high* by a factor $k_{sine\ wave}/k_{square\ wave} = 1.11$. The percentage error equals

$$\frac{1.11 - 1}{1} \times 100\% = 11\%$$

EXAMPLE 6-2

Repeat Example 6-1 if the voltage applied to the meter consists of a sawtooth waveform with a peak value of 150 V and a period of 3 s as shown in Fig. 6-7(b).

SOLUTION   (a) The analytical expression for the sawtooth waveform between the limits of $t = 0$ and $t = T = 3$ s is $e = 50t$ V. Therefore

$$E_{rms} = \sqrt{\frac{1}{T} \int_0^T e^2 \, dt} = \sqrt{\frac{1}{3} \int_0^3 (50t)^2 \, dt} = 50 \sqrt{3} \text{ V}$$

$$E_{av} = \frac{1}{T} \int_0^T e \, dt = \frac{1}{3} \int_0^3 50t \, dt = 75 \text{ V}$$

Form factor, $k = \dfrac{50\sqrt{3}}{75} = 1.155$

(b) The ratio of the two form factors is

$$\frac{k_{\text{sine wave}}}{k_{\text{sawtooth}}} = \frac{1.11}{1.155} = 0.961$$

The meter indication is *low* by a factor of 0.961. The percentage error equals

$$\frac{0.961 - 1}{1} \times 100\% = -3.9\%$$

Examples 6-1 and 6-2 point out that any departure from a true sinusoidal waveform may cause an appreciable error in the result of the measurement.

6-4 TRUE RMS-RESPONDING VOLTMETER

Complex waveforms are most accurately measured with an *rms-responding volt-meter*. This instrument produces a meter indication by sensing waveform *heating power*, which is proportional to the square of the rms value of the voltage. This heating power can be measured by feeding an amplified version of the input waveform to the heater element of a *thermocouple* whose output voltage is then proportional to E_{rms}^2.

One difficulty with this technique is that the thermocouple is often nonlinear in its behavior. This difficulty is overcome in some instruments by placing two thermocouples in the same thermal environment, as shown in the block diagram of the true rms-responding voltmeter of Fig. 6-8. The effect of the nonlinear behav-

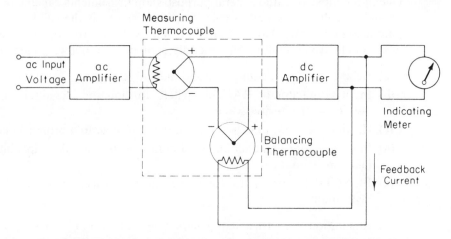

Figure 6-8 Block diagram of a true rms-reading voltmeter. The measuring and balancing thermocouples are located in the same thermal environment.

ior of the couple in the input circuit (the *measuring* thermocouple) is canceled by similar nonlinear effects of the couple in the feedback circuit (the *balancing* thermocouple). The two couple elements form part of a bridge in the input circuit of a dc amplifier. The unknown ac input voltage is amplified and applied to the heating element of the measuring thermocouple. The application of heat produces an output voltage that upsets the balance of the bridge. The unbalance voltage is amplified by the dc amplifier and fed back to the heating element of the balancing thermocouple. Bridge balance will be reestablished when the feedback current delivers sufficient heat to the balancing thermocouple, so that the voltage outputs of both couples are the same. At this point the dc current in the heating element of the feedback couple is equal to the ac current in the input couple. This dc current is therefore directly proportional to the effective, or rms, value of the input voltage and is indicated on the meter movement in the output circuit of the dc amplifier. The true rms value is measured independently of the waveform of the ac signal, provided that the peak excursions of the waveform do not exceed the dynamic range of the ac amplifier.

A typical laboratory-type rms-responding voltmeter provides accurate rms readings of complex waveforms having a *crest factor* (ratio of peak value to rms value) of 10/1. At 10 per cent of full-scale meter deflection, where there is less chance of amplifier saturation, waveforms with crest factors as high as 100/1 could be accommodated. Voltages throughout a range of 100 μV to 300 V within a frequency range of 10 Hz to 10 MHz may be measured with most good instruments.

6-5 ELECTRONIC MULTIMETER

6-5.1 Basic Circuit

One of the most versatile general-purpose shop instruments capable of measuring dc and ac voltages as well as current and resistance is the solid-state electronic multimeter or VOM. Although circuit details will vary from one instrument to the next, an electronic multimeter generally contains the following elements:

(a) Balanced-bridge dc amplifier and indicating meter
(b) Input attenuator or RANGE switch, to limit the magnitude of the input voltage to the desired value
(c) Rectifier section, to convert an ac input voltage to a proportional dc value
(d) Internal battery and additional circuitry, to provide the capability of resistance measurement
(e) FUNCTION switch, to select the various measurement functions of the instrument

In addition, the instrument generally has a built-in power supply for ac line operation and, in most cases, one or more batteries for operation as a portable test instrument.

Figure 6-9 shows the schematic diagram of a balanced-bridge dc amplifier using field effect transistors or FETs. This circuit also applies to a bridge amplifier with ordinary bipolar transistors or BJTs. The circuit shown here consists of two FETs which sould be reasonably well matched for current gain to ensure thermal stability of the circuit. The two FETs form the upper arms of a bridge circuit. Source resistors R_1 and R_2, together with ZERO adjust resistor R_3, form the lower bridge arms. The meter movement is connected between the source terminals of the FETs, representing two opposite corners of the bridge.

Without an input signal, the gate terminals of the FETs are at ground potential and the transistors operate under identical quiescent conditions. In this case, the bridge is balanced and the meter indication is zero. In practice, however, small differences in the operating characteristics of the transistors, and slight tolerance differences in the various resistors, cause a certain amount of unbalance in the drain currents, and the meter shows a small deflection from zero. To return the meter to zero, the circuit is balanced by ZERO adjust control R_3 for a true null indication.

When a positive voltage is applied to the gate of input transistor Q_1, its drain current increases which causes the voltage at the source terminal to rise. The resulting unbalance between the Q_1 and Q_2 source voltages is indicated by the meter movement, whose scale is calibrated to agree with the magnitude of the applied input voltage.

The maximum voltage that can be applied to the gate of Q_1 is determined by the operating range of FET and is usually on the order of a few volts. The range of input voltages can easily be extended by an input attenuator or RANGE switch, as shown in Fig. 6-10. The unknown dc input voltage is applied through a large resistor in the probe body to a resistive voltage divider. Thus, with the RANGE switch in the 3-V position as shown, the voltage at the gate of the input FET is

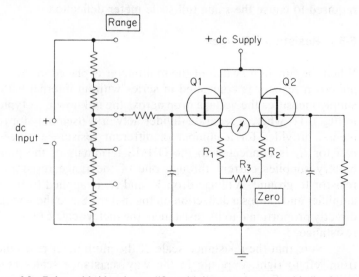

Figure 6-9 Balanced-bridge dc amplifier with input attenuator and indicating meter.

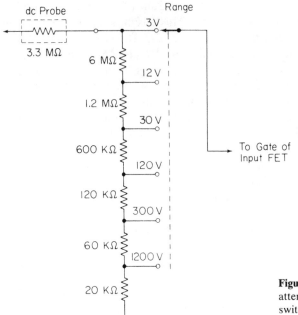

Figure 6-10 Typical input voltage attenuator for a VOM. The RANGE switch on the front of the panel of the VOM allows selection of the desired voltage range.

developed across 8 MΩ of the total resistance of 11.3 MΩ and the circuit is so arranged that the meter deflects full scale with 3 V applied to the tip of the probe. With the RANGE switch in the 12-V position, the gate voltage is developed across 2 MΩ of the total divider resistance of 11.3 MΩ and an input voltage of 12 V is required to cause the same full-scale meter deflection.

6-5.2 Resistance Ranges

When the function switch of the multimeter is placed in the OHMS position, the unknown resistor is connected in series with an internal battery, and the meter simply measures the voltage drop across the unknown. A typical circuit is shown in Fig. 6-11, where a separate divider network, used only for resistance measurements, provides for a number of different resistance ranges. When unknown resistor R_x is connected to the OHMS terminals of the multimeter, the 1.5-V battery supplies current through one of the range resistors and the unknown resistor to ground. Voltage drop V_x and R_x is applied to the input of the bridge amplifier and causes a deflection on the meter. Since the voltage drop across R_x is directly proportional to its resistance, the meter scale can be calibrated in terms of resistance.

Note that the resistance scale of the multimeter reads increasing resistance from left to right, opposite to the way resistance scales read on conventional multimeters (Sec. 4-9). This can be expected because the electronic multimeter

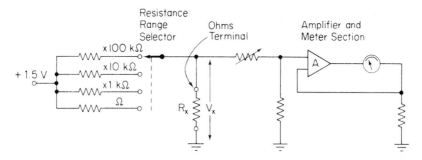

Figure 6-11 Resistance range selector circuit of a VOM.

reads a larger resistance as a higher voltage, whereas the ordinary multimeter indicates a higher resistance as a smaller current.

6-5.3 Commercial Multimeter

The simplified metering circuit of a commercial solid-state VOM is given in Fig. 6-12. The dc voltage from the input voltage divider (Fig. 6-9) is applied to the

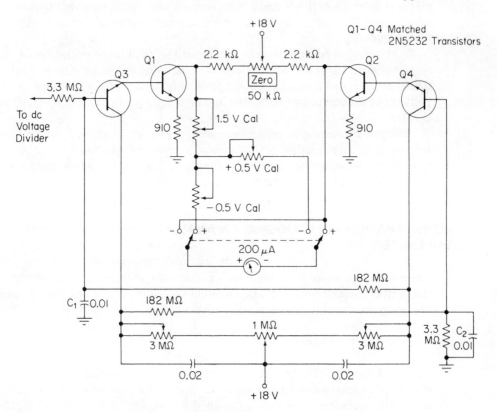

Figure 6-12 Typical metering circuit of a solid-state VOM.

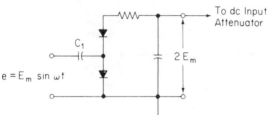

Figure 6-13 Full-wave peak-to-peak rectifier, also known as a voltage doubler.

bases of bridge preamplifier transistors Q_3 and Q_4. These emitter followers provide nearly infinite input impedence and therefore present a minimum load to the high-resistance input voltage divider. Preamplifier transistors Q_3 and Q_4 drive the bases of bridge amplifier transistors Q_1 and Q_2, respectively. The input impedances of Q_1 and Q_2 are very high because of their unbypassed emitter resistors, which prevent loading of the Q_3 and Q_4 emitters. The output voltage of the bridge amplifier is indicated on the 200-μA meter, connected between the collectors of Q_1 and Q_2. The front-panel ZERO control balances the meter amplifier output with zero input signal. Internal adjustments allow for meter calibration with two accurate test voltages of 0.5 V and 1.5 V, respectively. Also note that bypass capacitors C_1 and C_2 prevent ac signals from reaching the amplifier and affecting the meter reading.

Ac voltages being measured are applied to a full-wave peak-to-peak rectifier that charges a capacitor to the peak-to-peak value of the ac signal. A circuit of this type is also known as a *voltage doubler* and is shown in Fig. 6-13. The rectified ac voltage is then fed to the amplifier through the regular RANGE voltage divider.

When resistance is being measured, 1.5 V dc is applied to the unknown resistor through one of the resistance range resistors, as shown in Fig. 6-11. The known and the unknown resistances form a voltage divider whose output is fed to the amplifier and read on the meter in terms of resistance.

6-6 CONSIDERATIONS IN CHOOSING AN ANALOG VOLTMETER

The most appropriate instrument for a particular voltage measurement depends on the performance required in a given situation. Some important considerations in choosing a voltmeter are summarized below.

6-6.1 Input Impedance

To avoid loading effects, the input resistance or impedance of the voltmeter should be at least an order of magnitude higher than the impedance of the circuit under measurement. For example, when a voltmeter with a 10-MΩ input resistance is used to measure the voltage across a 100-kΩ resistor, the circuit is hardly disturbed and the loading effect of the meter on the circuit is negligible. The same meter placed across a 10-MΩ resistor, however, seriously loads the circuit and causes an error in measurement of approximately 50 per cent.

The input impedance of the voltmeter is a function of the inevitable shunt capacitance across the input terminals. The loading effect of the meter is particularly noticeable at the higher frequencies, when the input shunt capacitance greatly reduces the input impedance.

In some applications, a passive voltage-divider probe can be used to reduce the input capacitance at the point of measurement at the sacrifice of perhaps 20 dB of sensitivity. With such a probe, measurements can be easily made at random points without disturbing the circuit under test.

6-6.2 Voltage Ranges

The voltage ranges on the meter scale may be in the 1-3-10 sequence with 10 dB of separation, or in the 1.5-5-15 sequence, or in a single scale calibrated in decibels. In any case, the scale divisions should be compatible with the accuracy of the instrument. For example, a linear meter with a 1 per cent full-scale accuracy should have 100 divisions on the 1.0-V scale so that 1 per cent can be easily resolved. An instrument with an accuracy of 1 per cent or less should also have a mirror-backed scale to reduce parallax and improve accuracy.

6-6.3 Decibels

Use of the decibel scale can be very effective in measurements that cover a *wide range* of voltages. A measurement of this kind is found, for example, in the frequency response curve of an amplifier or a filter, where the output voltage is measured as a function of the frequency of the applied input voltage. Almost all voltmeters with dB scales are calibrated in dBm, referenced to some particular impedance. The 0-dBm reference for a 600-Ω system is 0.7746 V; for a 50-Ω system it is 0.2236 V. In many applications only a 0-dB reference is needed. In this case, 0 dBv (relative to 1 V) can be used for any impedance system.

6-6.4 Sensitivity Versus Bandwidth

Noise is a function of bandwidth. A voltmeter with a broad bandwidth will pick up and generate more noise than one operating over a narrow range of frequencies. In general, an instrument with a bandwidth of 10 Hz to 10 MHz has a sensitivity of 1 mV. A voltmeter whose bandwidth extends only to 5 MHz could have a sensitivity of 100 μV.

6-6.5 Battery Operation

For field work, a voltmeter powered by an internal battery is essential. If an area contains troublesome groundloops, a battery-powered instrument is preferred over a mains-powered voltmeter to remove the groundpaths.

6-6.6 AC Current Measurements

Current measurements can be made by a sensitive ac voltmeter and a series resistance. In the usual case, however, an ac *current probe* is used which enables the operator to measure an ac current without disturbing the circuit under test. The current probe simply clips around the wire carrying the unknown current and in effect makes the wire the one-turn primary of a transformer formed by a ferrite core and a many-turn secondary within the current-probe body. The signal induced in the secondary winding is amplified and the output voltage of the amplifier is applied to a suitable ac voltmeter for measurement. Normally, the amplifier is designed so that 1 mA in the wire being measured produces 1 mV at the amplifier output. The current is then read directly on the voltmeter, using the same scale as for voltage measurements.

In *summarizing* the preceding considerations, the following general guidelines can be stated:

(a) For measurements involving dc applications, select the meter with the broadest capability meeting the circuit's requirements.

(b) For ac measurements involving sine waves with only modest amounts of distortion ($<$10 per cent), the average-responding voltmeter provides the best accuracy and most sensitivity per dollar investment.

(c) For high-frequency measurements ($>$10 MHz), the peak-responding voltmeter with a diode-probe input is the most economical choice. Peak-responding circuits are acceptable if the inaccuracies caused by distortion in the input waveform can be tolerated.

(d) For measurements where it is important to determine the effective power of waveforms which depart from the true sinusoidal form, the rms-responding voltmeter is the appropriate choice.

6-7 DIGITAL VOLTMETERS

6-7.1 General Characteristics

The digital voltmeter (DVM) displays measurements of dc or ac voltages as discrete numerals instead of a pointer deflection on a continuous scale as in analog devices. Numerical readout is advantageous in many applications because it reduces human reading and interpolation errors, eliminates parallax error, increases reading speed, and often provides outputs in digital form suitable for further processing or recording.

The DVM is a versatile and accurate instrument that can be used in many laboratory measurement applications. Since the development and perfection of integrated circuit (IC) modules, the size, power requirements, and cost of the

DVM have been drastically reduced so that DVMs can actively compete with conventional analog instruments, both in portability and price.

The DVM's outstanding qualities can best be illustrated by quoting some typical operating and performance characteristics. The following specifications do not all apply to one particular instrument, but they do represent valid information on the present state of the art:

(a) *Input range:* from ± 1.000000 V to $\pm 1,000.000$ V, with automatic range selection and overload indication

(b) *Absolute accuracy:* as high as ± 0.005 per cent of the reading

(c) *Stability:* short-term, 0.002 per cent of the reading for a 24-hr period; long-term, 0.008 per cent of the reading for a 6-month period

(d) *Resolution:* 1 part in 10^6 (1 μV can be read on the 1-V input range)

(e) *Input characteristics:* input resistance typically 10 MΩ; input capacitance typically 40 pF

(f) *Calibration:* internal calibration standard allows calibration independent of the measuring circuit; derived from stabilized reference source

(g) *Output signals:* print command allows output to printer; BCD (binary-coded-decimal) output for digital processing or recording

Optional features may include additional circuitry to measure current, resistance, and voltage ratios. Other physical variables may be measured by using suitable transducers.

Digital voltmeters can be classified according to the following broad categories:

(a) Ramp-type DVM
(b) Integrating DVM
(c) Continuous-balance DVM
(d) Successive-approximation DVM

6-7.2 Ramp-Type DVM

The operating principle of the ramp-type DVM is based on the measurement of the time it takes for a linear ramp voltage to rise from 0 V to the level of the input voltage, or to decrease from the level of the input voltage to zero. This time interval is measured with an electronic time-interval counter, and the count is displayed as a number of digits on electronic indicating tubes.

Conversion from a voltage to a time interval is illustrated by the waveform diagram of Fig. 6-14. At the start of the measurement cycle, a ramp voltage is initiated; this voltage can be positive-going or negative-going. The negative-going ramp, shown in Fig. 6-14, is continuously compared with the unknown input voltage. At the instant that the ramp voltage equals the unknown voltage, a

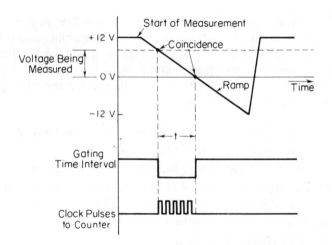

Figure 6-14 Voltage-to-time conversion using gated clock pulses.

coincidence circuit, or comparator, generates a pulse which opens a gate. This gate is shown in the block diagram of Fig. 6-15. The ramp voltage continues to decrease with time until it finally reaches 0 V (or ground potential) and a second comparator generates an output pulse which closes the gate.

An oscillator generates clock pulses which are allowed to pass through the gate to a number of decade counting units (DCUs) which totalize the number of pulses passed through the gate. The decimal number, displayed by the indicator tubes associated with the DCUs, is a measure of the magnitude of the input voltage.

The sample-rate multivibrator determines the rate at which the measurement cycles are initiated. The oscillation of this multivibrator can usually be adjusted by a front-panel control, marked *rate*, from a few cycles per second to as high as

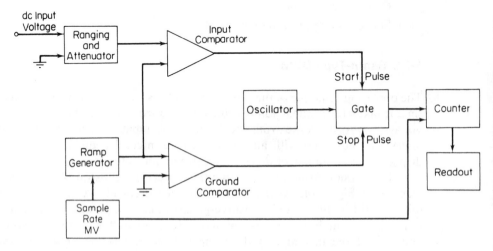

Figure 6-15 Block diagram of a ramp-type digital voltmeter.

1,000 or more. The sample-rate circuit provides an initiating pulse for the ramp generator to start its next ramp voltage. At the same time, a reset pulse is generated which returns all the DCUs to their 0 state, removing the display momentarily from the indicator tubes.

6-7.3 Staircase-Ramp DVM

The staircase-ramp DVM is given in block diagram form in Fig. 6-16. It is a variation of the ramp-type DVM but is somewhat simpler in overall design, resulting in a moderately priced general-purpose instrument that can be used in the laboratory, on production test-stands, in repair shops, and at inspection stations.

This DVM makes voltage measurements by comparing the input voltage to an internally generated staircase-ramp voltage. The instrument shown in Fig. 6-16 contains a 10-MΩ input attenuator, providing five input ranges from 100 mV to 1,000 V full scale. The dc amplifier, with a fixed gain of 100, delivers 10 V to the comparator at any of the full-scale voltage settings of the input divider. The comparator senses coincidence between the amplified input voltage and the staircase-ramp voltage which is generated as the measurement proceeds through its cycle.

When the measurement cycle is first initiated, the clock (a 4.5-kHz relaxation oscillator) provides pulses to three DCUs in cascade. The *units* counter provides a carry pulse to the *tens* decade at every tenth input pulse. The *tens* decade counts the carry pulses from the *units* decade and provides its own carry pulse after it has counted ten carry pulses. This carry pulse is fed to the *hundreds* decade which provides a carry pulse to an overrange circuit. The overrange circuit causes a front panel indicator to light up, warning the operator that the input capacity of the instrument has been exceeded. The operator should then switch to the next higher setting on the input attenuator.

Each decade counter unit is connected to a digital-to-analog (D/A) converter. The outputs of the D/A converters are connected in parallel and provide an output current proportional to the current count of the DCUs. The staircase amplifier converts the D/A current into a staircase voltage which is applied to the comparator. When the comparator senses coincidence of the input voltage and the staircase voltage, it provides a trigger pulse to stop the oscillator. The current content of the counter is then proportional to the magnitude of the input voltage.

The sample rate is controlled by a simple relaxation oscillator. This oscillator triggers and resets the transfer amplifier at a rate of two samples per second. The transfer amplifier provides a pulse that transfers the information stored in the decade counters to the front panel display unit. The trailing edge of this pulse triggers the reset amplifier which sets the three decade counters to zero and initiates a new measurement cycle by starting the master oscillator or clock.

The display circuits store each reading until a new reading is completed, eliminating any blinking or counting during the computation.

The ramp type of A/D converter requires a precision ramp to achieve accuracy. Maintaining the quality of the ramp requires a precision, stable capacitor and resistor in the integrator. In addition, the offset voltages and currents of the

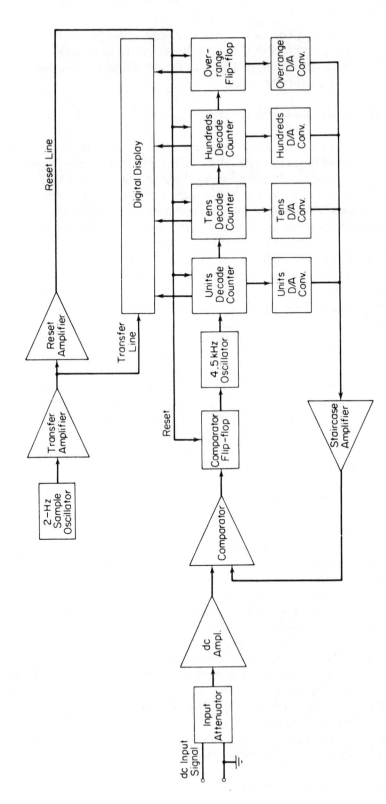

Figure 6-16 Block diagram of a staircase-ramp digital voltmeter.

operational amplifier used in the integrator are critical in the accurate ramp generator. One method of reducing the dependence of the accuracy of the conversion on the resistor, capacitor, and operational amplifier is to use a technique called the dual-slope converter.

In the dual-slope technique, an integrator is used to integrate an accurate voltage reference for a fixed period of time. The same integrator is then used to integrate with the reverse slope, the input voltage, and the time required to return to the starting voltage is measured.

It does not matter which of the two integrations occurs first, and for ease of understanding, the case where the unknown is used to integrate first and then the reference will be considered.

The output of an integrator shown in Fig. 6-17(a) is

$$V_{out} = - \frac{V_x t}{RC} \qquad (6\text{-}2)$$

where V_x = steady input voltage relative to ground

V_{out} = the output voltage from the integrator

R, C = integrator time-constant components

t = elapsed time from when the integration began

Equation (6-2) also assumes that the integrator capacitor started with no charge and thus the output of the integrator started at zero volts.

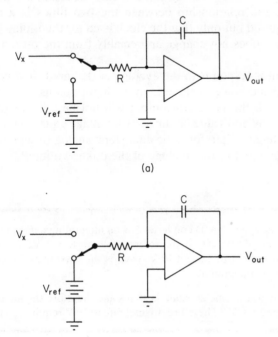

(a)

(b)

Figure 6-17 Functional schematic of the integrator of an integrating DVM.

If the integration were allowed to continue for a fixed period of time T_1, the output voltage would be

$$V_1 = -\frac{V_x}{RC} T_1 \qquad (6\text{-}3)$$

Notice that the integrator output has gone in the opposite polarity as the input. That is, a positive input voltage produces a negative integrator output.

If a reference voltage, V_{ref}, were substituted for the input voltage V_x, as shown in Fig. 6-17(b), the integrator would begin to ramp toward zero at a rate of V_{ref}/RC assuming that the reference voltage was of the opposite polarity as the unknown input voltage. The integrator for this situation does not start at zero but at an output voltage of V_1 and the output voltage can be represented as

$$V_{out} = V_1 + \frac{V_{ref}}{RC} t \qquad (6\text{-}4)$$

Notice that the second term in Eq. (6-4) has a negative sign due to its polarity.

Setting the output voltage of the integrator to zero and solving for V_x yields

$$V_x = \frac{T_x}{T_1} V_{ref} \qquad (6\text{-}5)$$

where T_x is the time required to ramp down from the output level of V_1 to zero volts.

Notice that the relationship between the reference voltage and the input voltage does not include R or C of the integrator but only the relationship between the two times. Because the relationship between the two times is a ratio, an accurate clock is not required but only that the clock used for the timing be stable enough that the frequency does not change appreciably from the up ramp to the down ramp.

Because the integrator responds to the average of the input, it is not necessary to provide a sample and hold, as changes in the input voltage will not cause significant errors. Although the integrator output will not be a linear ramp, the integration will represent the end value obtained by a voltage equal to the average of the unknown input voltage. Therefore, the dual-slope analog-to-digital conversion will produce a value equal to the average of the unknown input.

EXAMPLE 6-3

A dual-slope integrating type of A/D converter has an integrating capacitor of 0.1 μF and a resistance of 100 kΩ. If the reference voltage is 2 V, and the output of the integrator is not to exceed 10 V, what is the maximum time the reference voltage can be integrated?

SOLUTION   The integrator time constant is 10 ms and therefore the integrator output is 200 V/s or 5 ms/V. Therefore, to integrate to 10 V requires 50 ms.

The dual-slope type of A/D conversion is a very popular method for digital voltmeter applications. When compared to other types of analog-to-digital conversion techniques, the dual-slope method is slow but is quite adequate for a digital voltmeter used for laboratory measurements. For data acquisition applications, where a number of measurements are required, faster techniques are recommended. Many refinements have been made to the technique and many large-scale-integration (LSI) chips are available to simplify the construction of DVMs.

When a dual-slope A/D converter is used for a DVM the counters may be decade rather than binary and the segment and digit drivers may be contained in the chip. (Multiplexed counter displays are discussed in Chapter 10.) When the converter is to interface to a microprocessor, and many high-performance DVMs use microprocessors for data manipulation, the counters employed are binary.

One significant enhancement made to the dual-slope converter is automatic zero correction. As with any analog system, amplifier offset voltages, offset currents, and bias currents can cause errors. In addition, in the dual-slope A/D converter, the leakage current of the capacitor can cause errors in the integration and consequentially, an error. These effects, in the dual-slope A/D converter, will manifest themselves as a reading of the DVM when no input voltage is present. Figure 6-17 shows a method of counteracting these effects. The input to the converter is grounded and a capacitor, the auto zero capacitor, is connected via an electronic switch to the output of the integrator. The feedback of the circuitry is such that the voltage at the integrator output is zero. This effectively places an equivalent offset voltage on the automatic zero capacitor so that there is no integration. When the conversion is made, this offset voltage is present to counteract the effects of the input circuitry offset voltages. This automatic zero function is performed before each conversion, so that changes in the offset voltages and currents will be compensated.

Figure 6-18 shows a complete dual-slope A/D converter. Electronic

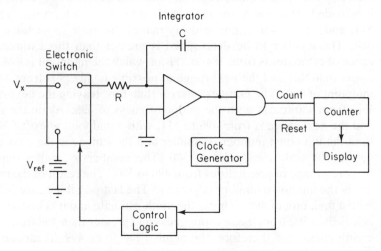

Figure 6-18 Block diagram of a dual-slope DVM.

switches, usually FET switches, are used to switch the input of the integrator alternately between the reference voltage and the unknown. Another pair of switches apply the integrator output to the automatic zero capacitor and ground the input for the automatic zero function.

All of the switch timing and the counting of the clock pulses to determine the unknown voltage are under control of the control logic. The output is made available to the external electronics after the conversion is complete.

If, in this example, the reference voltage were 1.000 V and the integrator were allowed to integrate the reference for 1,000 counts, the display would read 1 V full scale with a resolution of 1 mV.

The actual frequency of the clock is not critical, as previously explained, but has an effect on the speed of the conversion. As an example, a 10-kHz clock would allow a maximum conversion time of 0.2 s for the example described above.

6-7.4 Successive-Approximation Conversion

A very effective and relatively inexpensive method of analog-to-digital conversion is the method of successive approximation. This is an electronic implementation of a technique called binary regression.

Assume that one is to determine the value of a number and is allowed to make estimates. Each estimate would be evaluated and it would be known if the estimate was (1) equal to or less than or (2) greater than the number to be determined. The maximum and minimum value of the possible number is also known.

Consider, as an example, that a number to be determined is between 0 and 511. The best first guess would be some number midway between the extremes and, ideally, 256. To further the example, assume that the number to be determined is 499. The number is greater than the estimate of 256 and this information is provided. It is now known that the number to be determined is between 256 and 511, and, again, something midway makes the most sense for a guess, which is 384. The number to be determined is greater than this estimate, and the next range of estimates is from 384 to 511 for which the midpoint is 448. The number is larger than this and the next range of possible numbers is from 448 to 511, with a midpoint of 480. The number is larger than this, leaving the next range of possibilities from 480 to 511 with the midpoint guess of 496. Again the number is larger and the next range is from 496 to 511, with a midpoint guess of 504. For the first time the unknown number is smaller and the range for the next estimate is from 496 to 504 with a midpoint of 500. The number is smaller than this estimate, leaving a range of possibilities from 496 to 500. The result of the midpoint guess of 498 is the unknown number is greater. The last possible range is from 498 to 500, with a midpoint of 499. This is the ninth estimate and it is known that the number is less than 500 from the seventh guess and greater than 498 from the results of the eighth guess, and therefore, the number has to be 499. A tabular synopsis of the guesses and the results follows.

Estimate	Result
256	Equal to or less than
256 + 128 = 384	Equal to or less than
384 + 64 = 448	Equal to or less than
448 + 32 = 480	Equal to or less than
480 + 16 = 496	Equal to or less than
496 + 8 = 504	Greater than
496 + 4 = 500	Greater than
496 + 2 = 498	Equal to or less than
498 + 1 = 499	Correct

There are some interesting observations to be made from the tabulation. First, there were eight estimates set forth when the actual answer was known. After the eighth estimate the actual value was known to lie between 598 and 500, which is knowing the answer to an 8-bit accuracy plus or minus one bit.

Can all numbers between 0 and 512 be determined in eight guesses or less using this method? To determine the answer to this question, consider the following. The first estimate cannot be in error by more than 256. The second estimate cannot be in error by more than 128, the third by more than 64, and so on. A total of nine estimates are required to produce the final estimate, which is in error by no more than 1, which is the minimum possible error. Numbers from 0 to 511 can be represented by 9 binary bits. It is clear that this analysis can be extended to any number of binary bits, and the number of estimates required is exactly equal to the number of bits required of the analog-to-digital conversion.

A graphical representation of the estimates of the successive approximation conversion illustrates the converging nature of this technique. Figure 6-19 shows a graphical representation of the example of 499. As can be seen, the estimates approach the value from below, oscillating around the desired answer before settling on the correct answer. The oscillation is difficult to see as the error quickly becomes small and the amplitude of the oscillation is, likewise, small. As a comparison, Fig. 6-19 also shows a graphical representation of the estimates used to arrive at the value 320. There is more oscillation, but the actual value is arrived at within nine estimates.

The electronic implementation of the successive-approximation technique is relatively straightforward and is shown in Fig. 6-20. A D/A converter is used to provide the estimates. The "equal to or greater than" or "less than" decision is made by a comparator. The D/A converter provides the estimate and is compared to the input signal. A special shift register called a successive-approximation register (SAR) is used to control the D/A converter and consequentially the estimates. At the beginning of the conversion all the outputs from the SAR are at logic zero. If the estimate is greater than the input, the comparator output is high and the first SAR output reverses state and the second output changes to a logic "one." If the comparator output is low, indicating that the estimate is lower than the input signal, the first output remains in the logic one state and the second

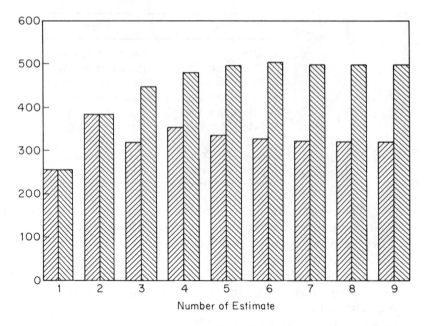

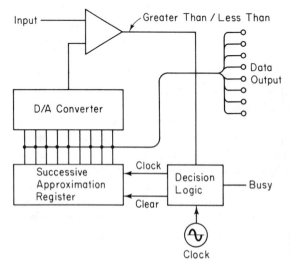

Figure 6-19 Graphical display of the values of estimates from a successive approximation type DVM.

output assumes the logic one state. This continues to all the states until the conversion is complete.

This sequence of events performs, electronically, the same estimating procedure that was outlined previously. An estimate is made on the edge of the SAR clock. For an N-bit conversion after N clocks, the actual value of the input is known. The least significant bit is the state of the comparator. In some systems

Figure 6-20 Block diagram of a successive-approximation DVM.

an additional clock is used to store the last bit in the SAR and thus $N + 1$ clocks are required for a conversion.

6-7.5 Quantizing Error

An electrical parameter, whether it is voltage, current, power, or something else, can assume any value within the possible range of that parameter. When the quantity is converted to a digital form, there are only a finite number of values that the quantity may assume. As an example, if a digital number consists of four bits, which has 16 different combinations, there are 16 levels that the analog quantity can be described.

Consider a voltage range of zero to 15 V to be digitized to a 4-bit number. There is a binary number for each volt of that range. What if the actual analog value is between the quantizing levels such as 2.25 V? The digitizing can produce a value equal to 2 V, which is represented as 0010, or 3 V, represented as 0011. The solution is simple. Round off the number to 2.0 and accept the digitized value of 0010. There is an error, however. The difference between the actual value of the analog quantity and the digitized value is 0.25. The number of bits of the quantized number could be increased by two, to 6 bits, and the number could be represented as 0010.01, which is exact with no error. What if the actual analog value were 2.27. Now the closest possible values of the binary number would be 2.25 or 2.50. Clearly, the closest is 2.25, which expresses the analog quantity with an error of only 0.02.

It would be clear, regardless of how many bits are used to express an analog quantity, that there is always a possibility of error when the quantities to be digitized fall between the exact digital values. The maximum error is equal to plus or minus one-half of the value of the least significant bit, which is called the quantizing error.

Older analog meters that used a meter scale as an indicator device required some sort of ranging circuitry so that the meter could be used over a large range of input parameters. As an example, if the full scale of a meter is 1,000 V, it would be nearly impossible to see the effects of a 1-V input. Therefore, a switched attenuator was used in the meter to provide a 1,000-, 100-, 10- and 1-V full-scale reading so that the desired meter deflections could be easily readable.

In the case of a digital meter, if a four-digit meter had a full-scale reading of 999.9 V, a 1-V reading would appear as 001.0. This represents two significant digits for the 1-V reading. The meter, however, is a four-digit meter, and 99 per cent of the meter capability is not used when reading 1 V. This is based on the fact that a four-digit meter can resolve 1 part in 10,000, while the two significant digits reflected by the 001.0-V display would represent 1 part in 100 or only 1 per cent of 1 part in 10,000. A switchable attenuator in the digital meter would achieve a similar effect. If an attenuator were used so that the full-scale readings of the digital meter were 999.9, 99.99, 9.999, and 0.9999 V, the 1-V reading would be 1.000, which is four significant digits and does not waste any of the capability of the meter.

Modern digital meters are capable of electronically switching the input attenuator, which makes the meter fully automatic. The electronics must determine if the present reading is less than the next-lower range of higher than full scale.

If the present reading is less than the full scale of the next-lower range, the attenuation is reduced. The attenuation continues to be reduced until the reading is between the next lower range and the full scale of the present range.

An opposite scenerio takes place when the present reading is more than full scale. In this case attenuation is increased until the present reading is less than full scale.

As an example, assume that the 1-V of the previous example was measured using a digital voltmeter that was presently on the 999.9-V range. The attenuator is in decade steps, which is 999.9, 99.99, 9.999, and 0.9999 V full scale. Since the reading is 001.0, this is less than the full scale of the next-lower range and the attenuation is decreased. This results in a reading of 01.00 V, which is still less than the full scale of the next-lower range and the attenuation is reduced automatically. The next reading is 1.000, which is greater than the next-lower-range full-scale reading of 0.9999 and the attenuation is reduced no further.

Because the variation in input voltage levels can be rather great, the input attenuator is often switched with relays rather than electronic switches. In addition, there are times when the input voltage is much greater than the full-scale reading and the input amplifiers and attenuators must be capable of withstanding a significant overload for a short period of time before the proper attenuation is found. This technique is called autoranging.

A similar technique is used to counteract the effects of various offset voltages within a digital meter. The input of the meter is electronically switched to ground while the input is disconnected. The meter will now read the results of the offset voltages, leakage currents, or other effects.

This offset reading is compensated by either subtracting the offset from the display or by feeding an analog offset in the opposite polarity. The offset check is performed at a regular rate to ensure that the change in offset is accounted for.

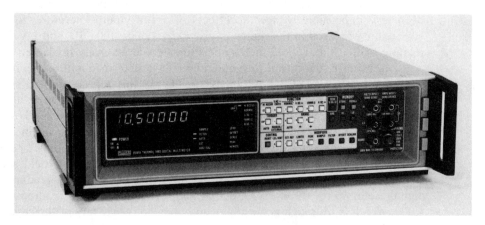

Figure 6-21 Example of a high performance digital multimeter. (Courtesy of John Fluke Mfg. Co., Inc.)

This technique is called auto zero and is necessary for instruments of high accuracy. Fig. 6-21 shows a high-performance auto-ranging multimeter with a true rms capability for ac measurements.

6-8 COMPONENT MEASURING INSTRUMENTS

Bridges for measuring component values of resistance, inductance, and capacitance were discussed in Chapter 5. Bridges are potentially very accurate and reliable for component measurements using measuring frequencies to the low megahertz region. They have some disadvantages in that they involve a variable inductor, resistor, or capacitor, depending on the type of bridge, and this usually involves an operator. This adjustment makes it difficult to automate or computerize the measurement since an actual mechanical movement is required. For manual measurements, this tends to slow down the measurements, but for computer interface, this tends to make the task nearly impossible.

6-8.1 All-Electronic Component Measurements

Chapter 5 discussed the Wheatstone bridge for resistance measurements, and the simple ohmmeter was discussed in Chapter 4. This is an example of a bridge and an all-electronic instrument for measuring resistance. (In the case of the moving-coil meter, the actual meter movement is mechanical, but this could be replaced with a digital readout, making the resistance measurement all-electronic.)

There are several methods of performing an all-electronic inductance or capacitance measurement where the measurement is not performed by a comparison, as is the case with a bridge. Fig. 6-22 shows one possible method of measuring the value of a capacitor, where a voltage is applied to the capacitor and the current through the capacitor can be measured. The relationship between the current through a capacitor and the voltage applied to the capacitor is

$$I_C = \frac{V}{X_C} = V(2\pi f C) \qquad (6-6)$$

where V = applied voltage

$\quad f$ = applied frequency

$\quad C$ = capacitance

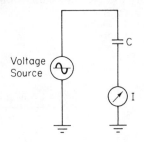

Figure 6-22 This circuit can measure the value of a capacitor by measuring the current through the capacitor with a known applied voltage.

The meter is simply calibrated in capacitance because of the linear relationship between the capacitance and the current. Although in theory this is a useful circuit, it is not practical because of the typical values of capacitors encountered in the electronics industry. Capacitors of a few picofarads are not unusual, and these capacitors typically could have working voltages of less than 25 V. RF current measuring devices, essentially thermocouple instruments, are not available for currents of less than a few hundred milliamperes, and thus the current expected must be greater than a few hundred milliamperes. If, as an example, a capacitor of 10 pF were to produce a current of 100 mA, with an applied voltage of 10 V rms, which would be safe for a 25-V capacitor, the frequency would have to be higher than 1,600 MHz. At this frequency, most capacitors have ceased to behave as capacitors and lead inductance, dissipation resistance, and other parasitic impedances will dominate the measurement. In addition, the accuracy of the measurement is dependent on the frequency of the generator, which would be difficult to control at 1,600 MHz. Therefore, smaller currents must be used for capacitance measurements.

An alternative method is shown in Fig. 6-23. In this example the current through the capacitor is sampled across a known resistance and the resultant voltage is amplified and measured. The amplifier provides the necessary gain so that the current through the capacitor can be quite small and within practicality. The voltage across the resistor can be expressed as

$$V = \frac{RV_{\text{in}}}{\sqrt{R^2 + \left(\dfrac{1}{2\pi fC}\right)^2}} \tag{6-7}$$

where R = resistance

V_{in} = generator voltage

V = voltage across the resistor

C = capacitance of the unknown capacitor

f = frequency of the generator.

If V_{in}, f, and R are kept constant, the voltage V is a function of the unknown capacitance. The scale would have to be calibrated in a nonlinear fashion because of the relationship of Eq. (6-7). An applied frequency of a few megahertz can provide a practical system using this technique. The actual movement of the meter depends on not only the constants mentioned above, but on the gain of the

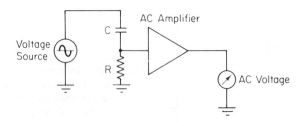

Voltage Source

AC Amplifier

C

R

AC Voltage

Figure 6-23 Measuring the current through a capacitor using a sampling resistor and an amplifier.

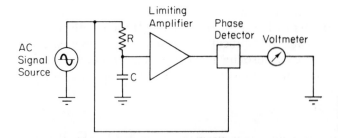

Figure 6-24 Capacitance measuring meter using the phase shift characteristics of RC circuit.

amplifier. It can be difficult to maintain a constant gain in an amplifier at several megahertz, especially for the large dynamic range encountered while measuring capacitance using this system. An alternative approach is shown in Fig. 6-24. In this example the phase angle between the applied voltage and the voltage across the capacitor is measured. An amplifier is used in this scheme except that the gain of the amplifier is not a factor in the measurement. Typically, a limiting amplifier such as that found in an FM receiver would be used. The phase angle can be expressed as

$$\theta = \tan^{-1} \frac{R}{X_C} = \tan^{-1} (2\pi fRC) \qquad (6\text{-}8)$$

The angle, θ, will be read by the meter in this circuit and the meter can be calibrated in capacitance since this angle is a function of the unknown capacitance. This would result in a nonlinear but useful display.

Using the Taylor expansion, the expression for the angle θ can be rewritten:

$$\theta = \tan^{-1} (2\pi fRC) = (2\pi fRC) - \tfrac{1}{3}(2\pi fRC)^3 + \tfrac{1}{5}(2\pi fRC)^5 \cdots \qquad (6\text{-}9)$$

As can be seen from the Taylor expansion, the value of the arctangent will approach the angle, in radians, if the value of $(2\pi fRC)$ is small. To gain an idea of how small the arctangent must be so that just one term of the Taylor expansion may be used, that is, the first term, consider an arctangent of less than 0.1. The value of the Taylor expansion using the first term only is, of course, 0.1. The actual value of the arctangent is 0.0996687, which is only 0.3% less than the actual angle, in radians. If the meter in this technique were calibrated directly in capacitance and the phase angle were restricted to less than 0.1 rad, the error due to this approximation would not exceed 0.3%. Therefore, $\theta = (2\pi fRC)$ for less than 0.1.

The capacitance meter based on the circuit of Fig. 6-24 could be configured to cover several ranges by changing the value of R, such that the full-scale reading is 0.1 rad. As an example, assume that it is desired that the lowest range cover from 0 to 100 pF full scale, with a source frequency of 1 MHz. Therefore, at 1 MHz the phase shift of the resistance, R, and 100 pF must be 0.1 rad or

$$0.1 = (2\pi R \times 100 \, pF) \qquad (6\text{-}10)$$

Solving for R gives

$$R = \frac{0.1}{6.28 \times 10^{-10}} \qquad (6\text{-}11)$$

To cover from 10 to 1,000 pF full-scale meter, the resistors could be 1,590 Ω for 10 pF full scale, 477 Ω for 30 pF, 159 Ω for 100 pF, 47.7 Ω for 300 pF full scale, and 15.9 Ω for 1,000 pF full scale. It is difficult to measure capacitors greater than 100 pF using the 1-MHz source because the impedance of a capacitor at 1 MHz is too low to achieve an accurate measurement with this type of instrument.

6-8.2 Sources of Error

The accuracy of low-capacitance measurements is limited by the distributed capacitance of the measuring circuits. Figure 6-25 shows the basic measuring circuit with the parasitic capacitances added. The series resistance, R, has some series inductance and the input of the amplifier will have a certain amount of input capacitance. Primarily, the amplifier input capacitance will have the greatest effect on the accuracy of the measurement. It would be difficult to design an amplifier with an input capacitance low enough to allow measurements of capacitors below 10 pF without some form of compensation. Figure 6-26 shows a modified measuring circuit allowing the effects of the input capacitance of the amplifier to be nulled out. In this example, the resistor has been placed at the amplifier input, and the signal source is applied to a transformer to create an out-of-phase component. The effects of the input capacitance are nulled out by injecting some of the out-of-phase signal through a variable capacitor. Except for the trimming circuits, the operation of this capacitance measuring system is similar to the previous example.

Another source of error is the harmonic distortion of the signal source. The phase shift of the RC circuit, which is the heart of the capacitance measuring system, will satisfy the equations presented only if the signal source is a pure sine function without any harmonic distortion. For an accuracy of 0.3 per cent, which was the theoretical limit for the linear approximation using the Taylor expansion, the harmonic content of the signal source must be better than 50 dB down from the nominal level. A crystal oscillator is capable of supplying a signal purity of this magnitude only if the output is carefully coupled from the oscillator. In addition to the coupling point, the signal should be passed through a low-pass filter.

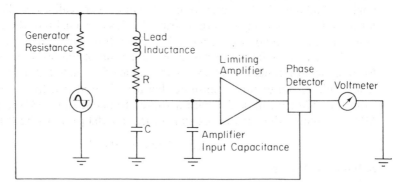

Figure 6-25 Capacitance measuring meter, showing the parasitic inductance, resistance, and capacitance.

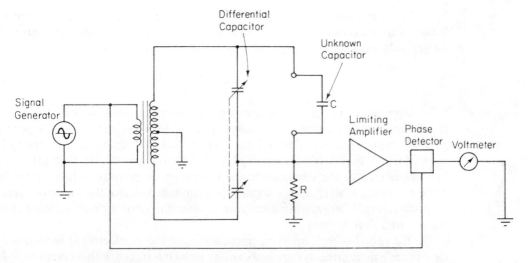

Figure 6-26 Nulling circuit for canceling the effects of the input stray capacitance of the amplifier.

By far the largest source of error is the equivalent series or parallel resistance. The series resistance called equivalent series resistance, or ESR, adds to the resistance of the circuit, but the phase measurement is made relative not to the capacitance but to the point where the ESR and the circuit resistance join, as shown in Fig. 6-27. This causes an error because the phase shift is not being measured accurately. Likewise, an equivalent parallel resistance, which is due to leakage resistance, will cause an erroneous reading because it changes the equivalent resistance as seen by the capacitor and hence changes the phase shift. This capacitance measuring method is not suitable for measuring capacitors with high dissipation factors or high ESR. Corrections can be made if the actual dissipation factor or ESR is known, but capacitance and dissipation factor can both be measured in a capacitance bridge. Generally, the quality of capacitors in the region of capacitance measured by this instrument is very good with insignificant ESR and dissipation factors, and the errors caused by these resistances are negligible.

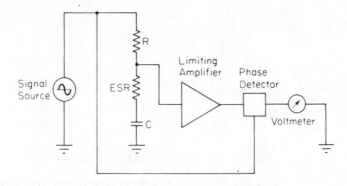

Figure 6-27 Effects of equivalent series resistance on capacitor measurement.

The same basic system can be used to measure inductance. Figure 6-28 shows a modification of the capacitance measuring instrument previously discussed to measure inductance. In this circuit the phase shift is

$$\theta = \tan^{-1} \frac{X_L}{R} \qquad (6\text{-}12)$$

As in the case of capacitance measuring, for phase shifts of less than 0.1 rad, the linearity of the output is sufficiently close to allow the same 0.3 per cent accurate measurements. The resistor value of 1 μH full scale is 62.8 Ω; for 3 μH it is 188.4 Ω; for 10 μH, 628 Ω; for 30 μH, 1,884 Ω; and for 100 μH, 6.28 kΩ. As the low impedance of the capacitor made it difficult to measure values larger than 1,000 pF with a 1-MHz source signal, the high impedance of the inductors greater than 100 μH makes measurements greater than this value difficult without changing the source frequency.

One of the more important applications of the all-electronic inductance or capacitance measuring system is its ability to be interfaced with a computer. This would require that the output of the phase detector be digitized and made available to the computer. Although there are digital phase detectors that supply a digital representation of the phase angle to within 1 part in 10,000 for a source frequency as high as 1 MHz, these devices are not practical. For the typical computer application, the output of the phase detector would be digitized using an analog-to-digital converter.

There are sources of error in this system, just as in the capacitor measuring system described previously, and, as in the capacitor measurements, they are primarily due to resistance. The equivalent series resistance of an inductor is expressed indirectly as the inductor Q. Mathematically, $Q = X_L/R$, where X_L is the capacitive reactance and R is the equivalent series resistance. From the equation it can be seen that, for smaller values of R, the Q or quality of an inductor goes up. It must be pointed out that the value of R is not that value of resistance that would be obtained if the inductor were measured with a dc bridge or ohmmeter. The value of R is due to losses in the inductor's core material and the variation of resistance due to the skin effect. Therefore, the ohmic resistance measured at dc would not be as great as the equivalent series resistance at 1 MHz, the

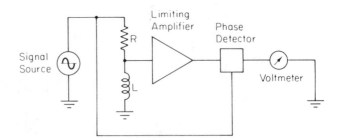

Figure 6-28 Modification of the capacitance measuring circuit described previously, allowing the measurement of inductance.

measuring frequency. Like the case with the capacitors, using this circuit to measure inductors depends on the Q of the inductor, and for components with low Q a bridge must be used for the measurement.

6-9 Q METER

6-9.1 Basic Q-Meter Circuit

The Q meter is an instrument designed to measure some of the electrical properties of coils and capacitors. The operation of this useful laboratory instrument is based on the familiar characteristics of a series-resonant circuit, namely, that the voltage across the coil or the capacitor is equal to the applied voltage times the Q of the circuit. If a fixed voltage is applied to the circuit, a voltmeter across the capacitor can be calibrated to read Q directly.

The voltage and current relationships of a series-resonant circuit are shown in Fig. 6-29. At resonance, the following conditions are valid:

$$X_C = X_L$$

$$E_C = IX_C = IX_L$$

$$E = IR$$

where E = applied voltage

I = circuit current

E_C = voltage across the capacitor

X_C = capacitive reactance

X_L = inductive reactance

R = coil resistance

The magnification of the *circuit*, by definition is Q, where

$$Q = \frac{X_L}{R} = \frac{X_C}{R} = \frac{E_C}{E} \tag{6-13}$$

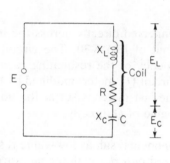
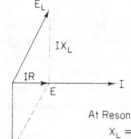

Figure 6-29 Series-resonant circuit.

Therefore if E is maintained at a constant and known level, a voltmeter connected across the capacitor can be calibrated directly in terms of the circuit Q.

A practical Q-meter circuit is shown in Fig. 6-30. The wide-range oscillator with a frequency range from 50 kHz to 50 MHz delivers current to a low-value shunt resistance R_{SH}. The value of this shunt is very low, typically on the order of 0.02 Ω. It introduces almost no resistance into the oscillatory circuit and it therefore represents a voltage source of magnitude E with a very small (in most cases negligible) internal resistance. The voltage E across the shunt, corresponding to E in Fig. 6-29, is measured with a thermocouple meter, marked "Multiply Q by." The voltage across the variable capacitor, corresponding to E_C in Fig. 6-29, is measured with an electronic voltmeter whose scale is calibrated directly in Q values.

To make a measurement, the unknown coil is connected to the test terminals of the instrument, and the circuit is tuned to resonance either by setting the oscillator to a given frequency and varying the internal resonating capacitor or by presetting the capacitor to a desired value and adjusting the frequency of the oscillator. The Q reading on the output meter must be multiplied by the index setting of the "Multiply Q by" meter to obtain the actual Q value.

The *indicated Q* (which is the resonant reading on the "Circuit Q" meter) is called the *circuit Q* because the losses of the resonating capacitor, voltmeter, and insertion resistor are all included in the measuring circuit. The *effective Q* of the measured coil will be somewhat greater than the indicated Q. This difference can generally be neglected, except in certain cases where the resistance of the coil is relatively small in comparison with the value of the insertion resistor. (This problem is discussed in Example 6-7.)

The inductance of the coil can be calculated from the known values of frequency (f) and resonating capacitance (C), since

$$X_L = X_C \quad \text{and} \quad L = \frac{1}{(2\pi f)^2 C} \quad \text{henry} \qquad (6\text{-}14)$$

6-9.2 Measurement Methods

There are three methods for connecting unknown components to the test terminals of a Q meter: direct, series, and parallel. The type of component and its size determine the method of connection.

Direct connection. Most coils can be connected directly across the test terminals, exactly as shown in the basic Q-circuit of Fig. 6-30. The circuit is resonated by adjusting either the oscillator frequency or the resonating capacitor. The indicated Q is read directly from the "Circuit Q" meter, modified by the setting of the "Multiply Q by" meter. When the last meter is set at the unity mark, the "Circuit Q" meter reads the correct value of Q directly.

Series connection. Low-impedance components, suh as low-value resistors, small coils, and large capacitors, are measured *in series* with the measuring

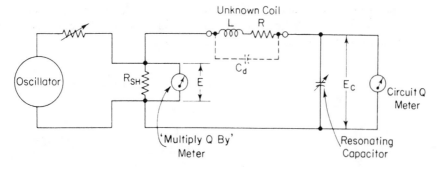

Figure 6-30 Basic *Q*-meter circuit.

circuit. Figure 6-31 shows the connections. The component to be measured, here indicated by [Z], is placed in series with a stable *work coil* across the test terminals. (The work coil is usually supplied with the instrument.) Two measurements are made: In the first measurement the unknown is short-circuited by a small *shorting strap* and the circuit is resonated, establishing a reference condition. The values of the tuning capacitor (C_1) and the indicated Q (Q_1) are noted. In the second measurement the shorting strap is removed and the circuit is returned, giving a new value for the tuning capacitor (C_2) and a change in the Q value from Q_1 to Q_2.

For the reference condition,

$$X_{C_1} = X_L \quad \text{or} \quad \frac{1}{\omega C_1} = \omega L \tag{6-15}$$

and neglecting the resistance of the measuring circuit,

$$Q_1 = \frac{\omega L}{R} = \frac{1}{\omega C_1 R} \tag{6-16}$$

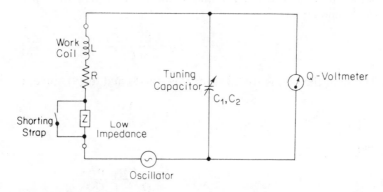

Figure 6-31 *Q*-meter measurement of a low-impedance component in the series connection.

For the second measurement, the reactance of the unknown can be expressed in terms of the *new* value of the tuning capacitor (C_2) and the *in-circuit* value of the inductor (L). This yields

$$X_S = X_{C2} - X_L \quad \text{or} \quad X_S = \frac{1}{\omega C_2} - \frac{1}{\omega C_1} \tag{6-17}$$

so that

$$X_S = \frac{C_1 - C_2}{\omega C_1 C_2} \tag{6-18}$$

X_S is inductive if $C_1 > C_2$ and capacitive if $C_1 < C_2$. The resistive component of the unknown impedance can be found in terms of reactance X_S and the indicated values of circuit Q, since

$$R_1 = \frac{X_1}{Q_1} \quad \text{and} \quad R_2 = \frac{X_2}{Q_2}$$

Also,

$$R_S = R_2 - R_1 = \frac{1}{\omega C_2 Q_2} - \frac{1}{\omega C_1 Q_1}$$

so that

$$R_S = \frac{C_1 Q_1 - C_2 Q_2}{\omega C_1 C_2 Q_1 Q_2} \tag{6-19}$$

If the unknown is *purely resistive*, the setting of the tuning capacitor would not have changed in the measuring process, and $C_1 = C_2$. The equation for resistance reduces to

$$R_S = \frac{Q_1 - Q_2}{\omega C_1 Q_1 Q_2} = \frac{\Delta Q}{\omega C_1 Q_1 Q_2} \tag{6-20}$$

If the unknown is a *small inductor*, the value of the inductance is found from Eq. (6-18) and equals

$$L_S = \frac{C_1 - C_2}{\omega^2 C_1 C_2} \tag{6-21}$$

The Q of the coil is found from Eqs. (6-18) and (6-19) since, by definition,

$$Q_S = \frac{X_S}{R_S}$$

and

$$Q_S = \frac{(C_1 - C_2)(Q_1 Q_2)}{C_1 Q_1 - C_2 Q_2} \tag{6-22}$$

If the unknown is a *large capacitor*, its value is determined from Eq. (6-18), and

$$C_S = \frac{C_1 C_2}{C_2 - C_1} \tag{6-23}$$

The Q of the capacitor may be found by using Eq. (6-22).

Parallel connection. High-impedance components, such as high-value resistors, certain inductors, and small capacitors, are measured by connecting them *in parallel* with the measuring circuit. Figure 6-32 shows the connections. Before the unknown is connected, the circuit is resonated, by using a suitable work coil, to establish reference values for Q and C (Q_1 and C_1). Then, when the component under test is connected to the circuit, the capacitor is readjusted for resonance, and a new value for the tuning capacitance (C_2) is obtained and a change in the value of circuit Q (ΔQ) from Q_1 to Q_2.

In a parallel circuit, computation of the unknown impedance is best approached in terms of its parallel components X_p and R_p, as indicated in Fig. 6-32. At the initial resonance condition, when the unknown is not yet connected into the circuit, the working coil (L) is tuned by the capacitor (C_1). Therefore

$$\omega L = \frac{1}{\omega C_1} \tag{6-24}$$

and

$$Q_1 = \frac{\omega L}{R} = \frac{1}{\omega C_1 R} \tag{6-25}$$

When the unknown impedance is not connected into the circuit and the capacitor is tuned for resonance, the reactance of the working coil (X_L) equals the parallel reactances of the tuning capacitor (X_{C_2}) and the unknown (X_p). Therefore

$$X_L = \frac{(X_{C_2})(X_p)}{X_{C_2} + X_p}$$

which reduces to

$$X_p = \frac{1}{\omega(C_1 - C_2)} \tag{6-26}$$

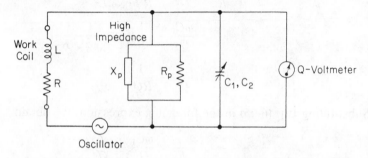

Figure 6-32 Work Coil L R — High Impedance — X_p — R_p — C_1, C_2 — Q-Voltmeter — Oscillator

Figure 6-32 Q-meter measurement of a high-impedance component in the parallel connection.

If the unknown is *inductive*, $X_p = \omega L_p$, and Eq. (6-26) yields the value of the unknown impedance:

$$L_p = \frac{1}{\omega^2(C_1 - C_2)} \tag{6-27}$$

If the unknown is *capacitive*, $X_p = 1/\omega C_p$ and Eq. (6-26) yields the value of the unknown capacitor:

$$C_p = C_1 - C_2 \tag{6-28}$$

In a parallel resonant circuit the total resistance at resonance is equal to the product of the circuit Q and the reactance of the coil. Therefore

$$R_T = Q_2 X_L$$

or by substitution of Eq. (6-24),

$$R_T = Q_2 X_{C_1} = \frac{Q_2}{\omega C_1} \tag{6-29}$$

The resistance (R_p) of the unknown impedance is most easily found by computing the *conductances* in the circuit of Fig. 6-32. Let

$$G_T = \text{total conductance of the resonant circuit}$$

$$G_p = \text{conductance of the unknown impedance}$$

$$G_L = \text{conductance of the working coil}$$

Then

$$G_T = G_p + G_L \quad \text{or} \quad G_p = G_T - G_L \tag{6-30}$$

From Eq. (6-29),

$$G_T = \frac{1}{R_T} = \frac{\omega C_1}{Q_2}$$

Therefore

$$\frac{1}{R_p} = \frac{\omega C_1}{Q_2} - \frac{R}{R^2 + \omega^2 L^2}$$

$$= \frac{\omega C_1}{Q_2} - \left(\frac{1}{R}\right)\left(\frac{1}{1 + \omega^2 L^2/R_2}\right)$$

$$= \frac{\omega C_1}{Q_2} - \frac{1}{R Q_1^2}$$

Substituting Eq. (6-25) in the foregoing expression, we obtain

$$\frac{1}{R_p} = \frac{\omega C_1}{Q_2} - \frac{\omega C_1}{Q_1}$$

and after simplifying, we obtain

$$R_p = \frac{Q_1 Q_2}{\omega C_1 (Q_1 - Q_2)} = \frac{Q_1 Q_2}{\omega C_1 \Delta Q} \tag{6-31}$$

The Q of the unknown is then found by using Eqs. (6-26) and (6-31) so that

$$Q_p = \frac{R_p}{X_p} = \frac{(C_1 - C_2)(Q_1 Q_2)}{C_1 (Q_1 - Q_2)} = \frac{(C_1 - C_2)(Q_1 Q_2)}{C_1 \Delta Q} \tag{6-32}$$

6-9.3 Sources of Error

Probably the most important factor affecting measurement accuracy, and the most often overlooked, is the *distributed capacitance* or *self-capacitance* of the measuring circuit. The presence of distributed capacitance in a coil modifies the *actual* or *effective Q* and the inductance of the coil. At the frequency at which the self-capacitance and the inductance of the coil are resonant, the circuit exhibits a purely resistive impedance. This characteristic may be used for measuring the distributed capacitance.

One simple method of finding the distributed capacitance (C_d) of a coil involves making two measurements at different frequencies. The coil under test is connected directly to the test terminals of the Q meter, as shown in the circuit of Fig. 6-33. The tuning capacitor is set to a high value, preferably to its maximum position, and the circuit is resonated by adjusting the oscillator frequency. Resonance is indicated by maximum deflection on the "Circuit Q" meter. The values of the tuning capacitor (C_1) and the oscillator frequency (f_1) are noted. The frequency is then increased to twice its original value ($f_2 = 2f_1$) and the circuit is returned by adjusting the resonating capacitor (C_2).

The resonant frequency of an LC circuit is given by the well-known equation

$$f = \frac{1}{2\pi\sqrt{LC}} \tag{6-33}$$

At the initial resonance condition, the capacitance of the circuit equals $C_1 + C_d$, and the resonant frequency equals

$$f_1 = \frac{1}{2\pi\sqrt{L(C_1 + C_d)}} \tag{6-34}$$

After the oscillator and the tuning capacitor are adjusted, the capacitance of the circuit is $C_2 + C_d$, and the resonant frequency equals

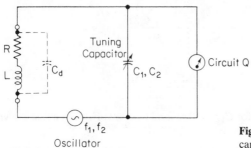

Figure 6-33 Determination of the distributed capacitance of an inductor.

$$f_2 = \frac{1}{2\pi\sqrt{L(C_2 + C_d)}} \tag{6-35}$$

Since $f_2 = 2f_1$, Eqs. (6-34) and (6-35) are related so that

$$\frac{1}{2\pi\sqrt{L(C_1 + C_d)}} = \frac{2}{2\pi\sqrt{L(C_1 + C_d)}}$$

and

$$\frac{1}{C_2 + C_d} = \frac{4}{C_1 + C_d}$$

Solving for the distributed capacitance yields

$$C_d = \frac{C_1 - 4C_2}{3} \tag{6-36}$$

EXAMPLE 6-4

The self-capacitance of a coil is to be measured by using the procedure just outlined. The first measurement is at $f_1 = 2$ MHz and $C_1 = 460$ pF. The second measurement, at $f_2 = 4$ MHz, yields a new value of tuning capacitor, $C_2 = 100$ pF. Find the distributed capacitance, C_d.

SOLUTION   Using Eq. (6-36), we obtain

$$C_d = \frac{C_1 - 4C_2}{3} = \frac{460 - 400}{3} = 20 \text{ pF}$$

EXAMPLE 6-5

Compute the value of self-capacitance of a coil when the following measurements are made: At frequency $f_1 = 2$ MHz, the tuning capacitor is set at 450 pF. When the frequency is increased to 5 MHz, the tuning capacitor is tuned at 60 pF.

SOLUTION   Since $f_2 = 2.5\,f_1$, Eqs. (6-34) and (6-35) are related as follows:

$$\frac{1}{2\pi\sqrt{L(C_2 + C_d)}} = \frac{2.5}{2\pi\sqrt{L(C_2 + C_d)}}$$

This reduces to

$$\frac{1}{C_2 + C_d} = \frac{6.25}{C_1 + C_d}$$

Solving for C_d, we obtain

$$C_d = \frac{C_1 - 6.25C_2}{5.25}$$

Substituting the values for $C_1 = 450$ pF and $C_2 = 60$ pF, we see that the value of the distributed capacitance is $C_d = 14.3$ pF.

The effective Q of a coil with distributed capacitance is less than the true Q by a factor that depends on the value of the self-capacitance and the resonating capacitor. It can be shown that

$$\text{true } Q = Q_e \left(\frac{C + C_d}{C} \right) \tag{6-37}$$

where Q_e = effective Q of the coil

C = resonating capacitance

C_d = distributed capacitance

The *effective Q* can usually be considered the *indicated Q*.

For many measurements, the *residual* or *insertion* resistance (R_{SH}) of the Q-meter circuit of Fig. 6-26 is sufficiently small to be considered negligible. Under certain circumstances, it can contribute an error to the measurement of Q. The effect of the insertion resistor on the measurement depends on the magnitude of the unknown impedance and, of course, on the size of the insertion resistor. For instance, the 0.02 Ω of insertion resistance may be neglected in comparison with a coil resistance of 10 Ω, but it assumes importance when compared to a coil resistance of 0.1 Ω. The effect of the 0.02-Ω insertion resistance is illustrated by Examples 6-6 and 6-7.

EXAMPLE 6-6

A coil with a resistance of 10 Ω is connected in the "direct-measurement" mode. Resonance occurs when the oscillator frequency is 1.0 MHz and the resonating capacitor is set at 65 pF. Calculate the percentage error introduced in the calculated value of Q by the 0.02-Ω insertion resistance.

SOLUTION The *effective Q* of the coil equals

$$Q_e = \frac{1}{\omega C R} = \frac{1}{(2\pi)(10^6)(65 \times 10^{-12})(10)} = 244.9$$

The *indicated Q* of the coil equals

$$Q_i = \frac{1}{\omega C(R + 0.02)} = 244.4$$

The percentage error is then

$$\frac{244.9 - 244.4}{244.9} \times 100\% = 0.2\%$$

EXAMPLE 6-7

Repeat the problem of Example 6-6 for the following conditions:

The coil resistance is 0.1 Ω.

The frequency at resonance is 40 MHz.

The tuning capacitor is set at 135 pF.

SOLUTION The *effective Q* of the coil is

$$Q_e = \frac{1}{\omega CR} = \frac{1}{2\pi \times 40 \times 10^6 \times 135 \times 10^{-12} \times 0.1} = 295$$

The *indicated Q* of the coil is

$$Q_i = \frac{1}{\omega C(R + 0.02)} = 246$$

The percentage error equals

$$\frac{295 - 246}{295} = 100\% = 17\%$$

Other sources of error include the *residual inductance* of the instrument, which is usually in the order of 0.015 μH and affects the measurement of only very small inductors (<0.5 μH). The *conductance* of the Q voltmeter has a slight shunting effect on the tuning capacitor at the higher frequencies, but this effect can usually be neglected.

6-10 VECTOR IMPEDANCE METER

Impedance measurements are concerned with both the magnitude (Z) and the phase angle (θ) of a component. At frequencies below 100 MHz, measurement of voltage and current is usually sufficient to determine the magnitude of the impedance. The phase difference between the voltage waveform and the current waveform indicates whether the component is inductive or capacitive. If the phase angle can be determined, for example, by using a CRO displaying a Lissajous pattern, the reactance can be calculated. If a component must be fully specified, its properties should be determined at several different frequencies, and many measurements may be required. Especially at the higher frequencies, these measurements become rather elaborate and time consuming, and many steps may be required to obtain the desired information.

The development of such instruments as the *vector impedance meter* makes impedance measurements over a wide frequency range possible. *Sweep-frequency plots* of impedance and phase angle versus frequency, providing complete coverage within the frequency band of interest, can also be made.

The vector impedance meter, shown in Fig. 6-34, makes simultaneous measurements of impedance and phase angle over a frequency range of from 400 kHz to 110 MHz. The unknown component is simply connected across the input terminals of the instrument, the desired frequency is selected by turning the front panel controls, and the two front panel readouts indicate the magnitude of the impedance and the phase angle.

The operation of the vector impedance meter is best understood by referring

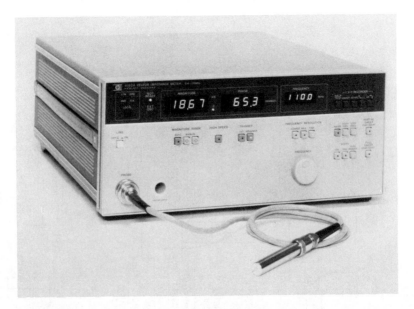

Figure 6-34 Vector impedance meter. (Courtesy of Hewlett-Packard Company.)

to the block diagram of Fig. 6-35 of a representative instrument. Two measurements take place: (1) The magnitude of the impedance is determined by measuring the current through the unknown component when a known voltage is applied across it, or by measuring the voltage across the component when a known current is passed through it; (2) the phase angle is found by determining the phase difference between the voltage across the component and the current through the component.

The block diagram of Fig. 6-35 shows that the instrument contains a *signal source* (Wien bridge oscillator) with two front panel controls to select the frequency range and to continuously adjust the selected frequency. The oscillator output is fed to an *AGC amplifier* which allows accurate gain adjustment by means of its feedback voltage. This gain adjustment is an internal control actuated by the setting of the *impedance range* switch, to which the AGC amplifier output is connected. The impedance range switch is a precision attenuator network controlling the oscillator output voltage and at the same time determining the manner in which the unknown component will be connected into the circuitry that follows the range switch.

The impedance range switch permits operation of the instrument in two modes: the *constant-current* mode and the *constant-voltage* mode. The three lower ranges (×1, ×10, and ×100) operate in the constant-current mode and the four higher ranges (×1k, ×10k, ×100k, and ×1M) operate in the constant-voltage mode.

In the constant-current mode the unknown component is connected across the input of the ac differential amplifier. The current supplied to the unknown depends on the setting of the impedance range switch. This current is held con-

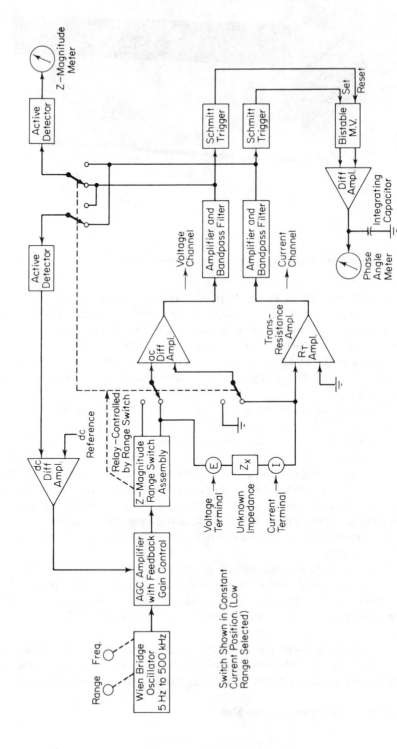

Figure 6-35 Block diagram of the vector impedance meter. (Courtesy of Hewlett-Packard Company.)

stant by the action of the transresistance or R_T amplifier, which converts the current through the unknown to a voltage output equal to the current times its feedback resistance. The R_T amplifier is an operational amplifier whose output voltage is proportional to its input current. The output of the R_T amplifier is fed to a detector circuit and compared to a dc reference voltage. The resulting control voltage regulates the gain of the AGC amplifier and hence the voltage applied to the impedance range switch. The output of the ac differential amplifier is applied to an amplifier and filter section consisting of high- and low-band filters that are changed with the frequency range to restrict the amplifier band-width. The output of the bandpass filter is connected, when selected, to a detector that drives the *Z-magnitude* meter. Since the current through the unknown is held constant by the R_T amplifier, the *Z*-magnitude meter, which measures the voltage across the unknown, deflects in proportion to the magnitude of the unknown impedance and is calibrated accordingly.

In the constant-voltage mode the two inputs to the differential amplifier are switched. The terminal that was connected to the input of the transresistance amplifier in the constant-current mode is now grounded. The other input of the differential amplifier that was connected to the voltage terminal of the unknown component is now connected to a point on the *Z*-magnitude range switch which is held at a constant potential. The voltage terminal of the unknown is connected to this same point of constant potential, or depending on the setting of the *Z*-magnitude range switch, to a decimal fraction of this voltage. In any case, the voltage across the unknown is held at a constant level. The current through the unknown is applied to the transresistance amplifier which again produces an output voltage proportional to its input current.

The roles of the ac differential amplifier and the transresistance amplifier are now reversed. The voltage output of the R_T amplifier is applied to the detector and then to the *Z*-magnitude meter. The output voltage of the differential amplifier controls the gain of the AGC amplifier in the same manner that the R_T amplifier did in the constant-current mode.

Phase-angle measurements are carried out simultaneously. The outputs of both the voltage channel and the current channel are amplified and each output is connected to a Schmitt trigger circuit. The Schmitt trigger circuits produce a positive-going spike every time the input sine wave goes through a zero crossing. These positive spikes are applied to a *binary phase detector* circuit. The phase detector consists of a bistable multivibrator, a differential amplifier, and an integrating capacitor. The positive-going pulse from the constant-current channel sets the multivibrator, and the pulse from the constant-voltage channel resets the multivibrator. The "set" time of the MV is therefore determined by the zero crossings of the voltage and current waveforms. The "set" and "reset" outputs of the MV are applied to the differential amplifier, which applies the difference voltage to an integrating capacitor. The capacitor voltage is directly proportional to the zero-crossing time interval and is applied to the *phase-angle* meter which then indicates the phase difference, in degrees, between the voltage and current waveforms.

Calibration of the vector impedance meter is usually performed by connecting standard components to the input terminals. These components may be standard resistors or capacitors. An electronic counter is needed to accurately determine the period of the applied test frequency. When the value of the component under test and the frequency of the test signal are both known accurately, the impedance or reactance can be calculated and compared to the indication on the Z-magnitude meter. With a standard resistor connected to the input terminals, the phase-angle meter should read 0°.

6-11 VECTOR VOLTMETER

A *vector voltmeter* measures the amplitude of a signal at two points in a circuit and simultaneously measures the phase difference between the voltage waveforms at these two points. This instrument can be used in a wide variety of applications, especially in situations where other methods are very difficult or time consuming. The vector voltmeter is useful in VHF applications and can be used successfully in such measurements as:

(a) Amplifier gain and phase shift
(b) Complex insertion loss
(c) Filter transfer functions
(d) Two-port network parameters

The vector voltmeter basically converts two RF signals of the same fundamental frequency (from 1 MHz to GHz) to two IF signals with 20-KHz fundamental frequencies. These IF signals have the same amplitudes, waveforms, and phase relationships as the original RF signals. Consequently, the fundamental components of the IF signals have the same amplitude and phase relationships as the fundamental components of the RF signals. These fundamental components are filtered from the IF signals and are measured by a voltmeter and a phase meter.

The block diagram of Fig. 6-36 shows that the instrument consists of five major sections as follows: two RF-to-IF converters, an automatic phase control section, a phase meter circuit, and a voltmeter circuit. The RF-to-IF converters and the phase control section produce two 20-kHz sine waves with the same amplitudes and the same phase relationship as the fundamental components of the RF signals applied to channels A and B. The phase meter section continuously monitors these two 20-kHz sine waves and indicates the phase angle between them. The voltmeter section can be switched to channel A or channel B to provide a meter display of the amplitude.

Each RF-to-IF converter consists of a sampler and a tuned amplifier. The sampler produces a 20-kHz replica of the RF input waveform, and the tuned amplifier extracts the 20-kHz fundamental component from this waveform replica. *Sampling* is a time-stretching process, with which a high-frequency repetitive signal is duplicated at a much lower frequency. The process is illustrated in

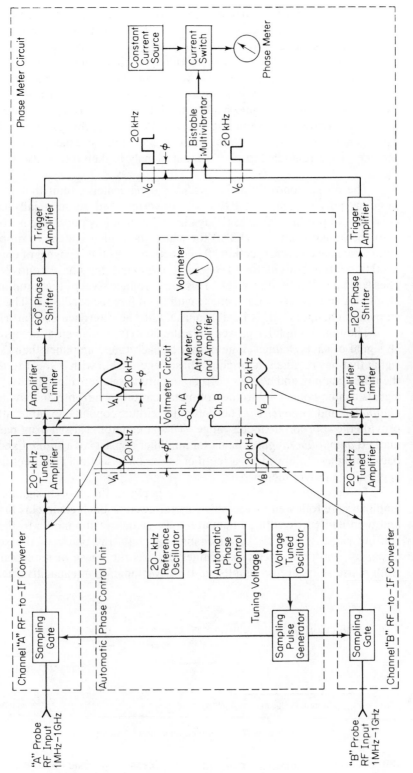

Figure 6-36 Block diagram of the vector voltmeter. (Courtesy of Hewlett-Packard Company.)

the diagram of Fig. 6-37. An electronic switch is connected between the RF input waveform and a storage capacitor. Each time the switch is momentarily closed, the capacitor is charged to the instantaneous value of the input voltage and holds this until the next switch closure. With appropriate timing, samples are taken at progressively later points on the RF waveform. Provided that the RF waveform is repetitive, the samples reconstruct the original waveform at a much lower frequency. Each input channel has a sampler consisting of a sampling gate and a storage capacitor. The sampling gates are controlled by pulses from the same pulse generator. Samples are taken in each channel at exactly the same instant, and the phase relationship of the input signals is therefore preserved in the IF signals.

The phase control unit is a rather sophisticated circuit that generates the sampling pulses from both RF-to-IF converters and automatically controls the pulse rate to produce 20-kHz IF signals. The sampling pulse rate is controlled by a voltage-tuned oscillator (VTO) for which the tuning voltage is supplied by the automatic phase control section. This section locks the IF signal of channel A to a 20-kHz reference oscillator. To get initial locking, the phase control section applies a ramp voltage to the VTO. This ramp voltage sweeps the sampling rate until channel A IF is 20 kHz and in phase with the reference oscillator. Then the sweep stops and channel A IF is held in phase with the reference oscillator.

The tuned amplifier passes only the 20-kHz fundamental component of the IF signal of each channel. The output of each tuned amplifier then consists of a signal that has retained its original phase relationship with respect to the signal in the other channel and also its correct amplitude relationship. The two filtered IF signals can be connected to the voltmeter circuit by a front panel switch, marked *channel A* and *channel B*. The voltmeter circuit contains an input attenuator to provide the appropriate meter range. This attenuator is also a front panel control, marked *amplitude range*. The meter amplifier consists of a stable fixed-gain feedback amplifier, followed by a rectifier and a filter section. The rectified signal is applied to a dc voltmeter.

To determine the *phase difference* between the two IF signals, the tuned amplifiers are followed by the phase meter circuit. Each channel is first amplified and then limited, resulting in square-wave signals at the inputs to the IF phase-shifting circuits. The circuit in channel A shifts the phase of the square-wave signal by +60°; the circuit in channel B shifts the phase of its signal by −120°. Both phase shifts are accomplished by a combination of capacitive networks and

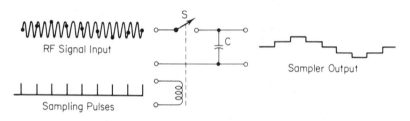

Figure 6-37 Simplified diagram of a sampling circuit.

inverting and noninverting amplifiers whose vector-sum outputs provide the desired phase shift. The outputs of the phase-shift circuits are amplified and clipped, producing square waveforms, and applied to the trigger amplifiers. These circuits convert the square-wave input signals to positive spikes with very fast rise times. The bistable multivibrator is triggered by pulses from both channels. Channel A is connected to the *set* input of the MV; channel B is connected to the *reset* input of the MV. If the initial phase shift between the RF signals at the probes was 0°, the trigger pulses into the multivibrator are 180° out of phase owing to the action of the phase-shift circuits. The MV then produces a square-wave output voltage which is symmetrical about zero. Any phase shift at the RF probes carries through the entire system and varies the trigger pulses from their 180° relationship, producing an asymmetrical waveform.

The (asymmetrical) square wave controls the current switch, which is a transistor switched into conduction by the negative portion of the square wave. The switch connects the constant current supply to the *phase meter*. At 0° phase shift at the RF input, the switch is turned off and on for equal amounts of time and the current supply is adjusted to cause the meter to read 0° or center scale. Any RF phase shift results in an asymmetrical waveform and allows either more or less current to the phase meter, depending on whether the phase shift caused the negative half-cycle of the square wave to be larger or smaller. An input phase shift of 180° would cause the square wave to collapse into either a positive or a negative dc voltage and the switch would then allow no current or maximum current to the phase meter. These maximum deviations from the center reading of 0° are marked on the meter face as +180° and −180°. The *phase range* can be selected by a front panel switch that places a shunt across the phase meter and changes its sensitivity.

The instrument contains a power supply section, which is not shown on the block diagram of Fig. 6-36. The power supply generates all the necessary supply voltages for the various sections of the instrument.

Calibration procedures and the testing of performance specifications vary from one instrument to the next. Complete descriptions of the various tests are given in the manual of the instrument and usually include the procedure and instrumentation needed for such tests.

6-12 RF POWER AND VOLTAGE MEASUREMENT

One example of the amplified meter is the RF voltmeter, such as the unit shown in Fig. 6-38. Radio-frequency energy is essentially ac voltage, except that the frequencies involved are much greater than that which would be experienced in power distribution, audiofrequency amplifiers, or control systems. Radio frequencies extend well into the gigahertz region, where it is difficult to amplify and great care must be taken because normal components are often useless.

Radio-frequency voltage is measured by rectifying the alternating voltage and amplifying the resulting dc output. Because of the difficulty in amplifying the RF signal itself, the RF voltage is first rectified and the dc output is amplified.

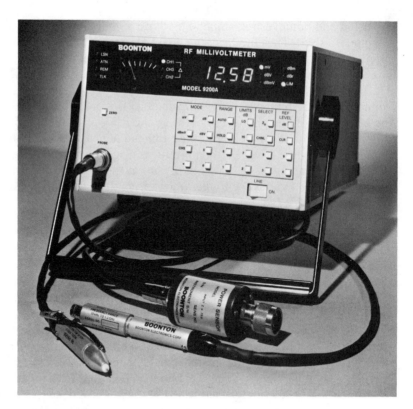

Figure 6-38 RF millivoltmeter for measuring RF power and voltage. (Courtesy of Boonton Electronics Corporation.)

The diodes used to rectify the RF waveform are not like the rectifiers used in a conventional ac meter, discussed in Chapter 4. The diodes used to rectify the RF signal are either Schottky barrier or point contact diodes. Conventional junction diodes with small geometries can be used for lower frequencies, but most detector diodes are not *PN* junction diodes. There are two significant problems with diodes used for RF rectification. First, most diodes have excessive capacitance for high-frequency RF rectification, and, second, most diodes have excessive reverse recovery time.

When diodes are operated at low forward-biased potentials, the rectified output does not equal the peak of the input. This means that for rather low amplitude RF voltages the resulting dc output is even lower, and a chopper-stabilized amplifier or other amplifier stabilized for dc drifts is required. Figure 6-39 shows a block diagram of a sensitive RF millivoltmeter. The actual RF rectifier or detector is usually mounted on a probe so that measurements can be made with the least amount of interconnecting RF cable, as even the losses of coaxial cable can cause significant errors at very high frequencies. The detected output is in the very low millivolt region, and often even lower, and is amplified via a chopper-stabilized amplifier, digitized and displayed on a digital readout.

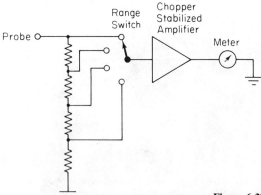

Figure 6-39 Block diagram of the RF millivoltmeter.

The type of measurement made by the RF millivoltmeter depends on the type of probe used. Voltage measurements are made with a probe similar to the one shown in Fig. 6-40(a). Voltage measurements are made with a relatively high impedance, but some capacitance is inescapable. This probe would be used within circuits where the impedances vary and the circuit cannot be isolated and terminated externally.

Many high-frequency circuits can be disconnected and terminated, usually in 50 Ω, externally, and the probe of Fig. 6-40(b) is used. This probe is more a power measuring probe rather than a voltage probe and can be used to measure powers to the nanowatt region. This power measurement is not a true rms measurement, and care must be taken in interpreting measurements, especially when the signal being measured has modulation applied.

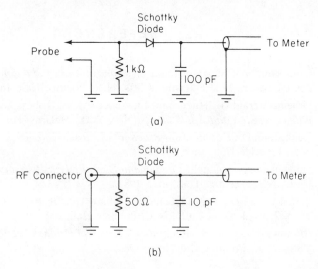

Figure 6-40 Two different RF probes for use with an RF millivoltmeter.

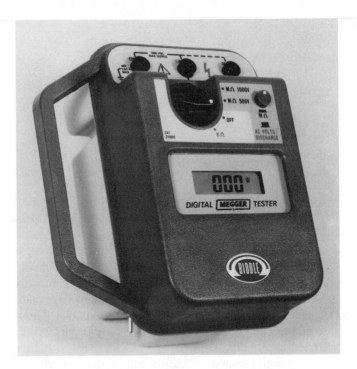

Figure 6-41 The Biddle Megger is a familiar test set for testing very high resistances. (Courtesy of Biddle Instruments.)

REFERENCES

6-1. Gothmann, William H., *Digital Electronics: An Introduction to Theory and Practice*, 2nd ed., chap. 11. Englewood Cliffs, N.J.: Prentice-Hall, Inc., 1982.

6-2. Graeme, Jerald G., Huelsman, Lawrence P., and Tobey, Gene E., *Operational Amplifiers: Design and Applications*. New York: McGraw-Hill Book Company, 1971.

6-3. Lenk, John D., *Handbook of Practical Electronic Circuits*, chap. 6. Englewood Cliffs, N.J.: Prentice-Hall, Inc., 1982.

6-4. Oppenheimer, Samuel, *Fundamentals of Electric Circuits*, chap. 23. Englewood Cliffs, N.J.: Prentice-Hall, Inc., 1984.

6-5. Prensky, Sol D., and Castellucis, Richard L., *Electronic Instrumentation*, 3rd ed., chap. 7. Englewood Cliffs, N.J.: Prentice-Hall, Inc., 1982.

6-6. Rutkowski, George B., *Integrated Circuit Operational Amplifiers*, 2nd ed. Englewood Cliffs, N.J.: Prentice-Hall, Inc., 1984.

PROBLEMS

6-1. What are the advantages of a chopper-stabilized amplifier?

6-2. What is the lowest full-scale voltage that could be displayed with a 100-μA meter movement with an internal resistance of 150 Ω? What would the sensitivity of this meter be in ohms per volt? Is there any way this meter could be used to construct a lower full-scale voltage reading?

6-3. A 25-mA full-scale current meter with an internal resistance of 100 Ω is available for constructing an ac voltmeter with a voltage range of 200 V rms. Using four diodes in a bridge arrangement, where each diode has a forward resistance of 500 Ω and inifinite reverse resistance, calculate the necessary series-limiting resistance for the 200-V rms range.

6-4. For measuring small values of capacitance, a 60-MHz signal source is to be used in a capacitance meter. What value of series resistance is required if the phase shift is to kept below 5.7 degrees for full-scale capacitance readings of 1, 10, and 100 pF?

6-5. What would a true-rms reading meter indicate if a pulse waveform of 5 V peak and a 25 per cent duty cycle were applied? What would the meter indicate if a 5-V dc input were applied (assume the meter has dc capability)?

6-6. To check the distributed capacitance of a coil, the coil is resonated at 10 MHz with 120 pF and then is resonated at 15 MHz with 40 pF. What is the inductance of the coil and what is the equivalent distributed capacitance?

6-7. A coil with a resistance of 3 Ω is connected to the terminals of the Q-meter of Fig. 6-34. Resonance occurs at an oscillator frequency of 5 MHz and resonating capacitance of 100 pF. Calculate the percentage of error introduced by the insertion resistance, $R_{\text{SH}} = 0.1$ Ω.

7

Oscilloscopes

7-1 INTRODUCTION

The cathode ray oscilloscope is probably the most versatile tool for the development of electronic circuits and systems, and has been one of the more important tools in the development of modern electronics. The cathode ray oscilloscope is a device that allows the amplitude of electrical signals, whether they be voltage, current, power, etc., to be displayed primarily as a function of time. The oscilloscope depends on the movement of an electron beam, which is then made visible by allowing the beam to impinge on a phosphor surface, which produces a visible spot. If the electron beam is deflected in either of two orthogonal axes, such as the familiar X and Y axes used in conventional graph construction, the luminous spot can be used to create two-dimensional displays. Typically, the X axis of the oscilloscope is deflected at a constant rate, relative to time, and the vertical or Y axis is deflected in response to an input stimulus such as a voltage. This produces the time-dependent variation of the input voltage, which is very important to the design and development of electronic circuits.

Time recording devices, such as pen and strip chart recorders, have existed for a long time; however, the oscilloscope is capable of much faster operation. Rather than the recording of events over a period of a few seconds, which is typical of the mechanical-type recorder, the oscilloscope is capable of displaying events that take place over periods of microseconds and nanoseconds.

7-2 OSCILLOSCOPE BLOCK DIAGRAM

The heart of the oscilloscope is the cathode ray tube, which generates the electron beam, accelerates the beam to a high velocity, deflects the beam to create the image, and contains the phosphor screen where the electron beam eventually becomes visible. To accomplish these tasks, various electrical signals and voltages are required, and these requirements dictate the remainder of the blocks of the oscilloscope outline as shown in Fig. 7-1. The power supply block provides the voltages required by the cathode ray tube to generate and accelerate the electron beam, as well as to supply the required operating voltages for the other circuits of the oscilloscope. Relatively high voltages are required by cathode ray tubes, on the order of a few thousand volts, for acceleration, as well as a low voltage for the heater of the electron gun, which emits the electrons. Supply voltages for the other circuits are various values, usually not more than a few hundred volts.

The laboratory oscilloscope has a time base which generates the correct voltage to supply the cathode ray tube to deflect the spot at a constant time-dependent rate. The signal to be viewed is fed to a vertical amplifier, which increases the potential of the input signal to a level that will provide a usable deflection of the electron beam. To synchronize the horizontal deflection with the vertical input, such that the horizontal deflection starts at the same point of the input vertical signal each time it sweeps, a synchronizing or triggering circuit is

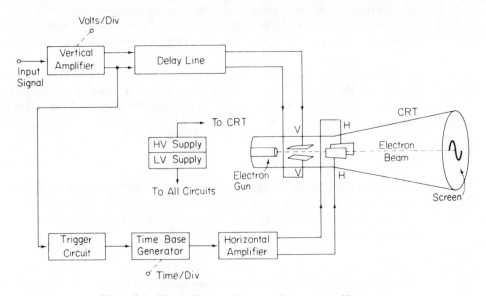

Figure 7-1 Block diagram of a general-purpose oscilloscope.

used. This circuit is the link between the vertical input and the horizontal time base.

7-3 CATHODE RAY TUBE

7-3.1 Early Cathode Ray Tubes

Figure 7-2 shows a cutaway view of an early cathode ray tube and is typical of some of the simpler tubes occasionally used in low-frequency oscilloscopes. Learning the basic operation of the modern cathode ray tube can be enhanced by understanding the simple cathode ray tube.

A heated cathode emits electrons, which are accelerated to the first accelerating anode, or the preaccelerating anode, through a small hole in the control grid. The amount of cathode current, which governs the intensity of the spot, can be controlled with the control grid in a manner similar to a conventional vacuum tube. The preaccelerating anode is a hollow cylinder that is at a potential a few hundred volts more positive than the cathode so that the electron beam will be accelerated in the electric field. A focusing anode is mounted just ahead of the preaccelerating anode and is also a cylinder. Following the focusing anode is the accelerating anode, which gives the electron beam its last addition of energy before its journey to the phosphor screen.

Although only one anode is referred to as a focusing anode, it actually takes three elements to perform the electron beam focusing. If the accelerated electrons were allowed to simply travel toward the phosphor screen, they would diverge owing to variations in energy and would produce a broad ill-defined spot on the phosphor. Therefore, the electron beam is focused with an electrostatic lens so that the electron beam converges on the phosphor screen as shown in Fig. 7-3. The electron lens requires three elements, with the center element at a lower potential than the two outer elements. Figure 7-4 shows two elements at two different potentials with the right-hand element at a higher potential than the left-hand element. Because of the potential difference, there would be an electric field generated as shown. The strength of the electric field is categorized by the amount of force a charged particle would experience in the field and is described by the following equation:

$$\varepsilon = \frac{f}{q} \tag{7-1}$$

Where ε is the electric field intensity in volts per meter, f is the force that would be experienced by a charged particle of charge q, in coulombs. An electron has a charge, e, of negative 1.60×10^{-19} C, and thus would experience a force in an electric field of ε of

$$f_e = e\varepsilon \tag{7-2}$$

The generated electric field is not uniform, and if equipotential lines were drawn as shown in Fig. 7-4, they would bulge at the center of the two cylinders. Only

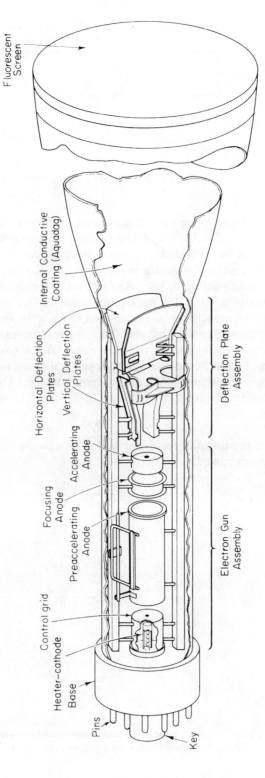

Figure 7-2 Internal structure of a cathode ray tube.

Fluorescent Screen

Internal Conductive Coating (Aquadag)

Horizontal Deflection Plates

Vertical Deflection Plates

Deflection Plate Assembly

Accelerating Anode

Focusing Anode

Preaccelerating Anode

Control grid

Heater–cathode

Base

Pins

Key

Electron Gun Assembly

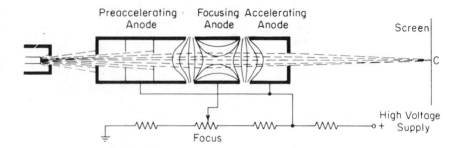

Figure 7-3 Electrostatic focusing system of a CRT.

electrons passing through the exact center of the two cylinders would experience no force. Electrons that are displaced from the center line will experience a force and thus will be deflected. To see this, consider an electron approaching an equipotential surface as shown in Fig. 7-5. The potential to the left of the surface, S, is $V-$, and to the right, $V+$. An electron, which is moving in a direction AB at an angle with the normal to the equipotential surface and entering the area to the left of S with a velocity v_1, experiences a force at the surface S. This force acts in a direction normal to the equipotential surface. Because of this force, the velocity of the electron increases to a new value, v_2, after it has passed S. The tangential component, v_t, of the velocity on both sides of S remains the same because there is no change of potential along the equipotential line. Only the normal component of the velocity, v_n, is increased; thus

$$v_t = v_1 \sin \theta_i = v_2 \sin \theta_r \tag{7-3}$$

where θ_i is the angle of incidence and θ_r, the angle of refraction of the electron beam. Rearranging Eq. (7-3),

$$\frac{\sin \theta_i}{\sin \theta_r} = \frac{v_2}{v_1} \tag{7-4}$$

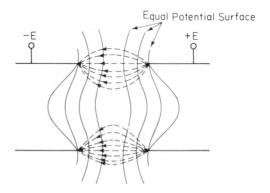

Figure 7-4 Equipotential surfaces for two cylinders placed end to end.

Oscilloscopes   Chap. 7

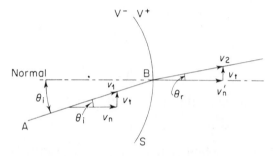

Figure 7-5 Refraction of an electron ray at an equipotential surface.

Equation (7-4) is identical to the equation relating the refraction of light passing through an area of different indexes of refraction, and thus the equipotential surfaces act as the surface of a lens in geometrical optics.

Each junction between two cylinders contains equipotential lines similar to a concave lens. The electron beam from the cathode passing through the first concave electrostatic lens tends to become more aligned toward the axis of the cathode ray tube and after passing through a second concave lens will become focused at the phosphor screen. Unlike the optical glass counterpart, the focal length of the lens can be adjusted by varying the potential difference between the two cylinders. Thus the electron beam can be made to focus precisely at the phosphor screen and produce a small bright spot.

7-3.2 Electrostatic Deflection

In discussing the electrostatic deflection method of an electron beam in an oscilloscope we return to the statement made in Sec. 7-3.1 regarding the force on the electron in a uniform electric field, as shown in Fig. 7-6. By definition of the electric field intensity, ε, the force on the electron is $f_e = -e\varepsilon$ Newton. The action of the force on the electron will accelerate it in the direction of the positive electrode, along the lines of the field flux. Newton's second law of motion allows us to calculate this acceleration since

$$f = ma \qquad (7\text{-}5)$$

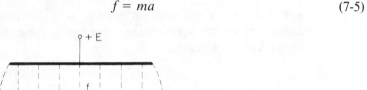

Figure 7-6 Force f an electron in a uniform electric field.

Substituting Eq. (7-2) into Eq. (7-5), we obtain

$$a = \frac{f}{m} = \frac{-e\varepsilon}{m} \quad \text{(m/s}^2)$$ (7-6)

where a = acceleration of the electron (m/s²)

f = force on the electron (N)

m = mass of the electron (kg)

When the motion of an electron in an electric field is discussed it is usually specified in respect to the customary Cartesian axes, as shown in Fig. 7-7. In discussing the concepts which follow, we shall use subscript notation for the *vector* components of velocity, field intensity, and acceleration. For example, the velocity component along the X axis will be written v_x (m/s); the component of the force along the Y axis is written f_y (N), etc. The motion of an electron in a given electric field cannot be determined unless the initial values of velocity and displacement are known. The term *initial* represents the value of velocity or displacement at the time of observation, or time $t = 0$. The subscript 0 will be used to indicate these initial values. For example, the initial velocity component along the X axis is written as v_{0x}.

Consider now an electric field of constant intensity with the lines of force pointing in the negative Y direction, shown in Fig. 7-8. An electron entering this field in the positive X direction with an initial velocity v_{0x} will experience a force. Since the field acts only along the Y axis, there will be no force along either the X or the Z axis, and the acceleration of the electron along these axes must be zero. Zero acceleration means constant velocity, and since the electron enters the field in the positive X direction with an initial velocity v_{0x}, it will continue to travel along the X axis at that velocity. Since the velocity along the Z axis was zero at time $t = 0$, there will be no movement of the electron along the Z axis.

Newton's second law of motion, applied to the force on the electron acting in the Y direction, yields

$$f = ma_y \quad \text{or} \quad a_y = \frac{f}{m} = \frac{-e\varepsilon_y}{m} = \text{constant}$$ (7-7)

Equation (7-7) indicates that the electron moves with a *constant acceleration* in the Y direction of the uniform electric field. To find the displacement of the electron due to this accelerating force, we use the well-known expressions for velocity and displacement:

$$v = v_0 + at \quad \text{(m/s)} \quad \text{(velocity)}$$ (7-8)

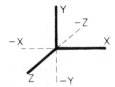

Figure 7-7   Cartesian coordinate system.

Figure 7-8 Path of a moving electron in a uniform electric field.

$$x = x_0 + v_0 t + \tfrac{1}{2}at^2 \quad \text{(m)} \quad \text{(displacement)} \tag{7-9}$$

Subject to the initial condition of zero velocity in the Y direction ($v_{0y} = 0$) Eq. (7-8) yields

$$v_y = a_y t \quad \text{(m/s)}$$

which, after substitution of Eq. (7-7), results in

$$v_y = \frac{-e\varepsilon_y t}{m} \quad \text{(m/s)} \tag{7-10}$$

The displacement of the electron in the Y direction follows from Eq. (7-9), which yields, applying the initial conditions of zero displacement ($y_0 = 0$) and zero velocity ($v_{0y} = 0$),

$$y = \tfrac{1}{2}a_y t^2 \quad \text{(m)}$$

which, after substitution of Eq. (7-7), results in

$$y = \frac{-e\varepsilon_y t^2}{2m} \quad \text{(m)} \tag{7-11}$$

The X distance, traveled by the electron in the time interval t, depends on the initial velocity v_{0x} and we can write, again using Eq. (7-9),

$$x = x_0 + v_{0x} t + \tfrac{1}{2}a_x t^2 \quad \text{(m)}$$

which, after applying the initial conditions for the X direction ($x_0 = 0$ and $a_x = 0$), becomes

$$x = v_{0x} t \quad \text{or} \quad t = \frac{x}{v_{0x}} \quad \text{(s)} \tag{7-12}$$

Substituting Eq. (7-12) into Eq. (7-11), we obtain an expression of the vertical deflection as a function of the horizontal distance traveled by the electron:

$$y = \left[\frac{-e\varepsilon_y}{2v_{0x}^2 m} \right] x^2 \quad \text{(m)} \tag{7-13}$$

Equation (7-13) shows that the path of an electron, traveling through an electric field of constant intensity and entering the field at right angles to the lines of flux, is *parabolic* in the X-Y plane.

In Fig. 7-9 two parallel plates, called *deflection plates*, are placed a distance d apart and are connected to a source of potential difference E_d, so that an electric field ε exists between the plates. The intensity of this electric field is given by

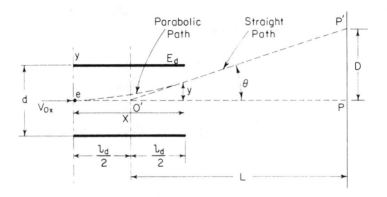

Figure 7-9 Deflection of the cathode ray beam.

$$\varepsilon = \frac{E_d}{d} \quad (\text{V/m}) \tag{7-14}$$

An electron entering the field with an initial velocity v_{0x} is deflected toward the positive plate following the parabolic path of Eq. (7-13), as indicated in Fig. 7-9. When the electron leaves the region of the deflection plates, the deflecting force no longer exists, and the electron travels in a straight line toward point P', a point on the fluorescent screen. The slope of the parabola at a distance $x = l_d$, where the electron leaves the influence of the electric field, is defined as

$$\tan \theta = \frac{dy}{dx} \tag{7-15}$$

where y is given by Eq. (7-13). Differentiating Eq. (7-13) with respect to x and substituting $x = l_d$ yields

$$\tan \theta = \frac{dy}{dx} = -\frac{e\varepsilon_y l_d}{mv_{0x}^2} \tag{7-16}$$

The straight line of travel of the electron is tangent to the parabola at $x = l_d$, and this tangent intersects the X axis at point O'. The location of this *apparent* origin O' is given by Eqs. (7-13) and (7-16) since

$$x - O' = \frac{y}{\tan \theta} = \frac{e\varepsilon_y l_d^2 / 2mv_{0x}^2}{e\varepsilon_y l_d / mv_{0x}^2} = \frac{l_d}{2} \quad (\text{m}) \tag{7-17}$$

The apparent origin O' is therefore at the center of the deflection plates and a distance L from the fluorescent screen.

The deflection on the screen is given by

$$D = L \tan \theta \quad (\text{m}) \tag{7-18}$$

Substituting Eq. (7-16) for $\tan \theta$, we obtain

$$D = L \frac{e\varepsilon_y l_d^2}{mv_{0x}^2} \quad (\text{m}) \tag{7-19}$$

The kinetic energy of the electron entering the area between the deflection plates with an initial velocity v_{0x} is

$$\tfrac{1}{2}mv_{0x}^2 = eE_a \qquad (7\text{-}20)$$

where E_a is the accelerating voltage in the electron gun. Rearranging Eq. (7-20), we obtain

$$v_{0x}^2 = \frac{2eE_a}{m} \qquad (7\text{-}21)$$

Substituting Eq. (7-14) for the field intensity ε_y, and Eq. (7-21) for the velocity of the electron in the X direction v_{0x} into Eq. (7-19), we obtain

$$D = L\,\frac{e\varepsilon_y l_d^2}{mv_{0x}^2} = \frac{Ll_d E_d}{2dE_a} \quad (\text{m}) \qquad (7\text{-}22)$$

where D = deflection on the fluorescent screen (meters)

L = distance from center of deflection plates to screen (meters)

l_d = effective length of the deflection plates (meters)

d = distance between the deflection plates (meters)

E_d = deflection voltage (volts)

E_a = accelerating voltage (volts)

Equation (7-22) indicates that for a given accelerating voltage E_a and for the particular dimensions of the CRT, the deflection of the electron beam on the screen is directly proportional to the deflection voltage E_d. This direct proportionality indicates that the CRT may be used as a *linear voltage-indicating device*. This discussion assumed that E_d was a fixed dc voltage. However, the deflection voltage usually is a varying quantity and the image on the screen follows the variations of the deflection voltage in a linear manner, according to Eq. (7-22).

The *deflection sensitivity* S of a CRT is defined as the deflection on the screen (in meters) per volt of deflection voltage. By definition, therefore

$$S = \frac{D}{E_d} = \frac{Ll_d}{2dE_a} \quad (\text{m/V}) \qquad (7\text{-}23)$$

where S is the deflection sensitivity (m/V). The *deflection factor G* of a CRT, by definition, is the reciprocal of the sensitivity S and is expressed as

$$G = \frac{1}{S} = \frac{2dE_a}{Ll_d} \quad (\text{V/m}) \qquad (7\text{-}24)$$

with all terms defined as for Eqs. (7-22) and (7-23). The expressions for deflection sensitivity S and deflection factor G indicate that the sensitivity of a CRT is independent of the deflection voltage but varies linearly with the accelerating potential. High accelerating voltages therefore produce an electron beam that requires a high deflection potential for a given excursion on the screen. A highly accelerated beam possesses more kinetic energy and therefore produces a brighter image on the CRT screen, but this beam is also more difficult to deflect and we sometimes speak of a *hard* beam. Typical values of deflection factors range from 10 V/cm to 100 V/cm, corresponding to sensitivities of 1.0 mm/V to 0.1 mm/V, respectively.

EXAMPLE 7-1

What is the minimum distance, L, that will allow full deflection of 4 cm at the oscilloscope screen with a deflection factor of 100 V/cm and with an accelerating potential of 2,000 V?

SOLUTION   To gain an insight into the physical restrictions of a cathode ray tube, refer to Fig. 7-9. The maximum deflection of the electron beam before it is shadowed by its own deflection plate can be calculated from the geometry of the cathode ray tube.

Rewriting Eq. (7-24) to solve for L, the following is obtained:

$$L = \frac{2dE_a}{Gl_d}$$

For a specific deflection factor, G, and accelerating voltage, the distance between the center of the deflection plates and the phosphor screen, L, is limited by the maximum deflection that produces a value of y equal to $d/2$. Any deflection greater than this produces a shadow on the CRT screen due to the electron beam striking its own deflection plates. The geometry of the electron beam produces two similar right triangles: one at the deflection plates consisting of the two sides, $d/2$ and $l_d/2$, and the second between the center of the deflection plates and the phosphor screen D and L. This geometry produces the following relationship:

$$\frac{L}{D} = \frac{l_d}{d}$$

Substituting this result into the preceding equation produces the relationship between the deflection factor, the accelerating potential, and the maximum deflection. Substituting the values from the example, the following is obtained:

$$L^2 = \frac{2DE_a}{G} = \frac{2 \times 4 \times 10^{-2} \times 2 \times 10^3}{10^4} = 0.016$$

$$L = 0.126$$

Thus the distance from the deflection plates to the oscilloscope tube screen is 12.6 cm. As a further example, if the accelerating potential is increased to 8,000 V and the deflection factor is to remain the same, the length of the oscilloscope tube will increase to 25.2 cm. Also, lower deflection factors, which are desirable to allow lower-voltage deflection amplifiers, would require even longer cathode ray tubes.

7-3.3 Postdeflection Acceleration

The amount of light given off by the phosphor screen depends on the amount of energy that is transferred to the phosphor by the electron beam. If the electron beam is to be deflected at a rapid rate, allowing the oscilloscope to respond to fast occurring events, the velocity of the electron beam must be great; otherwise, the light output will drop off. Thus, for a fast oscilloscope it is desirable to accelerate the electron beam to the greatest amount possible, while on the other hand the greater electron beam velocity will make it more difficult to deflect the beam.

It can be seen that the greater the accelerating potential the more difficult it is to deflect the electron beam. This would require higher deflection voltages, but, more important, because the voltage is higher the time change of voltage, that is, dV/dt, is also greater. This would require not only higher voltage for deflection but higher currents to charge the capacitance of the deflection plates. This becomes a very significant problem for high-frequency oscilloscopes with frequency responses greater than 100 MHz. Modern cathode ray tubes use a two-step acceleration to eliminate this problem. First, the electron beam is accelerated to a relatively low velocity through a potential of a few thousand volts. The beam is then deflected and, after deflection, is further accelerated to the desired final velocity. In this fashion the amount of acceleration after the deflection does not affect the deflection sensitivity. This type of cathode ray tube is called the *postdeflection acceleration tube*.

Figure 7-10 shows a diagram of a postdeflection acceleration cathode ray tube using a mesh that further increases the amount of the electron beam scan. In this example the electron beam is accelerated and deflected in a manner similar to the previous example of the simple tube. However, the beam is further acceler-

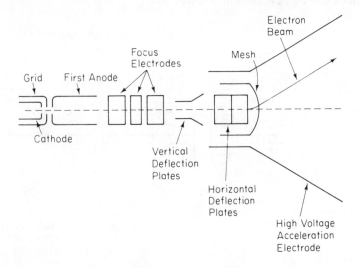

Figure 7-10 Postdeflection acceleration oscilloscope tube using a scan expansion mesh.

ated through a very high potential of 10,000 V or more, after the deflection, so it does not have an effect on the deflection sensitivity. A metallic mesh is suspended in the electron beam, and acts as a magnifying lens that causes the deflection to be further increased, which improves the deflection sensitivity. With this technique, deflection sensitivity can remain on the order of 5 to 50 V/cm even though the total electron beam acceleration is more than 10,000 V.

There are several disadvantages to the mesh type of postdeflection acceleration cathode ray tube. First, the mesh tends to defocus the electron beam and make the spot broader than it would be without the mesh interfering with the beam. Second, the mesh conducts some of the electron beam away from the screen. This results in a reduced beam current and thus reduced spot intensity. Another problem with the postdeflection acceleration cathode ray tube, and this problem is not unique to the mesh, is that the electron beam tends to be defocused in the vicinity of the deflection plates owing to repulsion from charge distributions within the tube. Several recent advances in cathode ray tube design have eliminated the mesh and alleviated these problems, thus producing a high-performance electron gun for use in high-frequency cathode ray tubes.

Figure 7-11 shows the electron gun for the meshless cathode ray tube. The electron beam is generated from a conventional heated cathode surrounded by the control grid. The first accelerating anode and two focus electrodes follow and provide focus, as well as the first accelerating voltage. These focus electrodes differ from the cylindrical elements used in the conventional tube in that they are constructed from individual metal wafers with noncylindrical holes in the center, as can be seen in Fig. 7-12. This allows for a different focusing characteristic in the horizontal plane and the vertical plane, typically divergent in one plane while being convergent in the other. The holes in the center of the metal wafers can be formed with greater precision than in a formed cylinder, and thus greater tolerances can be achieved at a lower cost.

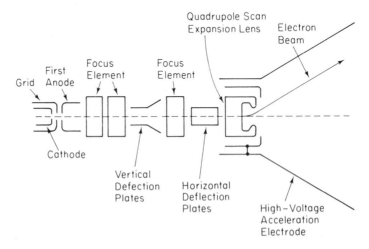

Figure 7-11 Diagram of a meshless scan expansion postdeflection acceleration cathode ray tube.

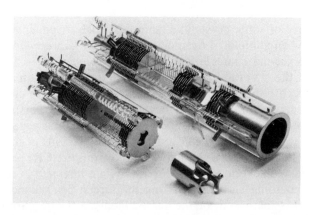

Figure 7-12 Modern oscilloscope tube electron gun showing the quadrupole electron lenses. (Courtesy of Tektronix, Inc.)

After the two focusing electrodes, the beam passes through the vertical deflection plates. The beam at this point is not fully focused, which decreases the amount of beam distortion due to the internal charge distributions. The beam will be further focused after deflection to provide a fine spot.

After vertical deflection, the beam passes through a scan expansion lens that increases the amount of beam bending in the vertical plane. The beam is then deflected in the horizontal direction and passed through another electron lens, which provides additional focusing.

The beam is accelerated to the final velocity by a *quadrupole* lens, which provides not only an increase in electron velocity, but adds to the scan angle (scan expansion, which is similar to the mesh in the previous example) without distorting or defocusing the electron beam.

The result of this design is an increased deflection sensitivity, typically 2.3 V/cm for the vertical deflection and 3.7 V/cm in the horizontal direction. The difference between the vertical and horizontal deflection sensitivities is due to the fact that the vertical deflection occurs at a lower beam velocity. Because the horizontal deflection of the oscilloscope involves only a time linear sweep, while the vertical deflection requires complex waveforms, the more sensitive deflection should be reserved for the vertical direction.

Using the meshless electron gun, 100-MHz plus oscilloscopes can be constructed with integrated circuits using only 40 or 50 V or even less for deflection. The meshless tube, being considerably shorter, results in smaller and lighter oscilloscopes for laboratory and portable use.

7-3.4 Screens for CRTs

When the electron beam strikes the screen of the CRT, a spot of light is produced. The screen material on the inner surface of the CRT that produces this effect is the *phosphor*. The phosphor absorbs the kinetic energy of the bombarding electrons and reemits energy at a lower frequency in the visual spectrum. The property of some crystalline materials, such as phosphor or zinc oxide, to emit light when stimulated by radiation is called *fluorescence*. Fluorescent materials have a second characteristic, called *phosphorescence,* which refers to the property of the material to continue light emission even after the source of excitation (in this case the electron beam) is cut off. The length of time during which phosphorescence, or afterglow, occurs is called the *persistence* of the phosphor. Persistence is usually measured in terms of the time required for the CRT image to decay to a certain percentage (usually 10 per cent) of the original light output.

The intensity of the light emitted from the CRT screen, called *luminance,* depends on several factors. First, the light intensity is controlled by the number of bombarding electrons striking the screen per second. If this so-called *beam current* is increased, or the same amount of beam current is concentrated in a smaller area by reducing the spot size, the luminance will increase. Second, luminance depends on the energy with which the bombarding electrons strike the screen, and this, in turn, is determined by the accelerating potential. An increase in accelerating potential will yield an increase in luminance. Third, luminance is a function of the time the beam strikes a given area of the phosphor; therefore sweep speed will affect the luminance. And finally, luminance is a function of the physical characteristics of the phosphor itself. Almost all manufacturers provide their customers with a choice of phosphor materials. Table 7-1 summarizes the characteristics of some of the commonly used phosphors.

As Table 7-1 shows, a number of factors must be considered in selecting a phosphor for a given application. For example, P11 phosphor, with its short persistence, is excellent for waveform photography but not at all suitable for visual observation of low-speed phenomena. P31 phosphor, with its high luminance and medium persistence, is the best compromise for general-purpose viewing and is therefore found in the majority of standard laboratory-type CROs.

It is possible to inflict serious damage to the CRT screen by incorrect handling of the front panel controls. When a phosphor is excited by an electron beam with excessive current density, permanent damage of the phosphor may occur

TABLE 7-1 Phosphor Data Chart

Phosphor type	Fluorescence	Phosphorescence	Relative luminance[a]	Decay to 0.1% (ms)	Comments
P1	Yellow-green	Yellow-green	50%	95	General-purpose; replaced by P31 in most applications
P2	Blue-green	Yellow-green	55%	120	Good compromise for high- and low-speed applications
P4	White	White	50%	20	Television displays
P7	Blue	Yellow-green	35%	1,500	Long decay; observation of low-speed phenomena
P11	Purple-blue	Purple-blue	15%	20	Photographic applications
P31	Yellow-green	Yellow-green	100%	32	General-purpose; brightest available phosphor

[a] Luminance is the photometric equivalent of brightness and is based on measurements made with a sensor having spectral sensitivity approximating the human eye. P31 is the reference phosphor.

through *burning,* and the light output will be reduced. Two factors control the occurrence of burning: beam density and duration of excitation. Beam density is controlled with the INTENSITY, FOCUS, and ASTIGMATISM controls on the oscilloscope front panel. The length of time that the beam excites a certain area on the phosphor can be adjusted with the sweep or TIME/DIV control. Burning, and possibly complete *destruction* of the phosphor, can be avoided by keeping the beam intensity down and the exposure time short.

The bombarding electrons striking the phosphor release secondary-emission electrons, thus keeping the screen in a state of electrical equilibrium. These secondary-emission low-velocity electrons are collected by a conductive coating, known as *aquadag,* on the inside surface of the glass tube, which is electrically connected to the second anode. In some tubes, particularly CRTs with magnetic focusing (such as TV picture tubes), the accelerating anode is dispensed with entirely and the conductive coating is used as the final accelerating anode.

7-3.5 Graticules

Calibrated horizontal and vertical marks are placed on the cathode ray tube screen to facilitate the use of the oscilloscope. The accuracy of these marks depends on how close the graticule marks can be placed to the actual phosphor to eliminate parallax. Early oscilloscope tubes used an external graticule to provide the necessary marks, but the distance between the marks on the graticule and the actual phosphor coating could be nearly 1 cm, which caused measurement errors if not used carefully. If the graticule lines are etched on the inner surface of the front glass of the cathode ray tube, the distance separating the phosphor and the graticule is nearly zero and parallax errors are practically nonexistent.

This internal graticule causes two problems. First, since the graticule cannot be aligned once the tube has been assembled, any misalignment between the deflection plates and the internal graticule must be corrected by electronic means. This is usually done by supplying a magnetic field by wrapping the cathode ray tube with wire carrying a current. The magnetic field rotates the electron beam and effectively rotates the cathode ray tube trace. A cathode ray tube using the external graticule could be aligned by simply rotating the graticule. Second, it is somewhat more difficult to illuminate the internal graticule lines for photographic purposes, and therefore some cathode ray tubes have special electron guns that flood the entire phosphor screen to enhance the internal graticule lines.

7-4 CRT CIRCUITS

The cathode ray tube must be supplied with several dc potentials to provide the proper control, acceleration, and focusing action. Figure 7-13 shows a cathode ray tube and the associated circuits that provide the required potentials for operation. The first requirement is a low voltage for the cathode heater. This is usually supplied from a well-insulated and separate winding on the power transformer so

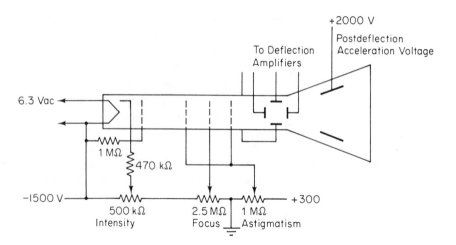

Figure 7-13 Cathode ray tube showing the development of the necessary electrode voltages and adjustments.

that the heater potential is relatively close to the cathode to prevent breakdown between those two elements. The total accelerating voltage will be applied to the cathode ray tube in two halves. First, a high negative potential will be applied to the cathode, grid, and focusing electrodes. Second, a high positive potential will be applied to the postdeflection acceleration electrode, resulting in the deflection plates being at approximately ground potential. This prevents the output of the deflection amplifier from being at a high potential and simplifies the design of that circuit.

Very few of the elements of the cathode ray tube require any significant power, and the required operating voltages are derived from simple voltage dividers, as shown in Fig. 7-13. Three controls are associated with the operating voltages of the cathode ray tube: intensity, focus, and astigmatism. The intensity control varies the potential between the cathode and the control grid and simply adjusts the beam current in the tube. The increased beam current increases the number of electrons landing on the phosphor and thus adjusts the light output. The focus control adjusts the focal length of the electrostatic lens. The astigmatism control adjusts the potential between the deflection plates and the first accelerating electrode and is used to produce a round spot.

The deflection sensitivity, and thus the accuracy of the oscilloscope, is dependent on the value of the accelerating voltage before the deflection plates, and usually this voltage is regulated. Deflection sensitivity is not a function of the postdeflection acceleration voltage, and this supply is not usually regulated.

Although not a part of the required tube voltages, a constant current, as adjusted by the *trace rotation* control, is supplied to the wire supplying the trace rotation magnetic field.

7-5 VERTICAL DEFLECTION SYSTEM

The function of the vertical deflection is rather straightforward; it must provide an amplified signal of the proper level to drive the vertical deflection plates without introducing any appreciable distortion into the system.

Although the oscilloscope can eventually be used to display practically any parameter, the input to the oscilloscope is voltage. The general laboratory oscilloscope can accept as low as a few millivolts per centimeter of deflection up to hundreds of volts using the built-in attenuator and external probes. Figure 7-14 shows the block diagram of a complete vertical deflection system. The input connector feeds an input attenuator, after which follows the vertical amplifier. The input impedance of an oscilloscope is rather high, being on the order of 1 MΩ, which is desirable for measuring voltages in high impedance circuits. The attenuator sets the sensitivity of the oscilloscope in the common 1-2-5 sequence. As an example, the input attenuator could provide for 10, 20, 50, 100, 200 mV, etc., per centimeter. The input attenuator must provide the correct 1-2-5 sequence attenuation while maintaining a constant input impedance, as well as maintaining both the input impedance and attenuation over the frequency range for which the oscilloscope was designed.

Figure 7-15 shows a resistive divider attenuator connected to an amplifier with a 10-pF input capacitance. If the input impedance of the amplifier is high, the input impedance to the attenuator is relatively constant regardless of the switch setting of the attenuator. The impedance as seen by the amplifier changes dramatically depending on the setting of the attenuator. Because of this, the *RC* time constant and thus the frequency response of the amplifier are dependent on the setting of the attenuator, which is highly undesirable. The attenuator of Fig. 7-15 would have a high-frequency roll-off due to the shunt capacitance of the vertical amplifier.

Figure 7-16 shows an attenuator with both resistive and capacitive voltage dividers. The capacitive voltage divider improves the high-frequency response of the attenuator. This combination of capacitive and resistive voltage dividers is

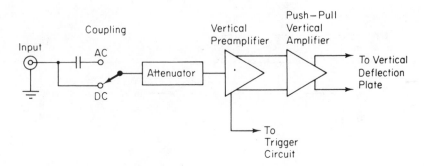

Figure 7-14 Block diagram of the vertical section of an oscilloscope.

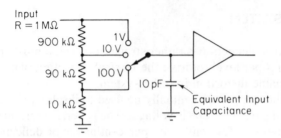

Figure 7-15 Uncompensated attenuator showing the input capacitance of the amplifier.

known as a *compensated attenuator*. For oscilloscopes where the frequency range extends to 100 MHz and beyond, even more complex input dividers are required as shown in Fig. 7-17. This example shows an attenuator divided between the input and output of the vertical deflection preamplifier. The input attenuator provides switching for powers of ten, while the attenuator at the output of the vertical preamplifier provides the 1-2-5 attenuation. This greatly reduces the number of steps the input attenuator has to provide and improves the frequency response.

Practically all oscilloscopes provide a switchable input coupling capacitor. This is provided so that measurements of ac signals may be viewed in the presence of high dc voltages by including the coupling capacitor. When dc measurements are to be made, the capacitor may be removed. The value of the capacitor is chosen so that the frequency response of the oscilloscope is preserved down to a few hertz.

The input impedance of an oscilloscope is 1 MΩ shunted with between 10 and 30 pF. If a probe were connected to the oscilloscope, the input impedance at the probe tip would have greater capacitance because of the added capacitance of the probe assembly and of the connecting shielded cable. It is desirable, especially for high-frequency oscilloscopes, to have an input capacitance much less than 20 or 30 pF, and this is achieved by the use of an attenuator probe. Figure 7-18 shows a schematic diagram of a 10-to-1 attenuator probe connected to the input of an oscilloscope. Within the probe tip is a 9.0-MΩ resistor, and shunted across this resistor is a capacitor. In the base of the probe at the oscilloscope

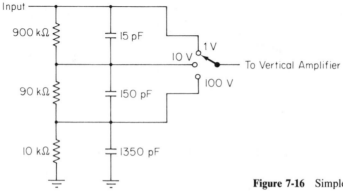

Figure 7-16 Simple compensated attenuator.

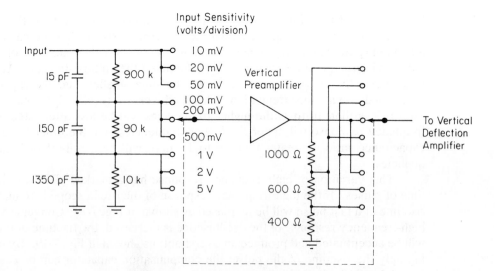

Figure 7-17 Two-stage attenuator for a high-frequency oscilloscope.

connector, there is an adjustable capacitor. This capacitor is adjusted so that the ratio of the shunt capacitance to the series capacitance is exactly 10 to 1. The attenuator probe, often called a 10-to-1 probe, provides approximately a 10-to-1 reduction in input capacitance, but also provides a 10-to-1 reduction in overall oscilloscope sensitivity.

Because the input capacitance of an oscilloscope cannot be guaranteed from unit to unit, the 10-to-1 probe is provided with an easily adjustable *compensating* capacitor. If the ratio of the series to shunt capacitance is not adjusted to be precisely 10 to 1, the frequency response of the oscilloscope will not be flat.

The effects of incorrect setting of an oscilloscope compensation can readily be seen by observing a fast risetime pulse. If the frequency response of the oscilloscope is not correct, the pulse will be distorted and the compensation can be adjusted to achieve the least amount of distortion. Referring to the schematic of the oscilloscope probe of Fig. 7-18, if the ratio of the capacitors is exactly 10 to 1, the voltage division at the oscilloscope input will be exactly 10 to 1 for all

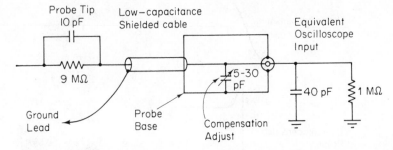

Figure 7-18 A 10-to-1 oscilloscope probe as it would appear connected to an oscilloscope input.

frequencies. If the ratio of the capacitance is different than 10 to 1, the voltage attenuation at higher frequencies will be incorrect. Notice that the capacitance affects only the higher frequencies. At low frequencies the reactance of the capacitors is great, and the resistance will dominate the voltage division. At higher frequencies the capacitive reactance will become smaller, and the capacitance ratio will dominate the attenuation. This produces two results. If the capacitance ratio is less than 10 to 1, the higher frequencies will be less attenuated and the frequency response will favor the higher frequencies. On the other hand, if the capacitance ratio is greater than 10 to 1, there will be a roll-off of higher frequencies.

This variation of high-frequency response has a marked effect on the risetime of pulses. If the high-frequency response of the oscilloscope is reduced, the risetime of a fast pulse will be increased as shown in Fig. 7-19. Conversely, if the high-frequency response of the oscilloscope is increased, the risetime of the pulse will be accentuated and produce an overshoot, as shown in Fig. 7-19. By observing only the risetime of the pulse, the compensation capacitor can be easily adjusted for perfectly flat frequency response.

The general effects of incorrect compensation can be seen in Fig. 7-20. Three waveforms are shown for a correctly compensated probe, an overcompensated probe, that is, favoring the higher frequencies, and an undercompensated probe, which favors lower frequencies. Figure 7-20(a) shows the proper representation of three waveforms, a square wave of 50 kHz, a single pulse with a sharp risetime and an exponential decay, and a 50-kHz sine wave. The overcompensated probe shown in Fig. 7-20 provides an overshoot on the leading and falling edges as previously explained. In the case of the single-pulse example, the risetime is not sufficiently fast to create a noticeable overshoot, but the amplitude of the pulse is distorted. In the case of a 50-kHz sine wave, the amplitude is in-

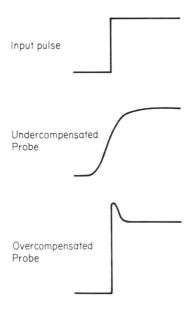

Input pulse

Undercompensated Probe

Overcompensated Probe

Figure 7-19 Effects of probe compensation.

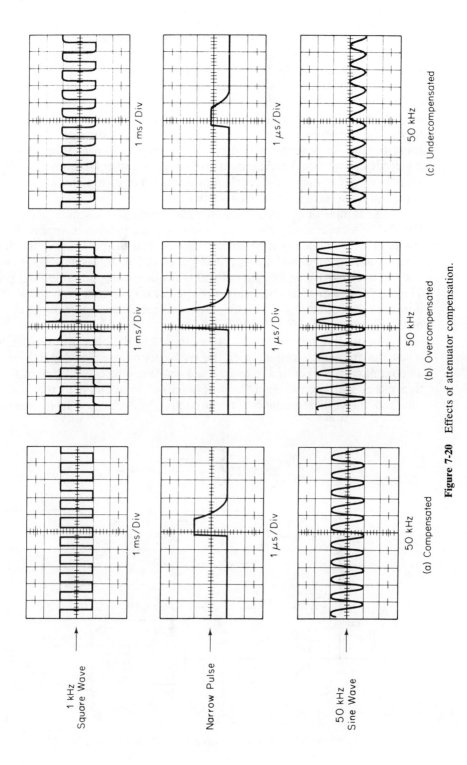

Figure 7-20 Effects of attenuator compensation.

207

creased because the frequency response of the probe favors high frequencies. In the case of the undercompensated probe, the risetime of the square wave is reduced and appears rounded, the single pulse is reduced in amplitude, and likewise the amplitude of the 50-kHz sine wave is reduced.

The voltage required to deflect the electron beam in the cathode ray tube varies from about 100 V peak to peak to about 500 V depending on the accelerating voltage and the construction of the tube. The input sensitivity of many laboratory oscilloscopes is on the order of a few millivolts per division, and the required gain from this low level to several hundred volts peak to peak is provided for with the vertical amplifier. In addition to providing this rather large amount of gain, the vertical amplifier must be direct coupled, must not distort the waveform in any way, and must have a broad frequency response.

An example of a vertical amplifier for a laboratory oscilloscope is shown in Fig. 7-21. As previously discussed, the operating potentials applied to the elements of the cathode ray tube are arranged so that the potential of the deflection plates is near ground, and the amplifier shown in Fig. 7-21 is designed to interface with an oscilloscope where the vertical deflection plates are at ground potential. The amplifier shown is completely push-pull or double ended, so 230 V of deflection can be supplied while operating the transistors at a supply voltage of only half of the peak-to-peak deflection voltage.

The oscilloscope deflection plates represent the plates of a capacitor, and when the frequency response of the oscilloscope exceeds 1 MHz or so, the amount of current required to charge and discharge the capacitance of the deflection plates can become significant. Therefore, in addition to the voltage gain required, the vertical amplifier is required to supply enough current gain to charge

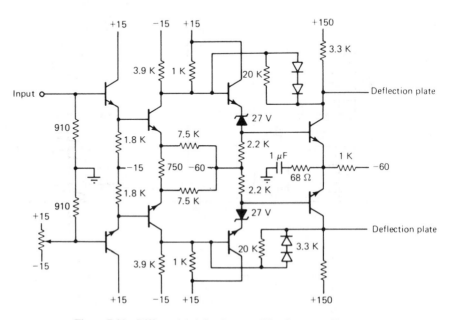

Figure 7-21 Differential deflection amplifier for an oscilloscope.

Oscilloscopes   Chap. 7

and discharge the deflection plate capacitance. The typical oscilloscope vertical amplifier operates with high-current class A amplifiers with feedback, as shown in Fig. 7-21.

7-6 DELAY LINE

7-6.1 Function of the Delay Line

All electronic circuitry in the oscilloscope (attenuators, amplifiers, pulse shapers, generators, and indeed the circuit wiring itself) causes a certain amount of time delay in the transmission of signal voltages to the deflection plates. Almost all of this delay is created in circuits that switch, shape, or generate. Comparing the vertical and horizontal deflection circuits in the oscilloscope block diagram of Fig. 7-22, we observe that the horizontal signal (time base, or sweep voltage) is initiated, or *triggered,* by a portion of the output signal applied to the vertical CRT plates. Signal processing in the horizontal channel consists of generating and shaping a trigger pulse (trigger pickoff) that starts the sweep generator, whose output is fed to the horizontal amplifier and then to the horizontal deflection plates. This whole process takes time: on the order of 80 ns or so.

To allow the operator to observe the leading edge of the signal waveform, the signal drive for the vertical CRT plates must therefore be *delayed* by at least the same amount of time. This is the function of the vertical delay line. We observe that in Fig. 7-22 a 200-ns delay line has been added to the vertical channel, so that the signal voltage to the CRT plates is delayed by 200 ns, and the horizontal sweep is started prior to the vertical deflection. Although the delay line can appear almost anywhere along the vertical signal path, the trigger pickoff *must* precede the delay line.

There are basically two kinds of delay line: the lumped-parameter delay line and the distributed-parameter delay line.

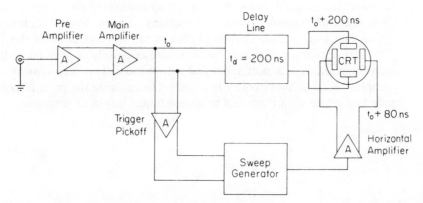

Figure 7-22   Delay of the vertical signal allows the horizontal sweep to start prior to vertical deflection.

7-6.2 Lumped-Parameter Delay Line

The lumped-parameter delay line consists of a number of cascaded symmetrical *LC* networks, such as the so-called *T-section* of Fig. 7-23.

If the T-section is terminated in its *characteristic impedance* Z_o, then, by definition, the impedance looking back into the input terminals is also Z_o. This condition of termination gives the T-section the characteristics of a low-pass filter whose attenuation and phase shift are a function of frequency, and whose passband is defined by the frequency range over which the attenuation is zero. The upper limit of the passband is called the *cutoff frequency* of the filter, given by

$$f_c = \frac{1}{\pi\sqrt{LC}} \tag{7-25}$$

If the spectrum of input signal v_i consists of frequencies much less than the cutoff frequency, output signal v_o will be a faithful reproduction of v_i, but delayed by a time

$$t_s \approx \frac{1}{\pi f_c} = \sqrt{LC} \tag{7-26}$$

where t_s is the time delay for a single T-section. A number of T-sections, cascaded into a so-called *lumped-parameter* delay line, increases the total delay time to

$$t_d = n t_s \tag{7-27}$$

where n is the number of cascaded T-sections.

Because of the sharp cutoff frequency of the lumped-parameter delay line, amplitude and phase distortion become a problem when the frequency of the input signal increases. The application of a step-voltage input, for example, which contains high-frequency components (odd harmonics), causes an output voltage that suffers from transient response distortion in the form of overshoot and ringing, as shown in Fig. 7-24. This kind of response can be improved to more closely resemble the original step-voltage input by modifying the design of the filter sections into, for example, m-derived sections. The m-derived section is a popular circuit that uses mutual coupling between the two inductors of the T-section.

It is important to *match* the delay line as closely as possible to its characteristic impedance Z_o, at both input and output ends. This requirement often leads to complex termination circuitry in an effort to optimize the balance between amplitude and phase distortion and to obtain better transient response.

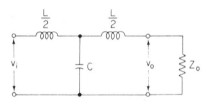

Figure 7-23   T-filter section.

Oscilloscopes   Chap. 7

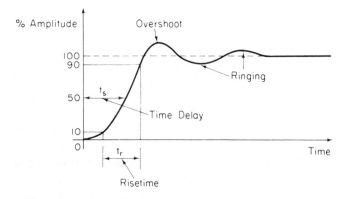

Figure 7-24 Step-voltage response of a T-section filter terminated in its characteristic impedance $Z_o\sqrt{L/C}$.

A practical delay line circuit in an oscilloscope is driven by a push-pull amplifier and then consists of a symmetrical arrangement of cascaded filter sections, as in Fig. 7-25. Optimum response of the delay line requires precise proportioning of the L and C components in each section; the variable capacitors must be carefully adjusted to be effective.

7-6.3 Distributed-Parameter Delay Line

The distributed-parameter delay line consists of a specially manufactured coaxial cable with a high value of inductance per unit length. For this type of delay line, the straight center conductor of the normal coaxial cable is replaced with a continuous coil of wire, wound in the form of a helix on a flexible inner core. To reduce eddy currents, the outer conductor is usually made of braided insulated wire, electrically connected at the ends of the cable. Construction details are shown schematically in Fig. 7-26.

The inductance of the delay line is produced by the inner coil, and it equals that of a solenoid with n turns per meter. The inductance can be increased by winding the helical inner conductor on a ferromagnetic core, which has the effect of increasing the delay time t_d and the characteristic impedance Z_o. The capacitance of the delay line is that of two coaxial cylinders separated by a polyethylene dielectric. The capacitance can be increased by using a thinner dielectric spacing between the inner and outer conductors.

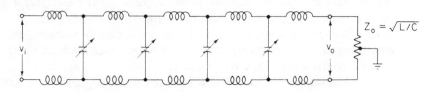

Figure 7-25 Push-pull transmission line with single termination.

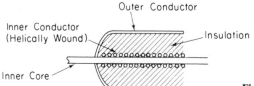

Figure 7-26 Helical high-impedance delay line.

Typical parameters for a helical, high-impedance delay line are $Z_o = 1,000 \ \Omega$ and $t_d = 180$ ns/m. The coaxial delay line is advantageous because it does not require the careful adjustment of a lumped-parameter line and it occupies much less space.

7-7 MULTIPLE TRACE

In the analysis of electronic circuits and systems, it is very useful to be able to view the behavior of two or more voltages simultaneously. This could be accomplished by using two oscilloscopes. Aside from the expense, it is difficult to trigger the sweeps of each oscilloscope at precisely the same time and to insure that the sweep generators operate precisely the same. Even if all this should be achieved, the two traces to be viewed will be on two different oscilloscope tubes, and not one atop the other where the comparison can be made easily.

There is an elegant solution to this problem, which requires a special cathode ray tube that has two separate electron guns generating two separate beams. Each electron beam has its own vertical deflection plates, but the two beams are deflected in the horizontal direction by a common set of deflection plates and deflection generator. This is called a *dual-beam cathode ray tube* and is only used in systems where absolute independence of the vertical channels is required.

A more common and less costly method is to use the dual-trace, as opposed to dual-beam, method. In this method, the same electron beam is used to generate two traces that can be deflected from two independent vertical sources. Two methods may be employed to generate the two independent traces. One method is to deflect the oscilloscope and display the first or A vertical input. Then the oscilloscope is triggered and the B vertical input is displayed at a different position on the oscilloscope screen. A block diagram of this system is shown in Fig. 7-27. An electronic switch is used to switch between the two vertical sources, which are processed in separate vertical amplifiers that include separate position controls. Each time the sweep generator is triggered, the electronic switch is changed to the other channel. The disadvantage of this system is that the display is not actually a representation of two simultaneous events; the events were at two different times. If the events are cyclical, this may not pose a problem. However, if the events are a single occurrence or are different each cycle, this *alternate-sweep* method will not provide a true picture.

A second method is to switch from one vertical channel to the other at such a rapid rate that the display is created from small segments of the actual waveform. This requires that the chopping frequency be much greater than the input wave-

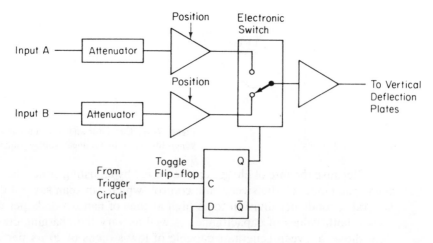

Figure 7-27 Block diagram of a dual-trace oscilloscope.

form in order to prevent the display from being unrecognizable. The practical switching speed of an electronic switch will limit the frequency capability of this method, and chopping frequencies are usually less than 500 kHz. The electronic circuits required to generate the *chop* method of dual-trace generation are precisely the same as those used to generate the alternate method, except the electronic switch is a high-frequency clock rather than the trigger generator.

Because there are significant advantages and disadvantages with each system, most oscilloscopes have a switch that is capable of selecting either method.

7-8 HORIZONTAL DEFLECTION SYSTEM

It is the purpose of most laboratory oscilloscopes to deflect the horizontal portion of the trace at a constant rate relative to time, which is often referred to as *linear sweep*. The horizontal deflection system consists of a time-base generator, a trigger circuit, and a horizontal amplifier as shown in Fig. 7-1. The time-base generator controls the rate at which the beam is scanned across the face of the cathode ray tube and is adjusted from the front panel. The trigger circuit, as previously described, insures that the horizontal sweep starts at the same point of the vertical input signal. The horizontal amplifier is similar to the vertical amplifier described previously and is required to increase the amplitude of the signals generated in the sweep generator to the level required by the horizontal deflection plates of the cathode ray tube.

The sweep generator uses the charging characteristics of a capacitor to generate linear risetime voltages to feed to the horizontal amplifier. Figure 7-28 shows a capacitor being charged from a constant-current source. The rate of voltage rise is given as

$$\frac{\text{change of voltage}}{\text{time}} = \frac{I}{C}$$

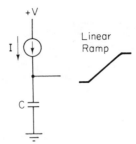

Figure 7-28   Capacitor and constant-current generator generating a linear voltage ramp.

Because the rate of charge can be varied by adjusting either the current, *I*, or the capacitance, *C*, the sweep rate control, which can span several decades from several seconds per division to as high as tens of nanoseconds per division, can switch both values of capacitance, as well as vary the charging current. Figure 7-29 shows a sweep generator capable of low sweeps of 20 μs per division to a maximum of 50 ns per division using both variable currents and switched capacitors. The sweep generator in this example follows the same 1-2-5 sequence that was used in the input attenuator in the vertical system. The resistors in the constant-current generator are switched to provide currents in a 1-2-5 sequence, which involves switching resistors in the reciprocal relationship, that is, $1\text{-}\frac{1}{2}\text{-}\frac{1}{5}$ sequence, while capacitors are switched in a decade sequence. In this fashion the eight decades required for the sweep rate can be handled by 11 timing components.

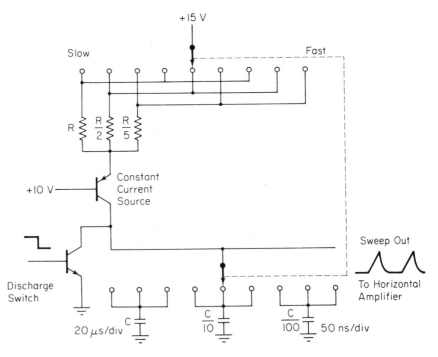

Figure 7-29   Simplified schematic of a triggered time base for an oscilloscope.

Oscilloscopes   Chap. 7

The sweep generator shown is used in a triggered scope, which means that the sweep does not start until a triggering pulse is received from the triggering circuit. This is accomplished very simply by shunting all the current from the constant-current generator around the capacitor with a transistor and thus preventing voltage buildup at the capacitor.

Once the sweep has been completed, the voltage at the capacitor is returned to zero by discharging the capacitor through the transistor, and after a period of time, called the *hold-off time,* the sweep is free to start again.

The relationship between the sweep generator and the trigger pulses, which represent the same point of the input waveform, is shown in Fig. 7-30. The sweep does not usually trigger for each cycle of the vertical input waveform unless the sweep plus the hold-off time is less than the period of the input. In some oscilloscopes the hold-off time can be adjusted from the front panel to facilitate stable triggering from complex waveforms.

When the oscilloscope has not been triggered, the electron beam in the cathode ray tube is turned off or *blanked.* Otherwise, a bright spot will appear at the left side of the screen and will in a short period of time destroy the phosphor coating at that point. In addition, the electron beam is turned off or blanked during the retrace. The image painted by the retrace is reversed in time and of a different rate. It therefore provides no useful information and clutters the desired trace. Generally, in an oscilloscope the trace is blanked and signals are applied to *unblank* the trace. When the triggering circuit supplies a negative-going pulse to allow the capacitor in the sweep circuit to charge and start the sweep, the same pulse is used to unblank the beam.

Without an input signal the beam is not unblanked, and no trace is visible on the screen. This can be a difficulty when it is necessary to set the vertical position control, as the beam cannot be seen. To facilitate locating the base line, most oscilloscopes have a built-in oscillator to trigger the beam when no input signal is present. When a signal of sufficient amplitude is available, the horizontal sweep is triggered by the vertical input signal.

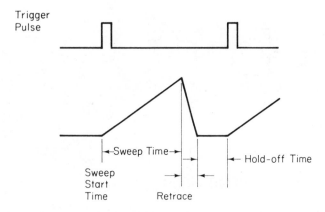

Figure 7-30 Relationship between the trigger pulse and the sweep in an oscilloscope.

Most laboratory oscilloscopes have two time bases that may interact in various ways. One popular method of interaction is to allow one time base to delay the triggering of a second time base. This would be useful if a signal with a long period were being viewed, but only a small portion of the signal were to be analyzed. In this case the triggering signal from the vertical amplifier would be applied to the first time base, and after a period of time, as set by the switches controlling the first time base, the second time base would be triggered. The significant advantage of this system is that the slow time base required by the long period of the input waveform could delay the second time base, which would be considerably faster, for a close inspection of the input waveform. In this example the oscilloscope would remain blanked until after the delay period, when the second time base would be triggered and the oscilloscope unblanked. Figure 7-31 shows a block diagram of this type of delayed sweep. Time base A supplies a linear voltage to a comparator, which triggers the second time base when the ramp voltage reaches the voltage supplied by the *time base multiplier* control on the front panel.

The delayed time base can either be triggered from the input signal or the sweep can begin immediately after the time delay. When the delayed portion of the waveform to be viewed may vary, in time, between the other parts of the waveform, it is usually desirable to trigger the time base. When no convenient edge of the signal is available after the delay time, it will be necessary to start the delayed time base after the delay time without a trigger.

There are significant disadvantages to this simple system. Once the delayed sweep is activated, the oscilloscope displays only the small portion of the waveform that is being investigated, and the picture of the remainder of the input waveform is lost. It would be useful to be able to view the entire waveform with the portion that is or will be expanded highlighted in some fashion. This can be accomplished by intensifying the portion of the waveform that will be displayed after the delay using the circuit in Fig. 7-32. In this sweep circuit, the horizontal deflection is supplied from the slower time base, but the second or faster time base supplies a pulse to the unblanking circuits to intensify the trace. This shows the

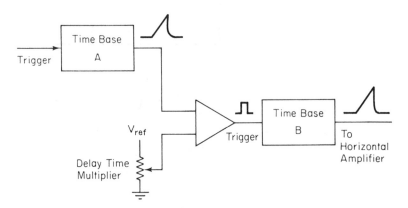

Figure 7-31 Block diagram of a dual time base for delayed trigger.

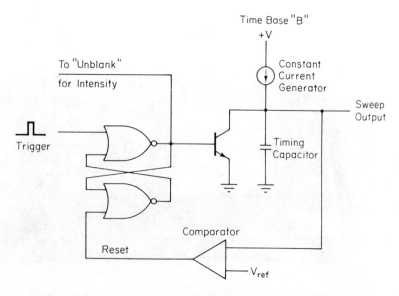

Figure 7-32 Time base schematic showing the origin of the intensify signal.

actual segment of the waveform that will be displayed when the oscilloscope is switched to the delayed mode. A more modern and versatile approach is the alternate sweep system. In this system the deflection is first supplied from the main or time base A generator. Then the trace is moved vertically and the delayed portion is displayed. This is equivalent to switching an intensified trace to a delayed trace while simultaneously changing the vertical position of the trace. When this switching is done at a rapid rate, two stable traces appear simultaneously.

Another method of alternate time-base operation is to switch the sweep speed after the delay time. In this method the beginning portion of the waveform is visible at the slow sweep speed, while at the delay time the sweep speed changes to the faster rate. The only significant disadvantages of this system are that only the beginning portion of the waveform is visible and that part of the screen has to be shared with the slower portion of the sweep, which reduces the amount of area for observing the fast portion of the sweep. There are some significant advantages, however, because only one sweep is required to display both the slow and delayed portions of the trace. This becomes significant when viewing complex waveforms where a stable triggering point is difficult or impossible to obtain.

There is another technique for obtaining an expanded view of the input waveform without using two independent time bases; it uses a magnifier. This circuit simply increases the gain of the horizontal amplifier and increases the sweep rate at the cathode ray tube screen by a factor of 5 or 10, depending on the gain increase. If, for example, the magnifier increased the amplifier gain by a factor of 10, the sweep speed would increase but 90 per cent of the trace would be invisible. The horizontal position control can be used to place that portion of the

Figure 7-33 Modern 100-MHz oscilloscope. (Courtesy of Phillips Test and Measurement.)

trace that is of interest on the tube, but this is not like a calibrated delay time. In addition there is no control over the delay other than the fixed magnification ratio. Nevertheless, the magnifier can be included in inexpensive oscilloscopes for very little cost. Figure 7-33 shows a typical 100-MHz oscilloscope.

7-9 OSCILLOSCOPE PROBES AND TRANSDUCERS

The primary function of the oscilloscope is to display voltage as a function of time, and the description of the attenuator probe was covered in Sec. 7-5. There are other probes and transducers that can make the oscilloscope more versatile. Other than the 10-to-1 probe, there are also other attenuation ratios such as a 1 to 1, which is nothing more than a cable with a probe tip and no other components. One useful probe is the active probe, which achieves a lower capacitance without the attenuation associated with the 10-to-1 probe. The final special transducer, since it is more than just a probe, is the current probe, which allows the oscilloscope to measure current without breaking the circuit under test.

A schematic of the active probe is shown in Fig. 7-34. In this example a field effect transistor is used as the active element to amplify the input signal. Although the voltage gain of the FET follower circuit shown in Fig. 7-34 is unity, the follower circuit provides a power gain so that the input impedance can be increased. To be effective, the FET must be mounted directly in the voltage probe tip so that the capacitance of an interconnecting cable can be eliminated. This requires that power for the FET be supplied from the oscilloscope to the FET in the probe tip. The FET voltage follower drives a coaxial cable, but instead of the cable connecting directly to the high-input impedance of the oscilloscope, the

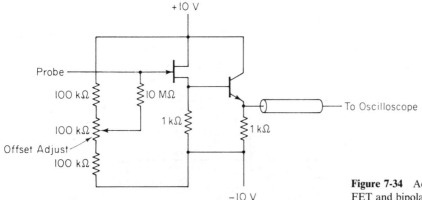

+10 V

Probe

100 kΩ   10 MΩ

1 kΩ

100 kΩ

Offset Adjust

100 kΩ

1 kΩ

To Oscilloscope

−10 V

Figure 7-34 Active probe using an FET and bipolar transistors.

cable is terminated in its characteristic impedance. In this fashion, there is no high-frequency roll-off of the frequency due to the capacitance of the cable.

There is a significant disadvantage with the FET probe. Because there is no signal attenuation between the FET amplifier and the probe tip, the range of signals that can be handled by the FET probe is limited to the dynamic range of the FET amplifier, and this is typically less than a few volts. Therefore, to handle a larger dynamic range, external attenuators are added at the probe tip. Adding the attenuator to the FET probe is effectively making the FET probe an attenuator probe; thus there is no real need to use the active probe, and a conventional probe could serve the purpose unless the extremely low capacitance available from an active probe with an attenuator is required. It is for this reason that active voltage probes have limited use.

Oscilloscopes are typically used with a 10-to-1 attenuator probe, as many circuits are affected by the capacitance of the 1-to-1 probe. For this reason most oscilloscopes have input sensitivities of 2 or 5 mV per division so that the versatility of the oscilloscope is not destroyed by the attenuation of the signal.

One very valuable probe is the current probe. This device can be clamped around a wire carrying an electrical current without any physical contact to the probe, allowing the oscilloscope to be used to measure the magnitude of the current with a frequency response from dc to 50 MHz. The current sensor consists of two parts, a conventional transformer for transforming alternating current to voltage, and a Hall effect device for converting direct current to a voltage, as shown in Fig. 7-35.

A magnetic core with a removable piece is used as the coupling element for the current probe. The wire carrying the current to be measured is inserted in the center of the magnetic core and is similar to a primary of a transformer. Alternating current in the wire will induce voltage in the secondary winding by conventional transformer action. Only alternating current will introduce a voltage in the secondary. Any direct currents will not appear at the current transformer secondary. In addition, the direct current passing through the wire will cause the magnetic flux in the core to increase and will affect the permeability of the material used for the core. This is undesirable, especially if the current in the wire should

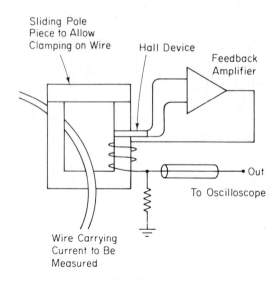

Figure 7-35 Current probe capable of measuring from dc to several megaherz.

cause the core material to become saturated. If this should happen, the transformer action of the current transformer will become severely affected and provide inaccurate measurements.

To provide a frequency response to zero or dc, a Hall effect sensor is also included in the current probe. The Hall effect occurs in many semiconductors as shown in Fig. 7-36. Current flow through the semiconductor shown is by the drift of electrons through the bulk material. The drift current occurs when an electric field is applied to the semiconductor material that causes the electrons to enter at the negative terminal and drift to the positive terminal. Generally, the motion is

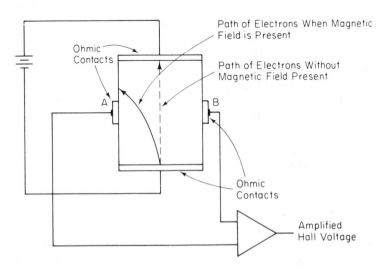

Figure 7-36 Hall-effect generator showing the path of electrons.

along a straight line, and if the potential were measured between points *A* and *B*, the potential difference would be zero. If the semiconductor material were subjected to a magnetic field, the direction of which is perpendicular to the motion of the drift electrons, the moving electrons would experience a force that would cause the paths of the electrons in the semiconductor material to move to one side, as shown in Fig. 7-36. Because the distribution of electrons is greater on one side than the other, there would exist a potential between the two sides, *A* and *B*.

The Hall effect sensor is included in the magnetic core structure of the current probe. A feedback system is arranged with an amplifier such that any magnetic field present in the Hall sensor will cause current to be induced into the secondary windings of the current transformer to counteract the magnetic field introduced by the wire being measured. Thus the Hall effect sensor assures that the static magnetic flux in the core is exactly zero. The amount of current required to counteract the magnetic field induced by the wire being measured is directly proportional to the magnitude and direction of the current passing through the wire being measured. Because the current required to counteract the static magnetic flux in the core also passes through the terminating resistor for the secondary of the current transformer, this applied counteracting current appears as a dc voltage at the oscilloscope and represents the magnitude of the current in the wire.

7-10 OSCILLOSCOPE TECHNIQUES

The oscilloscope is a very versatile instrument whose use is limited only by the skill of the operator. Although there are more sophisticated instruments, most oscillographic measurements are made with a dual-trace oscilloscope with delay sweep capabilities. Therefore, the discussion of oscilloscope techniques will assume the use of this class of instrument.

7-10.1 Determining Frequency

Determining the frequency of a waveform using the oscilloscope requires that the period be measured. To measure a frequency, the waveform viewed by the oscilloscope must be periodic, which sounds simple but can actually be somewhat difficult and lead to measurement errors. Consider as an example the simple sine function shown in Fig. 7-37. It is quite clear that the period of the sine function is between any alternate zero crossing. The period can also be measured between

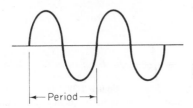

|← Period →|

Figure 7-37 Period of a simple sine function.

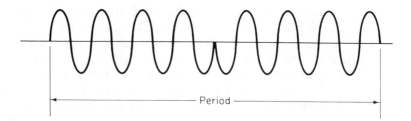

Figure 7-38 Complex waveform showing the correct period.

any two positive peaks or any two negative peaks. A more complex example is shown in Fig. 7-38. In this example the period cannot be determined from peak to peak or from every other zero crossing, because for every four cycles of the sine function the phase is changed by 180°. The complete waveform of this example is eight complete sine cycles, four of one phase and four additional of the opposite phase. When determining the period, be sure that the cycle is complete and that the next cycle is exactly the same. To determine the frequency, take the reciprocal of the period measurement:

$$\text{frequency} = \frac{1}{\text{period}}$$

The oscilloscope is not a precision frequency measuring tool because the accuracy of the frequency depends directly on the accuracy of the oscilloscope time base, which is, at best, a few per cent. The oscilloscope should be used for a rough estimate of frequency or when the waveform is so complex that a frequency counter would not operate reliably.

7-10.2 Phase Angle and Time Delay Measurement

The oscilloscope is well-suited and indispensable for time and phase measurements. As an example, assume the phase angle between two sine functions is to be measured as shown in Fig. 7-39. A very simple and effective method is to

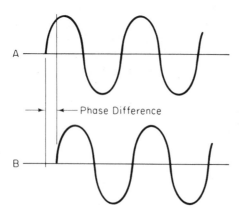

Figure 7-39 Two sine functions of the same frequency with a phase difference.

display the two sine functions as two separate traces on a dual-trace oscilloscope and measure the time delay between the two sine functions. When using this technique, it is imperative that the oscilloscope be triggered only from one of the two waveforms. This can be accomplished by setting the triggering source switch to either channel 1 or 2 or externally triggering the oscilloscope. *Mixed, composite,* or *both* modes of triggering allow the oscilloscope to be triggered by both of the two input signals and will not provide a suitable time reference.

The external trigger input allows the oscilloscope to be used for complex time and phase delay measurements. As an example, consider the digital circuit shown in Fig. 7-40. An input pulse initiates various output pulses that are timed to occur at specific times. Triggering the oscilloscope with the input clock, the oscilloscope probe can be moved to any one of the outputs and the time delay read. This technique allows a single-trace oscilloscope to be used for phase and time delay measurements because it is not necessary to view both the reference and the signal to be measured simultaneously. For this technique to be accurate, the triggering delay must be known. To determine this, trigger the oscilloscope using the external triggering input. One convenient method is to use an oscilloscope probe connected to the external triggering input. Most external triggering inputs have similar input impedances as the vertical inputs, which allows the same probes to be used. While the oscilloscope is being triggered from the external input, view the same signal with a second probe connected to the vertical input. Note the position on the oscilloscope screen of the edge of the input pulse used to

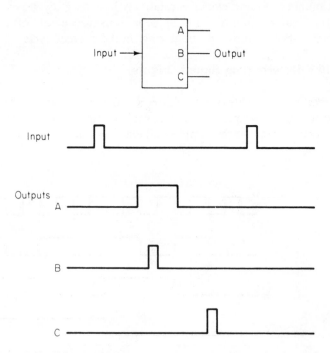

Figure 7-40   Triggered logic sequence generator and associated waveforms.

trigger the oscilloscope. This is the zero time reference, and all subsequent measurements will be made relative to this point. If the time base is changed, it will be necessary to repeat this calibration procedure for the new time-base setting.

In many situations, more than one input cycle will be received for each cycle of the output, such as the example of a 3-bit binary counter shown in Fig. 7-41. In this case the first output will occur every two input cycles, the second output will occur every four input cycles, and the third output will occur every eight input cycles. Therefore, if the time delay is to be measured between the input clock and the divide-by-eight, or third output, there will be eight positive edges of the input clock to trigger from. There are two methods of measuring the delay between the input clock and the third output, which is called *propagation delay,* a very important parameter in digital systems. The first method involves allowing the oscilloscope to trigger each positive edge of the input. Thus, for each output of the counter there will be eight oscilloscope traces, of which seven are undesired. This will produce a blurry picture, but it is possible to see, without much difficulty, the desired output transition and to make the necessary measurement.

A second method involves using the trigger hold-off control to eliminate the undesired traces. The trigger hold-off, as discussed previously, sets the time that the oscilloscope cannot be retriggered after a sweep. If the hold-off can be adjusted so that the oscilloscope triggers on the edge of the input clock and does not trigger on the next seven cycles, only one trace of the output will be displayed for each eight input cycles. The only difficulty with this method is that it is not possible to select the actual input cycle that the oscilloscope will trigger from. If the number of input cycles is relatively low, such as two or four, or even eight as in this example, a little bit of hit or miss using the hold-off control will eventually result in the oscilloscope triggering on the correct cycle.

7-10.3 Determining Signal Origins

In many electronic systems, failures will involve signals appearing where they do not belong. A simple example of such a situation is a power supply where ripple voltages appear on the supply voltages. A source of this ripple voltage could be

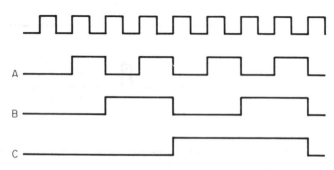

Figure 7-41   Waveforms associated with a 3-bit binary counter.

from the 60-Hz supply, or it could also be injected from a malfunctioning circuit. One method of determining the source of such extraneous signals is to trigger the oscilloscope from a suspected source and view the extraneous signal with the oscilloscope. If the triggering signal of the oscilloscope is the source of the offending signal, the viewed waveform will be stationary, but if the source is elsewhere, the viewed waveform will move in time. Returning to the power supply example, if the oscilloscope were triggered from the 60-Hz line, and many oscilloscopes have a switch for this purpose, the offending ripple, if it were due to the 60-Hz line, would be stationary; if it were developed elsewhere, it would move in time. Sixty-hertz pickup by electronics circuits is very common, and this technique allows quick verification of the source of the undesired signal.

7-10.4 Determining Modulation Characteristics

The oscilloscope can be used to measure the amount of amplitude modulation applied to a carrier for adjusting and troubleshooting amplitude-modulated transmitters, both full carrier and single sideband. Figure 7-42(a) shows a full-carrier

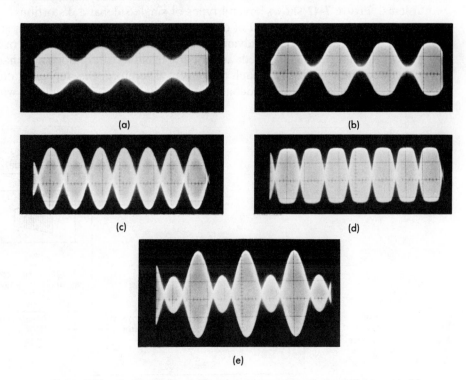

(a)

(b)

(c)

(d)

(e)

Figure 7-42 (a) Amplitude-modulated carrier modulated about 50 per cent; (b) carrier modulated about 80 per cent; (c) properly modulated two-tone single-sideband carrier; (d) clipped two-tone single-sideband signal; (e) double sideband single-tone signal with excessive carrier leakage.

amplitude-modulated signal. To display the carrier, the oscilloscope must be capable of covering the carrier frequency of the transmitter. The horizontal sweep, on the other hand, only has to cover the modulation frequencies, which in most cases is the voice-frequency band from 300 Hz to 3 kHz. The modulation percentage can be determined from the waveform and is calculated from the following relationship:

$$\text{modulation percentage} = \frac{A - B}{A + B} \times 100\% \qquad (7\text{-}28)$$

where A is the peak of the modulated envelope and B is the minimum.

If the oscilloscope is operating near the limits of its frequency response and obtaining a reliable trigger is difficult to achieve, the oscilloscope may be triggered from the audio modulating source through the external trigger input. Many oscilloscopes can be used well beyond their advertised frequency range from some tasks, such as modulation determination, with good results.

Single-sideband modulation can be observed in two ways. The first method requires connections identical to those used to observe full-carrier modulation. Because there is no such thing as percentage of modulation in the single-sideband signal, the waveform will be used to locate distortions and other system problems. Figure 7-42 shows several types of single-sideband distortions.

An alternative method of observing single-sideband modulation is shown in Fig. 7-43. The significant advantage of this arrangement is that the oscilloscope does not have to be triggered, and the trapezoidal pattern does not change shape with complex waveforms such as speech. Therefore, this representation can be used to evaluate a single-sideband transmitter while it is modulated with its normal form of modulation.

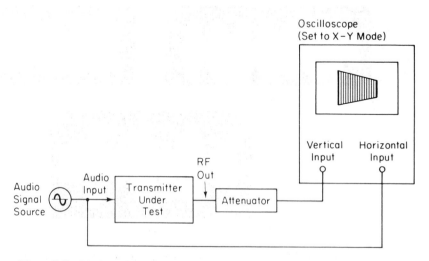

Figure 7-43 Method of observing the performance of a single-sideband transmitter.

7-11 SPECIAL OSCILLOSCOPES

7-11.1 Storage Oscilloscope

In the conventional CRT the persistence of the phosphor ranges from a few milliseconds to several seconds (see Table 7-1), so that an event that occurs only once will disappear from the screen after a relatively short period of time. A storage CRT can retain the display much longer, up to several hours after the image was first written on the phosphor. This *retention* feature can also be useful when displaying the waveform of a very low-frequency signal. In the conventional (non-storage) oscilloscope, the start of such a display would fade before the end is written.

Storage CRTs can be classified as bistable tubes and halftone tubes. The bistable tube will either store or not store an event and produces only one level of image brightness. The halftone tube can retain an image for varying lengths of time (variable persistence) and at different levels of image brightness. Both the bistable and halftone tubes use the phenomenon of secondary electron emission to build up and store electrostatic charges on the surface of an insulated target. The following discussion applies to either type of tube.

When a target is bombarded by a stream of primary electrons, an energy transfer takes place which separates other electrons from the surface of the target in a process known as *secondary emission*. The number of secondary electrons emitted from the target surface depends on the velocity of the primary electrons, the intensity of the electron beam, the chemical composition of the target, and the condition of its surface. These characteristics are reflected in the so-called *secondary-emission ratio*, defined as the ratio of secondary-emission current and primary beam current, or

$$\delta = \frac{I_s}{I_p} \tag{7-29}$$

The simple experimental circuit of Fig. 7-44 can be used to demonstrate how the secondary-emission ratio varies as a function of the target voltage V_t. The electron gun in Fig. 7-44 emits a focused beam of high-velocity electrons in much the same way as does the gun in a conventional CRT. This electron beam is directed at the surface of a metal target which will emit secondary electrons under favorable conditions. The collector, which completely surrounds the target except for a small aperture to pass the primary beam, collects all secondary-emission electrons. This constitutes the secondary current I_s. The target voltage is adjustable over a wide range (from 0 to +3,000 V), while the collector is held a few volts above the target by battery V_c.

The bombarding energy of a primary electron is directly related to the potential difference between the electron source (the cathode) and the target. When the target voltage is zero, the energy of the bombarding electron is zero and there is no secondary emission. Hence $\delta = 0$. When the target voltage is increased from

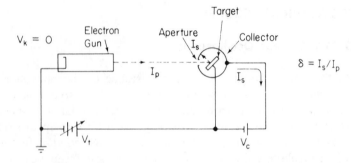

Figure 7-44 Experimental circuit used to demonstrate secondary-electron emission.

zero, the bombarding energy increases and causes some secondary-electron emission. Hence δ increases from zero, as shown in the secondary-emission curve of Fig. 7-45. At some positive target voltage (+50 V in Fig. 7-45), the number of secondary-emission electrons equals the number of primary beam electrons, so that $I_s = I_p$ and δ = 1. This point on the curve is known as the *first crossover point*. When the target voltage is increased beyond this crossover point, the secondary-emission ratio initially increases to some maximum value (δ = 2 in Fig. 7-45), and then it decreases again until $I_s = I_p$ and δ = 1. This point on the curve is the *second crossover point*.

Fig. 7-46(a) is a modification of the previous circuit and shows the collector voltage fixed at +200 V. The target voltage is adjustable over a wide range, as before. The fixed collector voltage drastically modifies the secondary-emission ratio, as indicated in Fig. 7-46(b). When the target voltage is larger than the collector voltage, the secondary electrons emitted from the target enter the retarding field of the collector and are reflected back into the target. Hence the target collects the total primary beam current I_p and the collector current I_s is zero. The *effective* secondary-emission ratio, defined by Eq. (7-29) as δ = I_s/I_p, is therefore zero, and the curve is modified as in Fig. 7-46(b). The other change occurs when the target voltage is approximately 0 V. When the target is slightly negative, the primary electrons cannot reach the target but are deflected to the collector. Although there can be no secondary-electron emission, the collector current

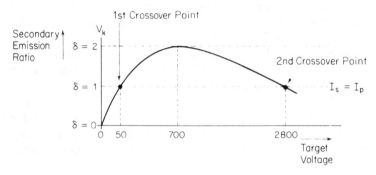

Figure 7-45 Typical secondary-emission curve.

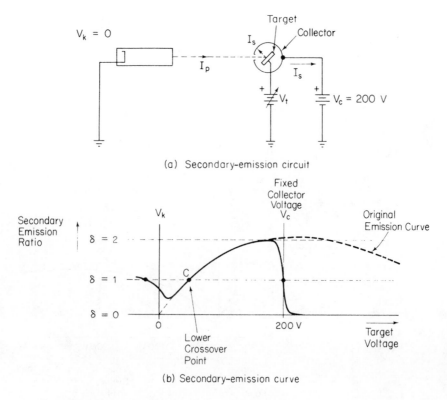

(a) Secondary-emission circuit

(b) Secondary-emission curve

Figure 7-46 Secondary-emission circuit with fixed collector voltage.

equals the primary beam current, and the target has an *apparent,* or *effective,* secondary-emission ratio of one. As the target voltage is increased from the negative side and approaches zero, the target no longer repels the primary beam, so that actual target bombardment takes place and real secondary emission results. These effects are shown in the modified curve of Fig. 7-46(b).

A further modification of the basic circuit is shown in Fig. 7-47(a). The collector voltage is again fixed at +200 V, but the target can be disconnected by switch S to become a so-called *floating target.* This CRT with floating target is capable of simple storage effects. Note that the secondary-emission curve for this tube, shown in Fig. 7-47(b), is similar to the one shown for the previous circuit.

Switch S is initially closed and the target voltage is set at some low value, say +20 V. At this point, the secondary-emission ratio is typically on the order of 0.5, so that the current in the collector circuit is one-half the primary beam current, or $I_s = \frac{1}{2} I_p$. The other half of the primary current is simply collected by the target and returned to the target battery. Hence target current $I_t = \frac{1}{2} I_p$. When switch S is now opened, the current in the target lead is interrupted and the primary beam current changes the target in the negative direction. The target voltage therefore decreases (becomes less positive), and the secondary-emission ratio changes, following the curve of Fig. 7-47(b). The rate of charge decreases as

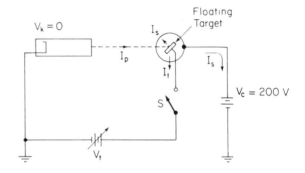

(a) Secondary-emission circuit with floating target

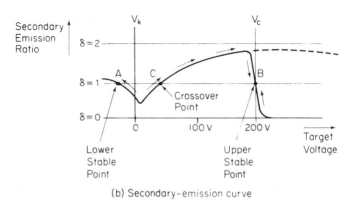

(b) Secondary-emission curve

Figure 7-47 Secondary-emission circuit with fixed collector voltage and floating target. The target voltage always assumes one of the stable conditions A or B.

the target voltage approaches point A on the curve. At this point, the secondary-emission current equals the primary beam current, and the net charging rate is zero. At point A, the target voltage is slightly negative, the secondary-emission ratio is one, and the target has reached a stable condition. Point A is called the *lower stable point,* and the target is considered to be in the *erased* condition.

If the initial or starting voltage of the target is to the right of crossover point C say at $+100$ V in Fig. 7-47(b), the secondary-emission ratio is greater than one. This means that I_s is greater than I_p and there must therefore be a net electron flow leaving the target surface. When switch S is now opened, the target continues to emit secondary electrons, so that it discharges and becomes more positive. Hence the secondary-emission ratio moves up along the curve to point B where the rate of discharge is once again zero and the target obtains a stable condition. At this so-called *upper stable point* the secondary-emission ratio is one, and the target is considered to be in the *written* condition.

As long as the primary gun is *on* and primary electrons bombard the target, the target will always be at a stable point, upper or lower, depending on the initial voltage of the target. Crossover point C on the curve is uniquely unstable in the sense that the target voltage will always move up to point *B* or down to point *A*, depending on which way the target voltage is first shifted by noise.

The CRT of Fig. 7-47 is an elementary bistable storage device. Its condition can be interrogated by measuring the target voltage. If the target voltage is "high," the target is written; if the target voltage is "low," the target is erased. The tube therefore has an electrical readout and its storage condition is not visible.

Figure 7-48(a) shows the principle of a bistable storage tube capable of writing, storing, and erasing an image. This storage tube differs from the one in Fig. 7-47(a) in two aspects: It has a multiple-target area, and it has a second electron gun. The second electron gun is called the *flood gun*; it emits low-velocity primary electrons that flood the entire target area. The distinguishing feature of the flood gun is that it floods the target at all times and not just intermittently as does the writing gun. The cathode of the flood gun is at ground potential, so that the target voltage will follow the secondary-emission curve indicated in Fig. 7-48(b). The lower stable point of the target is a few volts negative with

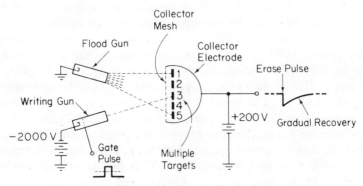

(a) Storage tube with multiple targets and two electron guns

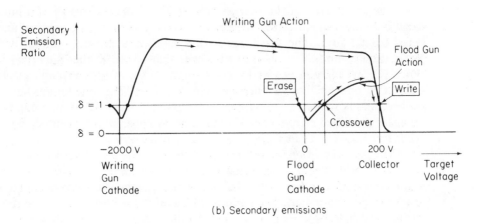

(b) Secondary emissions

Figure 7-48 Storage CRT with multiple targets and two electron guns.

respect to the flood gun cathode, and the upper stable point is at +200 V, the collector voltage. The cathode of the writing gun, however, is at −2,000 V, and its secondary-emission curve is superimposed on the flood gun curve. It is found that the combined effect of writing gun and flood gun is simply the sum of the individual effects of each electron beam by itself.

The flood gun is on at all times. Assume that the target is at its lower stable point, the erased condition. When the writing gun is gated on, its primary electrons arrive at the target with a potential of 2,000 V, which causes high secondary emission from the target. The target voltage therefore leaves the lower stable point and starts to increase. The flood gun electrons, however, attempt to maintain the target in its stable condition and oppose the increase in target voltage. If the writing gun is switched on long enough to carry the target past the crossover point, the flood gun electrons will aid the writing gun electrons and carry the target all the way to the upper stable point, so that the target is written. Even if the writing gun is now switched off, the target will be held in its upper stable condition by the flood gun electrons, thereby *storing* the information delivered by the writing gun. When the writing gun is not switched on long enough to carry the target past the crossover point, the flood gun electrons will simply move the target back to its lower stable condition, and storage does not occur.

Erasing the target simply means restoring the target voltage to the lower stable point. This can be accomplished by pulsing the collector negative, so that it momentarily repels the secondary-emission electrons and reflects them back into the target. This reduces the collector current I_s and the secondary-emission ratio drops below one. The target then collects primary electrons from the flood gun (remember that the writing gun is off) and charges negative. The target voltage decreases until it reaches the lower stable point where the charging ceases, and the target is in the erased condition. After erasure, the collector must be returned to its original positive voltage (+200 V in this case), and the erase pulse must therefore be returned to zero. As indicated in Fig. 7-48(a), this must happen gradually, so that the target is not accidentally driven past the crossover point and becomes written again.

The target area of the storage tube in Fig. 7-48(a) consists of a number of small individual metal targets electrically separated from one another and numbered from 1 to 5. The flood gun is of simple construction, without deflection plates, and it emits low-velocity electrons that cover all the individual targets. When the writing gun is gated on, a focused beam of high-velocity electrons is directed at one small target (number 3 in this case). This one target then charges positive and is written to the upper stable point. When the writing gun is turned off again, the flood electrons hold target 3 at its upper stable point (store). All the other targets are held at their lower stable points (erase).

The last step in our development of the bistable direct-viewing storage tube consists of replacing the individual metal targets with a single dielectric sheet, as in the typical tube of Fig. 7-49. This dielectric storage sheet consists of a layer of scattered phosphor particles capable of having any portion of its surface area written and held positive or erased and held negative without affecting the adjacent areas on the surface of the sheet. This dielectric sheet is deposited on a

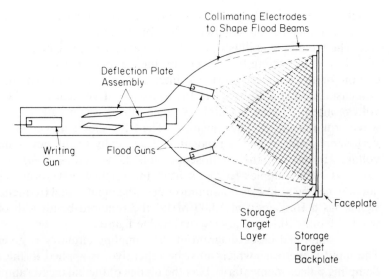

Figure 7-49 Schematic view of a bistable storage tube. (Courtesy of Tektronix, Inc.)

conductive-coated glass faceplate. The conductive coating is called the *storage target backplate,* and it is the collector of secondary-emission electrons. In addition to the writing gun and its deflection plate assembly, this storage CRT has two flood guns and a number of collimation electrodes that form an electron lens to distribute the flood electrons evenly over the entire surface area of the storage target.

After the write gun has written a charge image on the storage target, the flood guns will store the image. The written portions of the target are being bombarded by flood electrons that transfer energy to the phosphor layer in the form of visible light. This light pattern can be viewed through the glass faceplate. Since the storage target areas are either positive or negative, the light output produced by the flood electrons is either at full brightness or at minimum brightness. There is no gray scale in between.

7-11.2 Sampling Oscilloscope

When the frequency of the vertical deflection signal increases, the writing speed of the electron beam increases. The immediate result of higher writing speed is a reduction in image intensity on the CRT screen. In order to obtain sufficient image brilliance, the electron beam must be accelerated to a higher velocity so that more kinetic energy is available for transfer to the screen and normal image brightness is maintained. An increase in electron beam velocity is easily achieved by raising the voltage on the accelerating anodes. A beam with higher velocity also needs a greater deflection potential to maintain the deflection sensitivity. This immediately places higher demands on the vertical amplifier.

The *sampling* oscilloscope uses a different approach to improve high-frequency performance. In the sampling oscilloscope the input waveform is reconstructed from many samples taken during recurrent cycles of the input waveform and so circumvents the bandwidth limitations of conventional CRTs and amplifiers. The technique is illustrated by the waveforms indicated in Fig. 7-50.

In reconstructing the waveform, the sampling pulse turns the sampling circuit on for an extremely short time interval. The waveform voltage at that instant is measured. The CRT spot is then positioned vertically to the corresponding voltage input. The next sample is taken during a subsequent cycle of the input waveform at a slightly later position. The CRT spot is moved horizontally over a very short distance and is repositioned vertically to the new value of the input voltage. In this way the oscilloscope plots the waveform *point by point,* using as many as 1,000 samples to reconstruct the original waveform. The sample frequency may be as low as one-hundredth of the input signal frequency. If the input signal has a frequency of 1,000 MHz, the required bandwidth of the amplifier would be only 10 MHz, a very reasonable figure.

A simplified block diagram of the sampling circuitry is given in Fig. 7-51. The input waveform, which must be repetitive, is applied to the sampling gate. Sampling pulses momentarily bias the diodes of the balanced sampling gate in the forward direction, thereby briefly connecting the gate input capacitance to the test point. These capacitances are slightly charged toward the voltage level of the input circuit. The capacitor voltage is amplified by the vertical amplifier and applied to the vertical deflection plates. Since the sampling must be synchronized with the input signal frequency, the signal is delayed in the vertical amplifier, allowing the sweep triggering to be done by the input signal. When a trigger pulse is received, the avalanche blocking oscillator (so called because it uses avalanche transistors) starts an exactly linear ramp voltage, which is applied to a voltage comparator. The voltage comparator compares the ramp voltage to the output voltage of a staircase generator. When the two voltages are equal in amplitude, the staircase generator is allowed to advance one step and simultaneously a sampling pulse is applied to the sampling gate. At this moment, a sample of the input voltage is taken, amplified, and applied to the vertical deflection plates.

The real-time horizontal sweep is shown in Fig. 7-51, indicating the horizontal deflection rate of the beam. Notice that the horizontal displacement of the beam is *synchronized* with the trigger pulses which also determine the moment of sampling. The resolution of the final image on the CRT screen is determined by the size of the steps of the staircase generator. The larger these steps, the greater the horizontal distance between the CRT spots that reconstitute the trace.

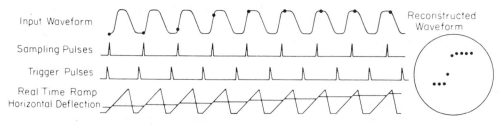

Figure 7-50 Waveforms pertinent to the operation of the sampling oscilloscope.

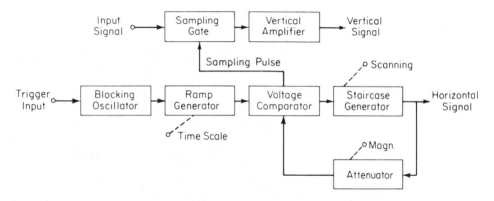

Figure 7-51 Simplified block diagram of the sampling circuitry. (Courtesy of Hewlett-Packard Company.)

7-11.3 Digital Storage Oscilloscopes

There are a number of distinct disadvantages of the storage cathode ray tube. First, there is a finite amount of time that the storage tube can preserve a stored waveform. Eventually, the waveform will be lost. The power to the storage tube must be present as long as the image is to be stored. Second, the trace of a storage tube is, generally, not as fine as a normal cathode ray tube. Thus, the stored trace is not as crisp as a conventional oscilloscope trace. Third, the writing rate of the storage tube is less than a conventional cathode ray tube, which limits the speed of the storage oscilloscope. Fourth, the storage cathode ray tube is considerably more expensive than a conventional tube and requires additional power supplies. Finally, only one image can be stored. If two traces are to be compared, they must be superimposed on the same screen and displayed together.

A superior method of trace storage is the digital storage oscilloscope. In this technique, the waveform to be stored is digitized, stored in a digital memory, and retrieved for display on the storage oscilloscope. The stored waveform is continually displayed by repeatedly scanning the stored waveform and, therefore, a conventional oscilloscope tube can be used for the display. The reduced cost of the conventional CRT relative to a storage oscilloscope tube can offset some of the cost of the additional circuitry for digitizing and storing the input waveform. The stored display can be displayed indefinitely as long as power is applied to the memory, which can be supplied with a small battery. The digitized waveform can be further analyzed by either the oscilloscope or by loading the contents of the memory into a computer.

Figure 7-52 shows the block diagram of a storage oscilloscope. The input is amplified and attenuated with input amplifiers as in any oscilloscope. The digital storage oscilloscope uses the same types of input circuitry and oscilloscope probes as a conventional oscilloscope, and many digital storage oscilloscopes can operate in a conventional mode, bypassing the digitizing and storing features. The output of the input signal amplifiers feeds an analog-to-digital (A/D) converter.

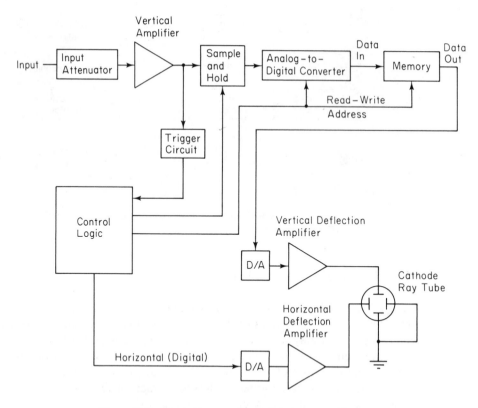

Figure 7-52 Block diagram of a digital storage oscilloscope.

The A/D converter can use any technique discussed in Chapter 6 relative to digital voltmeters or in Chapter 12 relative to data acquisition. However, in the storage oscilloscope application, the main requirement of the converter is speed. In a digital voltmeter application, accuracy and resolution were the main requirements of the converter, while speed was of secondary importance, as only slow-moving data were being digitized. In addition, in the oscilloscope, the digitized output need only be in binary form rather than BCD, which would be desirable for the display as digits on the digital voltmeter panel.

In an oscilloscope application, the typical resolution for the A/D conversion is 8 or 9 bits, which divides the input waveform into 256 parts for the 8-bit conversion and 512 parts for the 9-bit conversion. Any sort of ramp or integrating conversion is inherently too slow for digitizing oscilloscope use. The successive approximation type of converter is a useful method but is restricted to lower-frequency oscilloscopes. A discussion of the successive approximation A/D converter is found in Chapter 12, as this method of conversion is most often used in data acquisition. Another type of A/D converter known for its blazing speed is called the "flash" converter and is often found in digitizing oscilloscopes. The flash converter requires considerable circuitry for a fine-resolution converter, and is seldom used for such applications requiring accuracy and resolution as with the digital voltmeter. The significant characteristic of the flash converter is great speed. There are, also, A/D converters that use a combination of the flash and successive approximation techniques that have applications in digitizing oscilloscopes.

Before discussing the methods of A/D converters suited for digitizing oscilloscope applications, a feeling for the speed requirements will be developed. To accomplish this, the nature of the desired display will be investigated. An 8-bit A/D conversion supplies a useful display, as most conventional oscilloscopes are not capable of resolving signals to better than 1 part in 256, or roughly 0.4 per cent. An oscilloscope is for investigating waveshapes rather than for making precision measurements. If the oscilloscope screen were square, that is, the actual physical dimension of the screen, the display would be broken into 256 samples in both the horizontal as well as the vertical dimensions. This implies that the display is digitized 256 times for each display and the resolution of the digitizing is 8 bits, or a part in 256.

The speed of the A/D conversion can be determined in a simple fashion. As an example, if the oscilloscope is to display 100 μs for a full trace, which, typically, would be 10 μs per division, 256 conversions would be required for 100 μs of time. This requires a complete conversion each 390 ns.

For the case of an 8-bit successive approximation type of A/D converter, either N or $N + 1$ (8 or 9) clocks are required for each conversion and thus the clock must be 21 or 23 MHz. Although a successive approximation converter meeting this conversion rate is within the realm of conventional technology, the resulting 10-μs-per-division display does not represent a fast oscilloscope and will have limited application.

This derivation of the required speed for the A/D conversion for a storage oscilloscope assumed that each sampled point of the displayed trace was digitized consecutively. As quickly as the A/D converter could perform a conversion, the memory was accessed, the digitized value entered, and the next conversion made. If the waveform occurred only once, this would be necessary. If the waveform were repetitive, part of the conversion could be made each period of the input waveform. As an example, assume that the same 100-μs time was to be displayed on the oscilloscope. Rather than 256 conversions made at 390 ns per conversion as in the preceding example, a conversion is made each 3.12 μs. Of the required 256 conversions, only every eighth one was converted, which implies that the process must be repeated seven more times to fill in the missing conversions. Since the waveform is repetitive, the input waveform will be available to make the additional conversions.

It is necessary that the input to the A/D converter not change during the conversion. Since there are no guarantees on the input signal, it is sampled and the value of the input at a point in time is stored as a charge on a capacitor. This ensures that the input to the converter is steady during the conversion time.

For the example where every eighth sample was digitized, the obvious method of making the A/D conversions would be to start with sample number 1, then 9, 17, 25, 33, and so on. After the first pass of the waveform, the next pass would convert time slots 2, 10, 18, and so on. There are 256 time slots in this example and the A/D conversion would continued until all 256 time slots are converted. This conversion would last for a period of eight cycles of the input waveform. It is not necessary that every eighth sample be taken; the number could be every second, third, fourth, . . . , or tenth, eleventh, and so on. There are some undesirable effects when performing the digitizing in this fashion. Since the digitizing and waveform are both periodic, there can be undesirable interaction between the digitizing rate and the signal period. This is sometimes called aliasing or beating. To reduce this effect, a random sampling is used to smooth the displayed trace and eliminate aliasing or beating. The sample time slots may be of a random nature, such as 1, 10, 13, 20, and so on, on the first pass. The second pass may be 5, 11, 15, 26, and so on. The digitizing does not have a regularity to it and there is no beating.

It should be clear that the random sampling as well as the repetitive sampling will not function on a one-time waveform such as a transient. These one-time events are an important application of digital storage oscilloscopes, whether it be a digital storage type or an older type where the trace is stored on a special oscilloscope tube. A storage oscilloscope is also very useful for repetitive events when the event is very slow and the persistence of the oscilloscope tube is insufficient to allow easy viewing, or where further signal processing such as averaging, peak holding or spectral analysis is required.

The frequency restrictions applied to a sampled system follow the Nyquist rule, which states that if a bandwidth-limited waveform is sampled at a rate of least twice the bandwidth of the waveform, an accurate reconstruction could be made of the sampled waveform. For an example, a signal having a maximum spectral frequency component of 100 kHz could be sampled at a 200-kHz rate and completely reconstructed. This rule really does not apply to digital storage oscilloscopes. It is true that the waveform will be reconstructed to the bandwidth limits set by the Nyquist rate, but the visual appearance of the waveform will not provide a satisfactory display. The practice of oversampling, that is, providing many more samples than are theoretically required, is universally done.

Whether repetitive sampling is used or not, a fast A/D converter is an important part of a digital storage oscilloscope. Perhaps the fastest A/D conversion technology that can be applied to digitizing oscilloscopes is the flash converter. Figure 7-53 shows the block diagram of the flash converter. The input signal is applied to a string of comparators where one input of each comparator is connected to a reference voltage representing one of the possible quantizing levels. For an 8-bit conversion there are 255 different levels and, consequentially, 255 comparators. For the sake of instruction and clarity, the block diagram of Fig. 7-53 shows a 3-bit A/D converter which employs seven levels and seven comparators. All the comparators connected to reference voltages less than the input voltage changes state, while the comparators connected to reference voltages above the input voltage do not. The comparator outputs are decoded to change "greater than/less than" states to a binary number. In the case of the 3-bit example, the logic is moderately simple.

In the example flash converter shown in Fig. 7-53, the first reference level is $(0.5/8)V_{ref}$. The value $V_{ref}/8$ is the magnitude of the least significant bit. There-

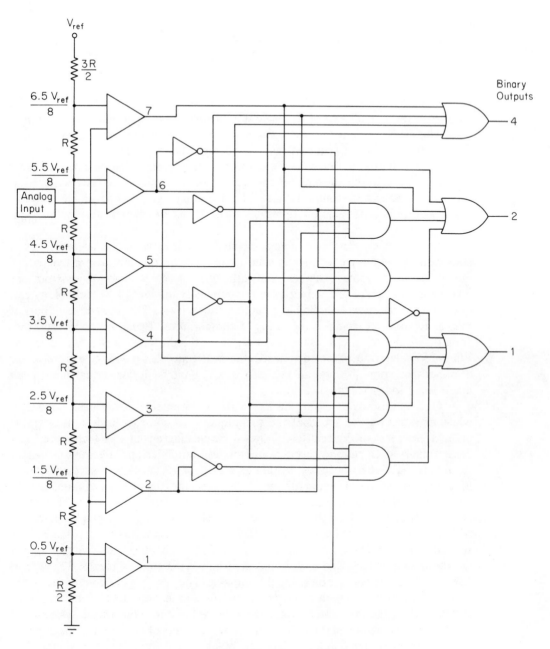

Figure 7-53 Three-bit flash converter.

fore, any input voltage between zero and one-half a least significant bit is quantized as zero. Values between 0.5 times the value of a significant bit and 1.5 times the value are quantized as 1, and so on. The reference levels for the comparators are $(N + 0.5)V_{ref}/8$, where N is an integer.

In the example converter the comparator outputs are encoded to a binary number with three OR gates. In the 3-bit example the logic is moderately simple. However, the complexity of the logic doubles for each additional bit.

The flash converter is inherently fast—hence the name, as it converts "in a flash." No clocks are required and as the input level changes, the comparator outputs will change and provide the correct binary output. The time delay from a change in input to the subsequent change in binary output is the time required to propagate through the comparator and the decoding logic. Conventional technology can produce conversion times of 20 to 50 ns or even faster for an 8-bit conversion.

Clearly, the major disadvantage of the flash converter is the large number of comparators and logic elements used. The resistor divider is constructed of mostly identical value resistors, and although a large number of resistors are involved, this part of the converter is not unusually difficult. LSI technology can integrate the required decoding logic onto a chip or two. The most difficult part of the flash converter design is the string of comparators. Because the comparators are a linear circuit, and a fast one as well, it is difficult to use low-power circuits. When 255 comparators are considered, the heat generated would be excessive for a monolithic chip. Despite all the difficulties, 8-bit flash converters have been made using hybrid technologies.

One method of reducing the hardware requirements of the flash converter while maintaining the speed of the technique is to use two flash converters to make a conversion. Figure 7-54 shows a block diagram of this technique. The input signal, after passing through an isolation amplifier and a sample-and-hold circuit, is divided by a factor, in this example, of 16. A 4-bit flash converter is used to convert the attenuated signal. Because the signal is attenuated by the factor of 16, the value of each bit in the conversion is multiplied by 2^4. Therefore, the least significant bit of the 4-bit conversion of the attenuated signal has a weight of 16 based on a full-scale value of 255/256 times the input full-scale voltage. The second least significant bit has a weight of 32. The third and fourth bits are weighted 64 and 128. The conversion determines the state of the 16, 32, 64, and 128 weight bits of the complete binary number. The result of the conversion is fed to a D/A converter, which provides an output that is subtracted from the input. What results from the subtraction is the input value minus the 16, 32, 64, and 128 weight or most significant bits. The difference consists of only the least significant bits, as the most significant bits have been subtracted from the input signal. The difference is converted with a second flash converter without any attenuation, and the result of the conversion represents the least significant bits of the entire binary number.

This arrangement of the two flash converters reduces the number of comparators. Using the 8-bit example, the reduction is from 256 to 32. The price to be paid for the reduction of hardware is an increase in conversion time. The time for

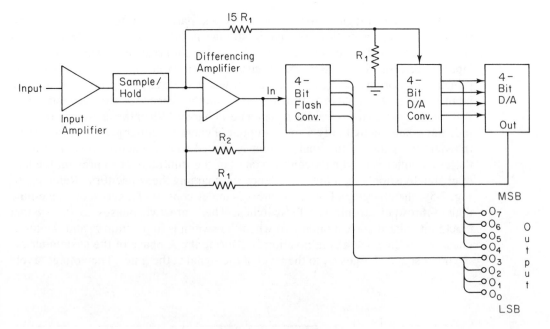

Figure 7-54 Eight-bit flash converter.

the conversion is the sum of the conversion time for two flash converters plus the conversion time of a D/A converter. Fast 4-bit D/A converters are common and the result is an 8-bit converter that is faster than practically any other technique other than an 8-bit flash converter.

EXAMPLE 7-2

If a digitizing oscilloscope is to have a 6-bit resolution in both the horizontal and vertical axes and is to display transients at a rate of 1 μs per division for a display of 10 divisions, what is the speed required of the input A/D converter?

SOLUTION A resolution of 6 bits requires a conversion of 1 part in 64. Since the example specified that the digitizing oscilloscope was to display transients, conversions must be made continuously rather over a period of several cycles. The total time for a trace is 10 μs, as the oscilloscope is to display 1 μs per division for 10 divisions. Therefore, 64 A/D conversions are required in 10 μs for 155 ns for each conversion. If a successive approximation converter were used, six or seven clocks would be required for a conversion that would allow for 22 ns per clock, or a clock frequency of 45 MHz. This would represent a very high performance successive approximation A/D converter. On the other hand, 6 bits is not beyond the realm of practicality for a flash converter or a dual flash converter, as described earlier.

For very fast digital storage oscilloscopes, particularly for storing one-time transient events, a method of analog storage is often used. This technique is shown as a block diagram in Fig. 7-55. An analog shift register is used to store the input waveform. The analog shift register, as shown in Fig. 7-55, works on the principle of storing a charge on a capacitor where the charge is proportional to the input voltage. The charge is transferred to another capacitor storage element so that the first capacitor is free to accept a new charge. When the charge is transferred, and assuming that only a small amount of charge is either gained or lost in the transfer, the value of the analog input is shifted from capacitor to capacitor. A buffer amplifier is used between the stages of the shift register to prevent the loss of charge by supplying a low impedance for charging the capacitors. Referring to Fig. 7-56, the charging of the capacitors is under control of a series of single-pole double-throw electronic (FET) switches. There are two phases to the system clock; one phase corresponding to when the switch is in position A and the other phase when the switch is in position B. During the A phase of the system clock, the input signal charges C_1 to the level of the signal at the time. The voltage levels

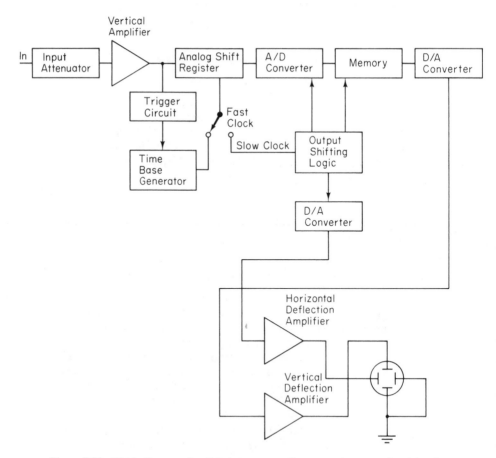

Figure 7-55 Block diagram of a digital storage oscilloscope using an analog delay line.

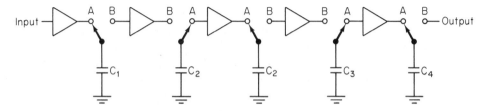

Figure 7-56 Schematic of a switched capacitor delay line.

of the other capacitors will not be considered at this time. When the phase changes to B, the voltage present on C_1 is transferred to C_2. In the next phase A, C_1 is again charged to the level of the signal at that time, which is now the first half of the second clock. In addition, the charge on C_2 has been transferred to C_3. During the B phase of the second clock, the value of C_1 is transferred to C_2. Thus C_2 is at the level of the input during the first half of the second clock. Also during this phase the charge of C_3 is transferred to C_4. At this point the sampled value of the input signal which is determined during the first half of the clock is distributed throughout the shift register. The first sample is contained in C_3 and C_4, while the second sample is contained in C_1 and C_2. Once the entire shift register has been loaded with the analog signal the input shifting is stopped and the shift register can be "unloaded." This unloading can occur at a much slower rate. The analog output is digitized using one of the techniques discussed previously. The results are stored in a memory and are treated as any other digitized waveform. The only function performed by the analog shift register is one of a slowing of the analog input signal.

Analog shift registers can be clocked in at frequencies above 50 MHz. The analog data cannot remain in the analog shift register indefinitely without degradation, but it is not unusual for a shift register to allow the data to be clocked out at rates below 100 kHz, giving sufficient time to make an analog-to-digital conversion. Many of the higher-frequency digital storage oscilloscopes use this technique.

Regardless of the technique, the result of each A/D conversion is stored in a random access memory for retrieval for display. To display the digitized waveform, the binary number representing each digitized sample is retrieved from the memory in sequence, fed to a D/A converter, and displayed as an analog deflection on the oscilloscope trace. The rate of selecting the digitized samples and displaying them on the oscilloscope is not critical, and the only criterion is that the display be updated often enough to prevent the display from appearing to flicker. A display rate of once per 10 ms is sufficiently fast to prevent flicker, which requires the 256 samples to be retrieved and displayed every 39 μs. This poses no burden on either the computer memory or the D/A conversion, and inexpensive circuits may be used.

One very important feature of a digital storage oscilloscope is its ability to provide a mode of operation called "pretrigger view." This means that the oscilloscope can display what happened before a trigger input is applied. This mode of operation is very useful when a failure occurs and is marked by the appearance of a signal. To determine what caused the failure, it would be necessary to see various waveforms before the failure. The digital storage oscilloscope will continuously record a selected waveform, and when the trigger waveform appears signifying a failure, the storage will stop and the waveform in the memory is available for viewing. Random or repetitive sampling type of storage oscilloscopes will not provide this function, as a failure is a one-time event and only the consecutive type of storage oscilloscope will provide this feature.

REFERENCES

7-1. Prensky, Sol D., and Castellucis, Richard L., *Electronic Instrumentation,* 3rd ed., chap. 10. Englewood Cliffs, N.J.: Prentice-Hall, Inc., 1982.

7-2. van Erk, Rien, *Oscilloscopes: Functional Operation and Measuring Examples.* New York: McGraw-Hill Book Company, 1983.

PROBLEMS

7-1. What are the major blocks of the oscilloscope, and what does each do?

7-2. What are the major components of a cathode ray tube?

7-3. How is the electron beam focused to a fine spot on the face of the cathode ray tube?

7-4. What effect does increasing the writing rate of an oscilloscope by increasing the accelerating potential have on the deflection sensitivity?

7-5. How much voltage is required across two deflection plates separated by 1 cm to deflect an electron beam 1° if the effective length of the deflection plates is 2 cm and the accelerating potential is 1,000 V?

7-6. What is the velocity of electrons that have been accelerated through a potential of 2,000 V?

7-7. Why are the operating voltages of a cathode ray tube arranged so that the deflection plates are nearly ground potential?

7-8. How is the vertical axis of an oscilloscope deflected? How does this differ from the horizontal axis?

7-9. What is oscilloscope probe *compensation*? How is this adjusted? What effects are noted when the compensation is not correctly adjusted?

7-10. Why is an attenuator probe used?

7-11. Why is a delay line used in the vertical section of the oscilloscope?

7-12. What are the advantages of dual trace over dual beam for multiple-trace oscilloscopes?

7-13. How does alternate sweep compare with chopped sweep? When would one method be chosen over the other?

7-14. What is delayed sweep? When is it used?

7-15. What are the advantages of using an active voltage probe?

7-16. How are the effects of direct current on the flux density of the current probe minimized?

7-17. What is the relationship between the period of a waveform and its frequency? How is an oscilloscope used to determine frequency?

7-18. How does the digital storage oscilloscope differ from the conventional storage oscilloscope using a storage cathode ray tube? What are the advantages of each?

7-19. How does the sampling oscilloscope increase the apparent frequency response of an oscilloscope?

7-20. What precautions must be taken when using a sampling oscilloscope?

─────── **8** ───────

Signal Generation

8-1 INTRODUCTION

The generation of signals is an important facet of electronic troubleshooting and development. The signal generator is used to provide known test conditions for the performance evaluation of various electronic systems and for replacing missing signals in systems being analyzed for repair. There are various types of signal generators, but several characteristics are common to all types. First, the frequency of the signal should be well known and stable. Second, the amplitude should be controllable from very small to relatively large values. Finally, the signal should be free of distortion.

There are many variations of these requirements, especially for specialized signal generators such as function generators, pulse and sweep generators, etc., and these requirements should be considered as generalizations.

8-2 THE SINE-WAVE GENERATOR

Because of the importance of the sine function, the sine-wave generator represents the largest single category of signal generators. This instrument covers the frequency range from a few hertz to many gigahertz, but in its simplest form the sine-wave generator is as shown in Fig. 8-1.

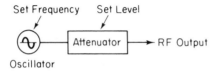

Set Frequency   Set Level

Attenuator   → RF Output

Oscillator

Figure 8-1   Block diagram of a simple sine-wave generator.

The simple sine-wave generator consists of two basic blocks, an oscillator and an attenuator. The generator's performance depends on the success of these two main parts. The frequency accuracy and stability and freedom from distortion depend on the design of the oscillator, while the amplitude accuracy depends on the design of the attenuator.

8-2.1 Inductor-Capacitor Tuned Oscillators

There is a broad class of oscillators that use the resonant characteristics of an inductor-capacitor, *LC,* circuit to generate a stable frequency. A block diagram of an oscillator is shown in Fig. 8-2. The oscillator consists of an amplifier and a feedback network such that the total gain of the loop, that is, the gain of the amplifier divided by the loss of the feedback network, is exactly equal to one, and the total phase shift around the loop is zero. Oscillators are designed such that these characteristics are met at only one frequency. This can be achieved by using various combinations of inductors, capacitors, and resistors.

    The resonant frequency of a circuit is given by

$$f = \frac{1}{2\pi\sqrt{LC}} \tag{8-1}$$

where L = circuit inductance (henrys)

    C = circuit capacitance (farads)

    f = resonant frequency (hertz).

When a resonant circuit is used in the feedback of an oscillator, the oscillation frequency is the resonant frequency of the circuit.

    Figure 8-3 shows the actual circuit of a *Hartley oscillator* and the equivalent circuit showing the amplifier and feedback components. Because a common-emitter amplifier is used as the active element of the oscillator, it is apparent that the circuit has a phase shift due to the amplifier of 180° regardless of the operating frequency. The feedback network, that is, the resonant circuit, has a phase shift

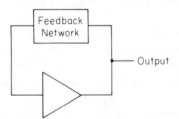

Feedback Network

Output

Figure 8-2   Block diagram of an oscillator, showing the amplifier and feedback network.

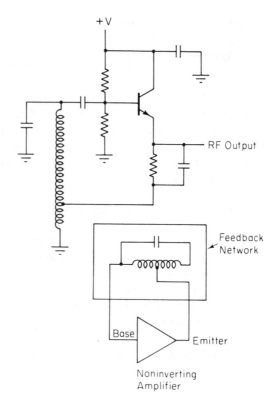

Figure 8-3 Hartley oscillator using a bipolar junction transistor.

of 180° at resonance. Therefore, the phase shift requirement for the oscillator can be met at the resonant frequency of the tuned circuit. It may not be clear how the loop gain can be equal to one, especially since the gain of a transistor amplifier can be quite high, and there is no loss in the tuned circuit. For an oscillator to sustain oscillations, the gain of the active element must be reduced, and this is accomplished by automatic adjusting of the operating characteristics of the transistor through self-bias. The amplitude of the ac voltages in the oscillator build until the effective gain of the transistor is reduced so that the total loop gain is equal to 1. This is accomplished in most oscillator circuits by increasing the transistor bias voltages so that the gain of the device is reduced. This usually results in large amplitude and distorted voltages and currents associated with the active device, which suggests that care should be taken when choosing the point from which to couple the oscillator output.

A circuit similar to the Hartley oscillator is shown in Fig. 8-4; it is called the *Colpitts oscillator*. Instead of the tapped inductor, the Colpitts oscillator uses a tapped capacitance to achieve the required 180° phase shift. Otherwise, the operation is identical. In fact, all simple transistor *LC* oscillators are practically identical.

These two basic circuits, as well as other simple oscillator circuits, are used

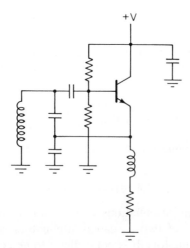

+V

Figure 8-4 Colpitts oscillator using a bipolar junction transistor.

as the signal source for most radio-frequency, RF, generators from tens of kilohertz to 1 GHz and greater. There are practical problems with constructing an oscillator of the simple sort for frequencies above 1 GHz using these circuits, and most signal generators for microwave frequencies use specialized oscillators. Likewise, for lower frequencies the size of the inductors required for the tuned circuit become prohibitive, and oscillators using other than LC tuned circuits are used.

Because both the inductance and capacitance have a similar control on the operating frequency of the oscillator, both elements can be used to set the frequency of the oscillator. In practice, the inductor is changed with a switch, while the capacitor is used for the tuning of the oscillator. This is usually accomplished by switching the inductor in bands while the capacitor is connected to the signal generator dial.

The second part of the sine generator is the attenuator. The signal generator is to supply signals of known amplitude as well as known frequencies. If a signal of a known fixed amplitude were applied to the input of an attenuator, the output signal level would be known as long as the attenuator were accurate. Signal generators are often used to supply known signal levels at very low levels for testing and evaluating receivers. It is not possible to measure and calibrate a signal at a very low level, and thus low-level signals are generated by feeding an attenuator with a higher-level signal for which the amplitude is easily measured and calibrating the attenuator steps. An attenuator is a device that will reduce the power level of a signal by a fixed amount. The attenuator should terminate with a fixed impedance, relative to either the input or output, regardless of the value of attenuation.

The attenuator reduces the power of an input such that the ratio of the input power to the output power is a constant. The reduction in power can be expressed as the log of the input power ratio by the following relationship:

$$A \text{ (dB)} = 10 \log \frac{P_i}{P_o} \qquad (8\text{-}2)$$

where A (dB) is the attenuation in decibels, P_o is the power output, and P_i is the power input of the attenuator. If a signal is passed through two attenuators in cascade as shown in Fig. 8-5, the output of the first attenuator is reduced by the ratio P_i/P_o, while the signal is further reduced by the ratio of the second attenuator, P_i'/P_o'. The total reduction is the product of the two attenuations, or

$$A \text{ (dB)} = 10 \log \left(\frac{P_i}{P_o}\right)\left(\frac{P_i'}{P_o'}\right) = 10 \log \frac{P_i}{P_o} + 10 \log \frac{P_i'}{P_o'}$$

Replacing each attenuation ratio with the corresponding decibel representation yields

$$A \text{ (dB)} = A_1 + A_2 \qquad (8\text{-}3)$$

where A_1 and A_2 are the attenuations of each attenuator. Therefore, the total attenuation, in decibels, of two cascaded attenuators is the sum of the decibel attenuation of each attenuator. It is not difficult to be convinced that this can be extended to more than two attenuators to derive the general rule that the attenuation, in decibels, of any number of cascaded attenuators is the sum of the decibel attenuations of all the attenuators. It should also be clear that the order of the cascaded attenuators will not affect the end result.

The decibel notation is convenient for a variety of reasons but needs a slight modification so that it can represent an absolute level. If the decibel equation were written as

$$\text{dBr} = 10 \log \frac{P}{P_r} \qquad (8\text{-}4)$$

where dBr is a decibel notation referenced to P_r, then dBr represents the number of decibels above or below some reference power, P_r. For example, if P_r were 1 W, the equation would read

$$\text{dBw} = 10 \log \frac{P}{1 \text{ W}} \qquad (8\text{-}5)$$

dBw, which is a standard notation, describes an absolute power level referenced to 1 W. Another important power level is the dBm, which is referenced to 1 mW across 50 Ω. For example, +3 dBm is 2 mW, while −3 dBm is $\frac{1}{2}$ mW or 500 μW. dBm is convenient for a 50-Ω system impedance, which includes the vast majority of equipment operating at frequencies greater than 1 MHz.

Various attenuator types can be used in signal generators. The *pi attenuator*, named for the Greek letter, which the schematic representation resembles, is one of the more common and versatile types. Three resistors are required for the pi attenuator, as shown in Fig. 8-6. The pi attenuator can be fabricated with standard components up to about 20 dB and for frequencies to about 100 MHz.

Figure 8-5 Two attenuators cascaded for increased attenuation.

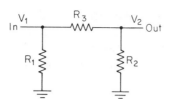

Figure 8-6 Schematic representation of the pi attenuator.

The capacitive reactances alter the attenuation for higher frequencies, and this can be eliminated to a certain extent by creating the higher-value attenuations by cascading lower-value attenuators. Standard components have two types of parasitic reactances: shunt capacitance and lead inductance. A pi attenuator constructed with standard components in actuality would look like the schematic in Fig. 8-7. The errors caused by the parasitic reactances only become a problem when the operating frequency becomes high. Special resistors called *rod* and *disk* are used for the series and shunt resistors, respectively, to minimize the parasitic reactances. Attenuators made with these components can be used to several gigahertz.

Referring to Fig. 8-6, the values of the resistors required for the pi attenuator, having a power reduction of N, where the attenuation in decibels is 10 log N, can be calculated in the following manner. The input resistance of the attenuator must be equal to the system resistance Z. This is true only when the attenuator is terminated at either end by the same resistance, Z. Looking into the attenuator, the resistance is R_3 in series with the parallel combination of R_2 and the termination, Z. This combination of R_2, R_3, and Z is also in parallel with R_1. The output resistance, with the input terminated, should also be equal to Z. Looking into the output with the input terminated, the resistance is R_3 in series with the parallel combination of R_1 and Z. This combination is in parallel with R_2, which should also be equal to Z. Because R_3 in series with either R_1 or R_2 in parallel with Z should produce a resistance of Z implies that R_1 and R_2 are the same. Writing the equation for the resistance looking into either end yields

$$Z = \frac{R_1\left(R_3 + \dfrac{R_1 Z}{R_1 + Z}\right)}{R_1 + R_3 + \dfrac{R_1 Z}{R_1 + Z}} \tag{8-6}$$

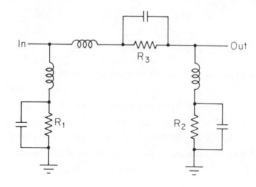

Figure 8-7 Schematic diagram of a pi attenuator showing the parasitic reactances of standard components.

This equation has two unknowns, R_1 and R_3 (Z is a constant of the system) and cannot be solved without the aid of, at least, a second equation.

The ratio of the input to output voltage, V_1/V_2, can be written using a simple voltage-divider equation, the voltage divider being R_3 and the parallel combination of R_1 and Z:

$$\sqrt{N} = \frac{V_1}{V_2} = \frac{R_3 + \dfrac{R_1 Z}{R_1 + Z}}{\dfrac{R_1 Z}{R_1 + Z}} \tag{8-7}$$

There are now two independent equations having only two unknowns (N in the second equation is also a constant), and there exists a solution.

For the convenience of mathematical manipulation, the second equation will be rewritten:

$$R_3 + \frac{R_1 Z}{R_1 + Z} = \frac{\sqrt{N}\, R_1 Z}{R_1 + Z} \tag{8-8}$$

The left side of this equation appears, in the identical form, in the first equation and a substitution can be made:

$$Z = \frac{\dfrac{R_1^2 \sqrt{N}\, Z}{R_1 + Z}}{R_1 + \dfrac{\sqrt{N}\, R_1 Z}{R_1 + Z}} = \frac{R_1^2 \sqrt{N}\, Z}{R_1^2 + R_1 Z + \sqrt{N}\, R_1 Z} \tag{8-9}$$

Because this equation contains only one variable, R_1, it can be solved for R_1:

$$R_1 = Z \left(\frac{\sqrt{N} + 1}{\sqrt{N} - 1} \right) \tag{8-10}$$

A relationship between R_3 and R_1 is contained in Eq. (8-8), which can be rewritten as

$$R_3 = \frac{\sqrt{N}\, R_1 Z}{R_1 + Z} - \frac{R_1 Z}{R_1 + Z} = \frac{Z R_1 (\sqrt{N} - 1)}{R_1 + Z} \tag{8-11}$$

Substituting the results for R_1 obtained in Eq. (8-10) into Eq. (8-11) results in the solution for R_3:

$$R_3 = \frac{Z(N - 1)}{2\sqrt{N}} \tag{8-12}$$

A double-pole, double-throw switch is required to switch a pi attenuator in or out of a cascade, as shown in Fig. 8-8, and this switch must have low parasitic reactances, as do the resistors. Various ingenious switch designs have made the switched pi attenuator useful to several gigahertz.

It was shown that the total attenuation of a cascade of attenuators was equal to the sum of the individual attenuations, in decibels, regardless of the order of the attenuators. For example, four switched, cascaded attenuators of 1, 2, 4, and 8

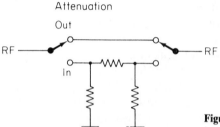

Figure 8-8 Double-pole, double-throw switch is required to switch attenuator sections.

dB could provide attenuations of 0 to 15 dB in 1-dB steps by manipulating the switches in a simple binary sequence. Using this technique, attenuators having up to 100-dB attenuation in 1-dB steps can be constructed using seven switched attenuators.

A type of attenuator that uses no resistors at all and has been very popular for moderately priced signal generators for many years is the *piston attenuator*, a schematic of which is shown in Fig. 8-9. When carefully made, it can provide excellent accuracy. The piston attenuator operates on the principle of a wave-guide beyond the cutoff wavelength. When a waveguide is operated below its intended frequency, the waveguide does not effectively transmit energy and thus there is a significant attenuation. Consider the attenuator in Fig. 8-9. An injection loop injects energy into the cylinder, which can be considered as a section of waveguide. At the opposite end of the waveguide section, a second loop is located, the pickup loop, which retrieves some of the energy launched into the waveguide. The attenuation is given by

$$A \text{ (dB)} = 32 \frac{L}{d} \tag{8-13}$$

where L is the distance between the loops in meters and d is the diameter of the cylinder, also in meters. Therefore, the power loss is proportional to the log of the distance separating the loops. It would be very convenient if the attenuator were calibrated in decibels, as that form of calibration would be linear along the attenuator scale.

This linear relationship between the length of the separation of the loops and attenuation in decibels is only valid when a certain minimum separation is observed. When the loops are close, other coupling mechanisms come into play

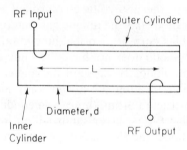

Figure 8-9 Piston-type attenuator showing injection loops and sliding cylinders.

other than the attenuation through a waveguide, and the result is an error. The practical piston-type attenuator cannot have an attenuation of less than about 20 dB without considerable error.

There are no resistive elements in the piston-type attenuator to dissipate the power that is attenuated, and the reduction in input power is accomplished by reflecting the unwanted power back into the source. This implies that the input impedance of the piston-type attenuator is not constant, and this can cause problems with certain types of oscillators. To stabilize the impedance as seen by the oscillator, a fixed resistive attenuator, called a *pad*, is inserted between the piston-type attenuator and the oscillator. This further increases the minimum loss between the oscillator and the signal generator output, which requires either more power from the oscillator or less power to the output. In spite of all the difficulties associated with the piston-type attenuator, it is used regularly, especially for signal generators for receiver testing where the largest signal needed is small.

There are several disadvantages to the simple signal generator shown in Fig. 8-1. First, there is no method of monitoring the voltage fed to the attenuator, and thus the calibration can be in error with no indication. A metering circuit could be applied to the input of the attenuator and the level set by a manual adjustment. This has been the technique employed by signal generators for many years, and it persists in low-cost generators. Rather than a manual adjustment, the level could be set automatically so that neither a meter nor a method of manual adjustment is required. This is the method, called *automatic level control* or ALC, employed in modern signal generators to set the level at the input of the attenuator. For this method to work, an accurate and broadband voltage measuring system is required to sense the voltage level. The voltage measuring system, ideally, would be a true rms type so that harmonic distortion would not affect the measurement. If the output of the signal generator has sufficiently low harmonic distortion, a simple diode detector may suffice. If the harmonic distortion is on the order of 20 dB or less down the carrier, the diode detector can cause considerable error. An alternative measuring method is the heated thermocouple. The rms value is the value of an ac voltage that would produce the same heating in a resistive load as an equivalent dc value. Therefore, the heating of the thermocouple automatically insures that the output would be the rms value.

Some method of varying the output voltage from the oscillator is required. In some early signal generators the level was set by varying the power supply voltage to the oscillator. This is not desirable, as this tends to vary the frequency of the oscillator, which will affect the calibration of the frequency dial.

To overcome problems such as these, a voltage-controlled attenuator using *PIN* diodes is employed. The *PIN* diode has the unique characteristic that the RF resistance is a function of the dc bias current. Normal junction diodes have this characteristic as well, but the *PIN* diode does not have the same order of magnitude of capacitance as the junction diode. Because of the higher capacitance of the junction diode, it is not possible to attenuate a signal any more than a few decibels. The *PIN* diode is constructed from three semiconductor layers. A heavily doped *N*-type and a similar *P*-type layer sandwich an intrinsic layer of pure silicon. Unlike the junction diode, the *P*- and *N*-type materials are separated by an intrinsic layer that is typically much thicker than the depletion region of a conventional junction diode. Because of the greater thickness, the resulting capacitance is much lower. When the diode is forward biased, carriers are injected from both the *P*- and *N*-type layers and provide the necessary carriers for conduction through the intrinsic layer. It is the introduction of impurities into silicon that makes the material conductive, and intrinsic silicon is essentially an insulator. Both electrons and holes are minority carriers in the intrinsic layer and will eventually combine, which makes them no longer available for conduction. If the recombination time is made long relative to the period of the RF energy to be conducted, even though the direction of the current has changed, sufficient carriers are available in the intrinsic layer to continue conduction. The amount of time that a carrier is typically available depends on the construction of the diode and can be made as long as several microseconds.

Figure 8-10 shows an attenuator fabricated using *PIN* diodes. In this attenuator, which uses the bridged T circuit, the current in the shunt diode is increased, while the current in the series diode is decreased for increased attenuation. The amount of amplitude variation from the signal generator oscillator is seldom more than a few decibels, and this attenuator is capable of supplying sufficient attenuation while maintaining an acceptable input and output impedance match.

Although an *LC*-tuned oscillator can be designed to have a stable and low distortion output, to prevent any interaction between the oscillator and the external load, some form of isolation is required. It may appear that it would be difficult to affect the frequency of the oscillator from external loads placed on the circuit, but even small variation in frequency is unacceptable for some signal generator applications. An oscillator operating at several hundreds of megahertz could require stabilities and frequency modulation characteristics that are better than a few kilohertz peak to peak. Depending on the type of oscillator circuit involved, this could require isolation of 20 dB or more.

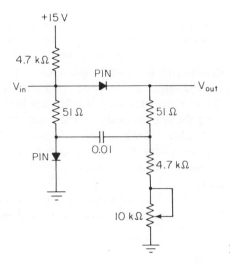

Figure 8-10   Bridged-T attenuator using *PIN* diodes.

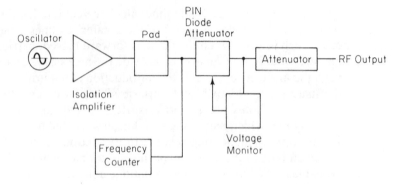

Figure 8-11 Sine-wave signal generator with frequency counter readout and automatic level control.

When the attenuator connected to the oscillator of the simple signal generator of Fig. 8-1 is set to values of 20 dB or more, the 20-dB requirement is easily met. However, when the signal is required to be a greater amplitude, the amount of attenuation must be reduced, and the amount of isolation is likewise reduced. If the output level of the oscillator is set to be 20 dB greater than the highest output of the signal generator, and 20 dB is the minimum attenuation, the required isolation is achieved. However, this would require that the oscillator provide a 20-dB-greater signal than is required, which will cause other problems.

Another solution to the problem is to provide an isolation amplifier between the oscillator and the attenuator. An amplifier having a gain of 10 dB can be made to have an isolation of 20 dB or greater. Applying the amplifier to the output of the oscillator will increase the signal level by 10 dB, while a 10-dB fixed attenuator can be inserted after the amplifier to return the signal level to the original level. This will cause no increase in signal level but will increase the isolation of the oscillator from the load to 30 dB.

Many applications of signal generators require a precise frequency readout. Early signal generators required precision dials and mechanical dial drives with hand-calibrated dial plates. Precision crystal calibrators were included in the more expensive generators to periodically check the dial calibrations. The introduction of the frequency counter made the entire task of frequency measurement simple and very accurate. The frequency counter was quickly applied to the signal generator as a method of checking the dial accuracy and to augment the dial when setting the frequency to an exact value. Many models of signal generators have the frequency counter built in as an electronic dial. Figure 8-11 shows a block diagram of a modern signal generator with a frequency counter display, an isolation amplifier, and an ALC system.

An obvious extension of the frequency counter as an electronic dial is the use of an electronic frequency setting system, which introduces the frequency synthesizer.

8-3 FREQUENCY-SYNTHESIZED SIGNAL GENERATOR

To understand the basic function of the frequency synthesizer, imagine that a technician, wishing to reduce the frequency drift of a signal generator, decides to set the frequency of the generator every few seconds by reading the counter and adjusting the generator accordingly to the correct frequency. This is the human equivalent of a phase-locked-loop frequency synthesizer. Although the imaginary technician accomplishes the task of adjusting the frequency every few seconds, the electronic equivalent can make these adjustments at a much greater rate. Actually, the technician stabilizing the generator manually would eventually be limited by the time required for the frequency counter to display the correct frequency, which depends on the resolution of the counter. The phase-locked loop avoids this problem by not requiring a frequency determination, but as its name implies, makes the frequency correction based on a phase measurement.

One every popular method of frequency synthesis is called the *indirect method*, or the phase-locked loop, and is shown in Fig. 8-12. Five main components are required: the VCO or voltage-controlled oscillator, the programmable divider, the phase detector, the phase reference, and the loop filter.

The *voltage-controlled oscillator* is the source of the output frequency and has the ability to be tuned electronically, usually by applying a variable voltage. Some oscillators are electronically tuned using a current, especially in the higher frequencies, but for the general discussion of a phase-locked-loop frequency synthesizer, the signal source will be considered a voltage-controlled device.

The *programmable divider* is a logic element that divides the frequency of the VCO by an integer that can be entered via programming switches, a microprocessor, or other method.

The *phase detector* provides an analog output that is a function of the phase angle between the two inputs, which in the case of a frequency synthesizer is the reference source and the output of the programmable divider.

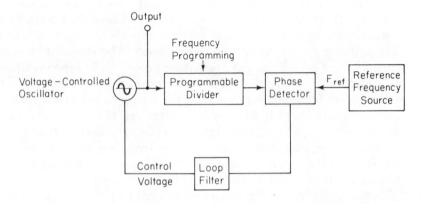

Figure 8-12 Block diagram of phase-locked loop.

The *reference source* is a very accurate and stable frequency source, which is typically a quartz crystal oscillator. The accuracy of the entire synthesizer is dependent on the accuracy of the reference source. The crystal oscillator operates in the region of 1 to 10 MHz, and that frequency is divided down using digital counters to provide the necessary clock and reference frequencies for the synthesizer.

The *loop filter* is an analog filter and is required to assure stable and noise-free operation of the synthesizer.

Assume that the VCO is to be electronically tuned to a multiple of the reference frequency of the example of Fig. 8-12. If an integer is entered into the programmable divider, we obtain

$$f_v = Nf_r$$

where f_v = desired frequency of the VCO

N = integer entered into the programmable divider

f_r = reference frequency applied to the phase detector.

Because the programmable divider divides the frequency of the VCO by N, the output frequency of the programmable divider is f_v/N or f_r.

The output of the programmable divider is fed to the phase detector and compared to the phase of the reference frequency. If the output of the phase detector were returned to the voltage-controlled oscillator, any variations in phase could be corrected so that the frequency of the VCO would be exactly N times the reference frequency. However, the phase determination can be made only once each reference frequency period, and thus the frequency of the VCO can only be corrected at this rate. This would cause the frequency of the VCO to be modulated and create spurious sidebands, called *reference sidebands*, in such a fashion as to make the VCO output useless for precision measurements. A filter is inserted between the phase detector and the VCO so that the periodic frequency changes are smoothed and the frequency modulation is reduced.

Although simple in theory, the phase-locked-loop frequency synthesizer has some significant disadvantages when used as the basis of a signal generator. First, even though the loop filter can remove a great deal of the frequency modulation caused by the output of the phase detector, it is never capable of removing all the sideband energy, and something must remain. The sideband level required for critical testing is very low and can make the phase-locked-loop synthesizer practically unsuitable. In addition, the loop filter does more to the characteristics of the phase-locked loop than just removing the reference sidebands; it affects the frequency slewing characteristics of the synthesizer. When it is necessary to change the frequency of the signal generator by a large amount, the time required for the change can be long. When the reference frequency becomes low, that is, the resolution of the frequency synthesizer is narrow, which is desirable for a signal generator, all the aforementioned problems become exacerbated.

Many of these problems can be mitigated by the use of complex, multiple phase-locked-loop synthesizers and other techniques; the discussion of these systems is beyond the scope of this text.

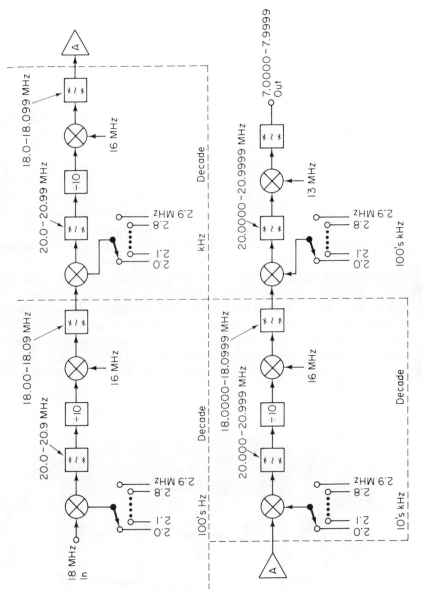

Figure 8-13 Example of direct synthesis.

There is a second method of frequency synthesis, not using the phase-locked loop and thus immune to some of the problems inherent in the phase-locked loop, called *direct synthesis*. Rather than stabilizing the frequency of a voltage-controlled oscillator by comparing the phase of a fraction of the VCO's frequency to a reference frequency, the direct method generates the desired frequency from a reference frequency.

Figure 8-13 shows an example of direct frequency synthesis. In this example a single 18-MHz reference frequency is used, which is divided, mixed multiplied, etc., to provide 10 outputs in 100-kHz steps from 2.0 to 2.9 MHz, as well as a 16-MHz output.

A selected frequency from the 2.0- to 2.9-MHz set is heterodyned with the 18-MHz reference, and the sum is filtered to produce 10 selectable frequencies from 20.0 to 20.9 MHz in the 100-kHz steps. This frequency is divided by 10 and mixed with 16 MHz to produce 10 selected frequencies from 18.0 to 18.09 MHz, but with 10-kHz steps. This is mixed with another selected frequency from the 2.0- to 2.9-MHz set to produce a sum from 20.00 to 20.99 MHz. This frequency is divided by 10 and mixed with 16 MHz, and so on. It can be seen from the block diagram that the circuits are repeated for each switch, and the repeated circuits are required whenever a decade is added. Although the synthesizer circuits appear complex and expensive, each decade is identical to the previous decade. Because there are no low-frequency circuits, the frequency can be changed practically instantaneously. In addition, because the entire synthesizer operates from a single reference frequency source, the close-in frequency spectrum can be quite pure without any undesired modulation. On the other hand, a multitude of intermediate frequencies is generated in the process of generating the ultimate output frequency, and great care must be taken in the design of direct synthesizers to shield and filter these undesired outputs. Figure 8-14 shows a signal generator using direct synthesis covering the range from 100 kHz to 160 MHz.

Figure 8-14 Example of direct synthesis: the Wavetek Model 5135A synthesized signal generator. (Courtesy of Wavetek, Inc.)

8-4 FREQUENCY DIVIDER GENERATOR

A type of signal generator that offers some of the advantages of a synthesized generator, yet avoids some of the pitfalls, is shown in Fig. 8-15. The frequency generation element of this signal generator is a very stable high-Q cavity oscillator operating in the 500-MHz region. The 500-MHz oscillator must cover a frequency range of 2 to 1, and a typical frequency range would be 256 to 512 MHz with a bit of overtravel. The oscillator is mechanically tuned, and if any type of phase-locked stabilization is included, the oscillator must be manually set to the desired frequency. The phase-locked loop provides only a small amount of frequency correction, primarily to compensate for any slow frequency drift. Because the phase-locked loop provides only a small amount of stabilization, the amount of reference sidebands generated is small.

Generally, the stability of the basic oscillator of the frequency divider signal generator is sufficient for most applications without the use of the phase lock. This is accomplished because the oscillator is required to cover only one range of frequencies and is optimized for that range. Typically, a signal generator is required to cover a large range of frequencies, which requires that the inductor of the oscillator be switched while the capacitor provides the necessary tuning. This usually means that inductor coils are mounted in a turret and a rather large capacitor, which is required for the lower-frequency ranges, is used to tune the oscillator. Aside from making the oscillator unit in the signal generator a large assembly, the mechanical components can generate mechanically induced frequency modulation because of their large areas. A cavity oscillator can be constructed with excellent immunity from residual frequency modulation and excellent frequency stability. The cavity becomes prohibitively large at lower frequencies and would be practically impossible to construct as a bandswitching unit. Essentially, the cavity is a single-frequency-range UHF oscillator. Outputs for other frequency ranges can be generated by dividing the frequency output from the cavity oscillator by powers of two using digital flip-flops. To use this scheme, the frequency range of the oscillator must be at least 2 to 1. As an example, the signal generator shown in Fig. 8-15 divides the 256- to 512-MHz output from the cavity oscillator to 128 to 256 MHz with the first divide by 2 circuit, and 64 to 128 MHz with the second. By repeating the frequency division, the 256- to 512-MHz cavity oscillator output is extended down to 1 to 2 MHz.

A frequency counter is used to display the signal generator frequency. Like the RF output of the generator, the frequency counted by the frequency counter is divided by powers of two so that the correct frequency is displayed. The frequency counter could be simply connected to the RF output; however, this could inject noise from the frequency counter circuits into the signal generator output.

Because the signal generator output is derived from digital logic, where the waveform is more likely to be a pulse-type waveform rather than a sine wave, the harmonics from the digital logic must be removed to generate a pure sine function. Because a perfect square wave would be the theoretical output from a flip-flop, only the odd harmonics would be present from the flip-flop frequency dividers. This would require a low-pass filter with a cutoff frequency higher than

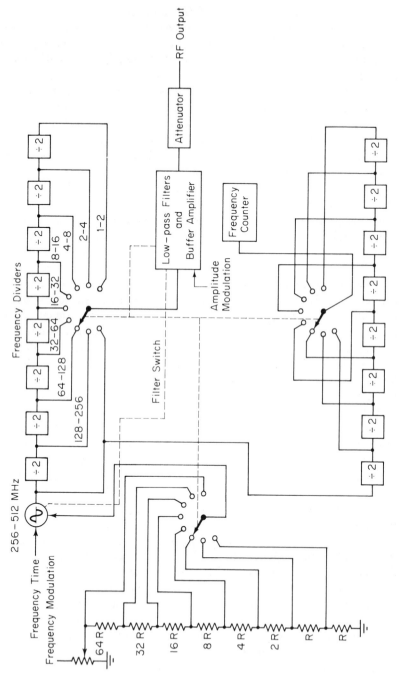

Figure 8-15 Block diagram of the frequency divider type of signal generator with frequency modulation.

the highest output frequency. As an example, the frequency output range of 4 to 8 MHz, which would be the output range after frequency division by 2, six times, would require a low-pass filter with a cutoff frequency slightly above 8 MHz. The lowest harmonic present in the output frequency range is the third harmonic of 4 MHz, or about 12 MHz.

The outputs of flip-flips contain only odd harmonics when the square wave is perfectly symmetrical. This is generally true at the lower frequencies; but as the frequency is increased, the logic becomes progressively less perfect and the amount of even harmonics starts to increase. The most troublesome of these is the second harmonic. To improve the spectrum of the higher frequencies, two low-pass filters are required, one for the lower portion of the output frequency range and a second for the upper portion of the frequency range, with the split occurring at the geometric mean of the output frequency range.

The geometric mean can be calculated by

$$\sqrt{f_l f_h} \tag{8-14}$$

where f_h is the high-frequency limit and f_l is the low-frequency limit. As an example, the geometric mean of the 64- to 128-MHz frequency range is 90.5 MHz. Some method must be employed to switch the low and high filters within a frequency band. This can be done by a control signal from the frequency counter or by a connection to the cavity oscillator. The point where the filter is switched is not critical, and a simple mechanical arrangement is more than sufficient.

One advantage of the divider signal generator is that the incident frequency modulation from the cavity is reduced by a factor of 2 each time the frequency is divided by the same factor. When spectral purity is a requirement and frequency modulation is not desired, this is an advantage. When frequency modulation is to be added to the signal generator output, this can be a problem. Frequency modulation is easily applied to the oscillator with a varactor diode. Because the amount of modulation is reduced each time the frequency is divided by 2, a correction circuit is required, which is controlled from the bandswitch. This circuit is shown in Fig. 8-15.

8-5 SIGNAL GENERATOR MODULATION

Most signal generators have the capability to be modulated with either frequency or amplitude modulation with a known modulation percentage or index. Amplitude modulation can be applied to the electronically leveled signal generator by modulating the *PIN* diode attenuator with the modulating signal. The serious problem with amplitude modulation is that the amplitude will vary from twice the carrier to zero for 100 per cent modulation, which implies that the voltage-controlled attenuator must have at least a nominal attenuation of 6 dB so that the amplitude can be increased to twice the carrier and must provide, theoretically, an infinite attenuation to provide the zero required by 100 per cent modulation. Regardless of the modulation technique, most signal generators provide amplitude modulation near but not equal to 100 per cent.

Frequency modulation does not suffer from a percentage of modulation problem, and there is no such thing as 100 per cent modulation. To frequency modulate the signal generator requires a method of electronically changing the frequency of the oscillator, and this is usually provided by a varactor diode in the oscillator-tuned circuit. The amount of modulation supplied by the varactor diode depends on the frequency of the oscillator and can vary over the tuning range of the oscillator. This requires that the signal generator have a method of correcting for this change in frequency modulation index. Applying modulation in a signal generator can be a complex problem when the signal generator is a synthesized generator. Each synthesized signal generator is a unique case, and numerous methods are used to supply an accurate source of modulation.

8-6 SWEEP-FREQUENCY GENERATOR

The previous discussion of sine-wave signal generators has concentrated on generators where a single output frequency was generated at a known and stable frequency. There are applications where a sweeping source of frequency is required, such as measuring the frequency response of amplifiers, filters, and other networks.

Compared to single-frequency signal generators, the sweep-frequency generator is a relatively new system. The difficulty, in the early days of electronics, was to find a method of electronically varying frequency so that a rapidly swept frequency output was available. Reactance tube modulators provided very little frequency variation, and usually a sweep generator resorted to electromechanical methods such as motor-driven capacitors. There were significant disadvantages to these early mechanical monsters, and most frequently response measurements were made by point-to-point techniques using conventional single-frequency signal generators. The development of broad-band communications systems brought with it the need for high-performance broad-band sweep-frequency generators.

The development of the solid-state variable capacitance diode did more for the development of sweep-frequency generators than any other electronic device. This diode provided the method of electronically tuning an oscillator and made the sweep generator a valuable instrument.

Figure 8-16 shows the block diagram of a simple sweep generator. The similarity to the single-frequency generator is apparent except that the sweep generator oscillator is capable of being electronically tuned, and a sweep voltage generator is supplied within the generator to provide the frequency sweep.

Because the relationship between the sweep voltage and the frequency of the oscillator is not linear, a compensating circuit is provided between the sweep frequency voltage and the oscillator tuning voltage. The amount of nonlinearity, and thus the amount of correction, required depends on the type of oscillator used and to a great extent on the frequency range covered by the oscillator. The narrower the frequency range of the sweep, the more linear the voltage-to-frequency relationship will be. Generally, there is a limit of 2 to 1 of the maximum-to-minimum frequency of any sweeping oscillator. Many modern systems, such

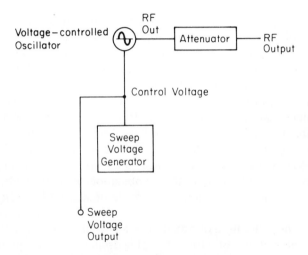

Figure 8-16 Example of a simple sweeping oscillator signal generator.

as those used for the transmission of cable and satellite television, have band-widths approaching hundreds of megahertz and require sweep techniques for evaluation and troubleshooting.

A typical *linearizing* circuit is shown in Fig. 8-17. As with most linearizers, the transfer characteristics can be adjusted to suit the oscillator. The nonlinear transfer is generated by using a piecewise-linear approximation. Linear slopes, where the slope and the break point can be adjusted by resistors in the circuit, provide an approximation to the desired transfer function. As an example, the

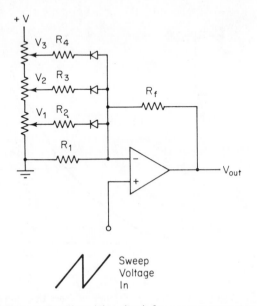

Figure 8-17 Linearizing circuit for a sweep generator.

gain the circuit shown is a function of the feedback resistor R_f and the net resistance of R_1 through R_4. The adjustable voltage divider is constructed from resistors considerably smaller than R_f or R_1 through R_4. The gain of the amplifier is primarily a function of the values of R_f and R_1 through R_4. When the input sweep voltage is low, none of the diodes are conducting and the gain of the op-amp circuit is $(R_f/R_1) + 1$. When the sweep voltage approaches V_1, the first diode conducts and the gain of the amplifier is increased to $(R_f/R_a) + 1$, where R_a is the combination of R_1 and R_2 in parallel. When the sweep voltage reaches V_2, the gain is increased again to $(R_f/R_b) + 1$, where R_b is the parallel combination of R_1, R_2, and R_3. When the sweep voltage reaches V_3, the gain is increased a third time to $(R_f/R_c) + 1$, where R_c is the combination of R_1 through R_4 in parallel. The net result is a nonlinear relationship made of straight line segments as shown in Fig. 8-18.

The generation of broad-band sweeping signals is handled by mixing a fixed-frequency oscillator with a sweeping oscillator at a frequency well above the band of frequencies to be generated. To remain below the 2-to-1 ratio for the sweeping oscillator, the operating frequency of the sweeping oscillator should be above the widest sweep width. Figure 8-19 shows an example of a modern wideband sweep generator. In this example 0- to 300-MHz signal is generated by mixing a 400- 700-MHz oscillator with a fixed 400-MHz oscillator. The output frequency range covers from literally zero to 300 MHz, while the maximum-to-minimum frequency ratio of the sweeping oscillator is less than two.

A broad-band sweep generator must have some sort of automatic amplitude adjusting circuit as shown in Fig. 8-19. It is not possible to mix two signals at several hundred megahertz, filter the difference, amplify the result, and maintain the resultant amplitude to within a few decibels. The automatic level control used

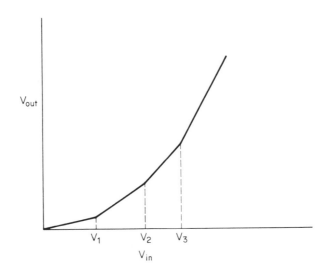

Figure 8-18 Transfer function of linearizing circuit such as the circuit shown in Fig. 8-17.

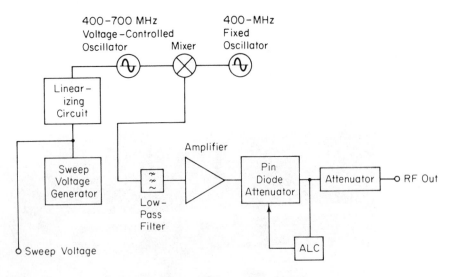

Figure 8-19 Wideband sweep generator.

in the broad-band sweep generator is similar to the one described previously except that the system must operate over a larger span of frequencies.

The sweep generator is almost invariably used to determine the frequency response of amplifiers or other systems and is not normally used to simulate normal operating signals. To determine the frequency response of a system requires two auxiliary devices, a detector and an oscilloscope. Figure 8-20 shows a typical setup to display the frequency response of an amplifier using a sweep generator. The sweep-generator output is fed to the amplifier input, and the output is fed to a crystal detector. The crystal detector is nothing more than a rectifier diode and a capacitor to remove the rectified ripple voltage, as shown in

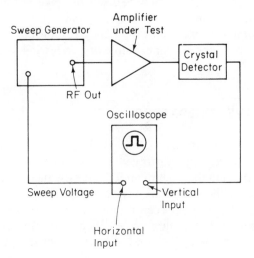

Figure 8-20 Typical test equipment hookup for measuring the frequency response of an amplifier.

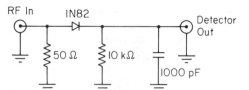

Figure 8-21 Schematic diagram of a crystal detector.

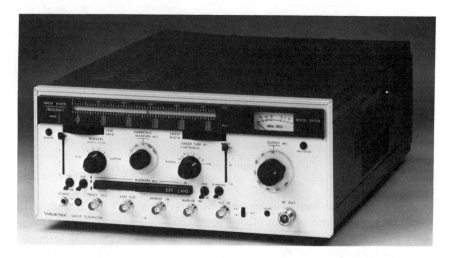

Figure 8-22 Wideband sweeping generator. (Courtesy of Wavetek RF Products, Inc.)

Fig. 8-21. Like the sweep generator, it is necessary that the frequency response of the detector be flat from the lowest frequency to be measured to the highest. One easy and accurate method of determining the frequency response of the sweep system is to connect the crystal detector directly to the sweep generator and display the results on the oscilloscope. This will display the frequency response of the measuring system, which can then be used to make any corrections required for the lack of perfect amplitude flatness. Usually, in a good sweep measuring system the maximum to minimum frequency response is less than 1 dB, which in most cases can be ignored. An example of a wideband sweeping generator is shown in Fig. 8-22.

8-7 PULSE AND SQUARE-WAVE GENERATORS

Pulse and square-wave generators are often used with an oscilloscope as the measuring device. The waveforms portrayed by the oscilloscope either at the output or at pertinent points in the system under test provide both qualitative and quantitative information of the system or device under test.

The fundamental difference between a pulse generator and a square-wave generator concerns the *duty cycle*. Duty cycle is defined as the ratio of the average value of the pulse over one cycle to the peak value of the pulse. Since the average value and the peak value are inversely related to their time duration, the duty cycle can be defined in terms of the *pulsewidth* and the *period* or *pulse repetition time*:

$$\text{duty cycle} = \frac{\text{pulsewidth}}{\text{period}}$$

Square-wave generators produce an output voltage with equal *on* and *off* times, so that their duty cycle equals 0.5, or 50 per cent. The duty cycle remains at 50 per cent as the frequency of oscillation is varied.

The duty cycle of a *pulse generator* may vary; very short duration pulses give a low duty cycle, and the pulse generator generally can supply more power during its *on* period than a square-wave generator can. Short duration pulses reduce the power dissipation in the component under test. For instance, measurements of transistor gain can be made with pulses short enough to prevent junction heating, and the resulting effect of heat on the gain of the transistor is then greatly minimized.

Square-wave generators are used whenever the low-frequency characteristics of a system are being investigated: testing audio systems, for instance. Square waves are also preferable to short-duration pulses if the transient response of a system requires some time to settle down.

8-7.1 Pulse Characteristics and Terminology

In selecting a pulse generator or square-wave generator, the quality of the pulse is of primary importance. A test pulse of high quality ensures that any degradation of the displayed pulse may be attributed to the circuit under test and not to the test instrument itself.

The pertinent characteristics of a pulse are shown in Fig. 8-23. The specifications describing these characteristics are usually given in the instrument manual or manufacturer's specification sheets.

The time required for the pulse to increase from 10 per cent to 90 per cent of its normal amplitude is called the *risetime* (t_r). Similarly, the time required for the pulse to decrease from 90 per cent to 10 per cent of its maximum amplitude is called the *falltime* (t_f). In general, the risetime and the falltime of the pulse should be significantly faster than the circuit or component under test.

When the initial amplitude rise exceeds the correct value, *overshoot* occurs. The overshoot may be visible as a single *pip*, or *ringing* may occur. When the maximum amplitude of the pulse is not constant but decreases slowly, the pulse is said to *droop* or *sag*. Any overshoot, ringing, or sag in the test pulse should be known to avoid confusion with similar phenomena caused by the test circuit.

The maximum pulse *amplitude* is of prime concern if appreciable input power is required by the tested circuit, such as, for example, a magnetic core memory. At the same time, the attenuation range of the instrument should be adequate to prevent overdriving the test circuit as well as to simulate actual operating conditions.

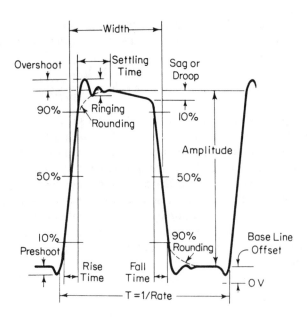

Figure 8-23 Characteristics of a pulse.

The range of the frequency control or *pulse repetition rate* (PRR) is of concern if the tested circuit can operate only within a certain range of pulse rates or if a variation in the rate is needed. Some of the more sophisticated pulse generators are capable of repetition rates of up to 100 MHz for testing "fast" circuits; other have a *pulse-burst* feature that allows a train of pulses rather than a continuous output to be used to check a system.

Some pulse generators can be *triggered* by externally applied trigger signals, similar to the trigger features found in laboratory oscilloscopes. Conversely, the output of the pulse generator or square-wave generator may be used to provide trigger pulses for operating external circuits. The output trigger circuitry of the pulse generator then allows the trigger pulse to occur either before or after the main output pulse.

The *output impedance* of the pulse generator is another important consideration in fast pulse systems. This is so because the generator, which has a source impedance matched to the connecting cable, will absorb reflections resulting from impedance mismatches in the external circuitry. Without this generator-to-cable match, the reflections would be re-reflected by the generator, resulting in spurious pulses or perturbations on the main pulse.

Dc coupling of the output circuit is necessary when retention of the dc bias levels in the test circuit is desired, in spite of variations in pulsewidth, pulse amplitude, or PRR.

Circuits used in pulse generation generally fall into two categories: *passive*, or pulse-shaping, and *active*, or pulse-generating circuits. In passive-type circuits, a sine-wave oscillator is used as the basic generator and its output is passed through a pulse-shaping circuit to obtain the desired waveform. For instance, an approximate square waveform may be obtained by first amplifying and then clipping a sine wave. Active generators are usually of the *relaxation* type. The relaxation oscillator uses the charge-and-discharge action of capacitor to control the conduction of a vacuum tube or a transistor. Some common forms of relaxation oscillator are the *multivibrators* and the *blocking oscillators*.

8-7.2 Astable Multivibrator

The *astable* or *free-running multivibrator* is widely used for the generation of pulses. It can be made to produce either square waves or pulses, depending on the choice of circuit components. A typical free-running multivibrator is shown in Fig. 8-24. Essentially, the circuit consists of a two-stage RC-coupled amplifier, with the output of the second state (Q_2) coupled back to the input of the first stage (Q_1) via capacitor C_1. Similarly, the output of Q_1 is coupled via C_2 to the input of Q_2. Since the coupling between the two transistors is taken from the collectors, the circuit is known as a *collector-coupled* astable multivibrator.

   The usual qualitative analysis of the circuit proceeds as follows: When the power is first applied to the circuit, both transistors start conducting. Because of small differences in their operating characteristics, one of the transistors will conduct slightly more than the other. This starts a series of events. Assume that Q_1 initially conducts more than Q_2. This means that the collector voltage of $Q_1(e_{c1})$ drops more rapidly than the collector voltage of $Q_2(e_{c2})$. The decrease in e_{c1} is applied to the R_cC_2 network, and because the charge on C_2 cannot change instantaneously, the full negative-going change appears across R_2. This decreases the forward bias on Q_2 which in turn decreases the collector current of $Q_2(i_{c2})$, and the collector voltage of Q_2 rises. This rise in Q_2 collector voltage is applied via the R_1C_1 network to the base of Q_1, increasing its forward bias. Q_1 therefore conducts even more heavily and its collector voltage drops still more rapidly. This negative-going change is coupled to the base of Q_2, further decreasing its collector current. The entire process is cumulative until Q_2 is entirely cut off and Q_1 conducts heavily (*bottoms*).

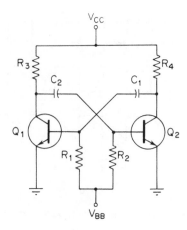

Figure 8-24 Astable or free-running multivibrator.

With Q_2 cut off, its collector voltage practically equals the supply voltage, V_{CC}, and the capacitor C_1 charges rapidly to V_{CC} through the low-resistance path from emitter to base of the conducting transitor Q_1. When the circuit action turns Q_1 fully on, its collector potential drops to approximately 0 V, and since the charge on C_2 cannot change instantaneously, the base of Q_2 is at least $-V_{CC}$ potential, driving Q_2 deep into cutoff.

The switching action now begins, C_2 begins to discharge exponentially through R_2. When the charge on C_2 reaches 0 V, C_2 attempts to charge up to the value of $+V_{BB}$, the base supply voltage. But this action immediately places a forward bias on Q_2 and this transistor starts to conduct. As soon as Q_2 starts conducting, its collector current causes a decrease in collector voltage e_{c2}. This negative-going change is coupled to the base of Q_1 which starts to conduct less, i.e., it comes out of saturation. This cumulative action repeats until Q_1 finally cuts off and Q_2 conducts heavily. At this instant, the collector voltage of Q_1 reaches its maximum value of V_{CC}. Capacitor C_2 charges to the full value of V_{CC}, and a full cycle of operation has been completed.

The waveforms appearing at the base and the collector of each transistor are the result of a *symmetrical* or *balanced* operation: The time constants R_1C_1 and R_2C_2, the transistors themselves, and the supply voltages are all identical. The conducting and nonconducting periods are therefore of almost the same duration. The waveforms for each of the two transistors are given in the waveform diagram of Fig. 8-25.

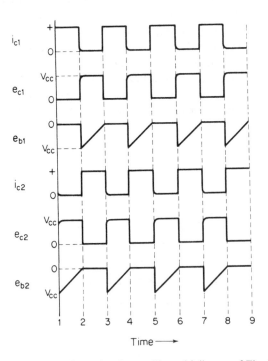

Figure 8-25 Waveforms for the astable multivibrator of Fig. 8-24.

Signal Generation   Chap. 8

Assume that at time $t = 1$, transistor Q_1 is fully turned on and transistor Q_2 is cut off. This makes the collector voltage e_{c1} of Q_1 a minimum (practically 0 V) and the collector voltage e_{c2} of Q_2 a maximum (V_{CC}). Capacitor C_1 is charging through the emitter-to-base resistance of Q_1 toward the supply voltage V_{CC} and reaches its full charge rapidly (low emitter-to-base resistance). Since e_{c1} is 0 V, capacitor C_2 begins to charge exponentially through R_2 toward the base supply voltage V_{BB} with a time constant equal to R_2C_2. Since the early part of the exponential charging curve is almost linear the increase in Q_2 base voltage (e_{b2}) is indicated by a linear slope on the graph of Fig. 8-25.

At time $t = 2$, e_{c2} reaches a value of approximately 0 V, placing a forward bias on the base of Q_2 which then starts to conduct. Within a very short time, the collector current of Q_2 reaches it maximum and collector voltage e_{c2} drops to 0 V. When Q_2 starts to draw current, the base of Q_1 becomes negative and Q_1 is quickly driven into cutoff. Its collector voltage, e_{c1}, reaches the V_{CC} value and collector current i_{c1} becomes zero. Within a very small fraction of the total Q_2 conducting time, capacitor C_2 is fully charged to V_{CC} through the low-resistance emitter-to-base path of Q_2.

Between times $t = 2$ and $t = 3$, transistor Q_1 is cut off, and its collector current and voltage remain constant. Similarly, the collector voltage and current for Q_2 remain constant. Only capacitor C_1 is charging and the base voltage e_{b1} of Q_1 is rising exponentially toward V_{BB}. At time $t = 3$, the base voltage of Q_1 exceeds the cutoff value (approximately 0 V) and Q_1 starts conducting again. Obviously, one complete cycle of operation, from time $t = 1$ to time $t = 3$, depends on the time required for the base voltage of the cutoff transistor to reach the forward bias value. This time depends on two things: the magnitude of the reverse bias ($-V_{CC}$) and the time constant of the capacitor charging circuit involved, namely, R_1C_1 or R_2C_2.

The *analytical* evaluation of circuit operation proceeds as follows: During its nonconducting period, the collector voltage of Q_1 equals

$$e_{c1} = V_{CC}(1 - e^{-t/\tau_3}) \tag{8-15}$$

where $\tau_3 = R_3C_2$.

When Q_1 switches on, its collector voltage is at ground potential and the base voltage of Q_2 becomes $-V_{CC}$ with respect to ground. The subsequent rise in Q_2 base voltage, through the R_2C_2 charging circuit, is described by

$$e_{b2} = (V_{BB} + V_{CC})(1 - e^{-t/\tau_2}) - V_{CC} \tag{8-16}$$

where $\tau_2 = R_2C_2$.

Q_2 remains cut off until e_{b2} reaches the value of 0 V (by good approximation) and the *off* time interval T_2 of Q_2 can be determined by setting e_{b2} in Eq. (8-16) to zero and solving for t, so that

$$0 = (V_{BB} + V_{CC})(1 - e^{-t/\tau_2}) - V_{CC} \tag{8-17}$$

and

$$T_2 = \tau_2 \ln \left(\frac{V_{BB}}{V_{BB} + V_{CC}} \right) \tag{8-18}$$

Similarly, when Q_2 is off and Q_1 is bottomed, the collector voltage of Q_2 can be described by

$$e_{c_2} = V_{CC}(1 - e^{-t/\tau_4}) \qquad (8\text{-}19)$$

where $\tau_4 = R_4 C_1$.

When now Q_2 is switched on, its collector voltage drops to 0 V and the base voltage of Q_1 is given by

$$e_{b1} = (V_{BB} + V_{CC})(1 - e^{-t/\tau_1}) \qquad (8\text{-}20)$$

where $\tau_1 = R_1 C_1$.

Solving for the *off* time interval T_1 of Q_1 by setting e_{b1} in Eq. (8-20) to zero, we obtain

$$0 = (V_{BB} + V_{CC})(1 - e^{-t/\tau_1}) - V_{CC} \qquad (8\text{-}21)$$

and

$$T_1 = \tau_1 \ln \left(\frac{V_{BB}}{V_{BB} + V_{CC}} \right) \qquad (8\text{-}22)$$

The total period of oscillation is given by

$$T = T_1 + T_2 \qquad (8\text{-}23)$$

In the case of *symmetrical* operation, when time constants $R_1 C_1$ and $R_2 C_2$ are equal, the waveform is a symmetrical square wave. By making time constant $R_1 C_1$ larger than time constant $R_2 C_2$, the output waveform becomes a pulse train because the *off* time of Q_1 will be larger than the *off* time of Q_2.

8-7.3 Laboratory Square-Wave and Pulse Generator

The block diagram of a typical general-purpose generator providing negative pulses of variable frequency, duty cycle, and amplitude is given in Fig. 8-26. The frequency range of the instrument is covered in seven decade steps from 1 Hz to 10 MHz, with a linearly calibrated dial for continuous adjustment on all ranges. The duty cycle can be varied from 25 per cent to 75 per cent. Two independent outputs are available: a 50-Ω source that supplies pulses with rise- and falltimes of 5 ns at 5-V peak amplitude, and a 600-Ω source that supplies pulses with rise- and falltimes of 70 ns at 30-V peak amplitude. The instrument can be operated as a free-running generator or it can be synchronized with external signals. Trigger output pulses for synchronization of external circuits are also available.

The basic generating loop, which is redrawn for greater clarity in Fig. 8-27 consists of two current sources, the ramp capacitor, the Schmitt trigger circuit, and the current-switching circuit (indicated by a simple switch). The two current sources provide a constant current for charging and discharging the ramp capacitor. The ratio of these two currents is determined by the setting of the *symmetry* control which then determines the duty cycle of the output waveform. The *fre-*

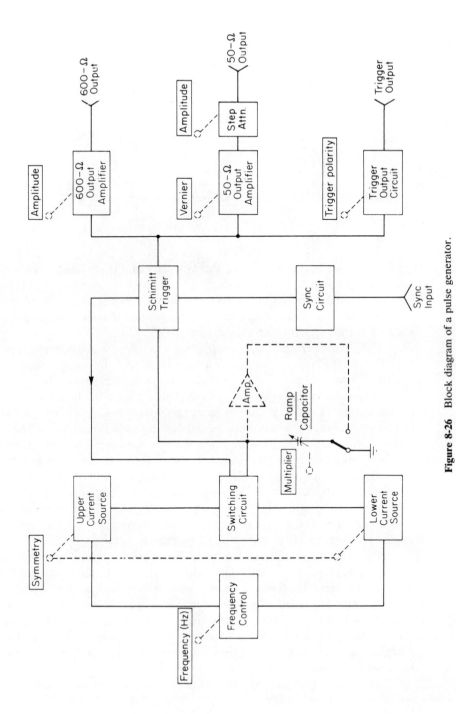

Figure 8-26 Block diagram of a pulse generator.

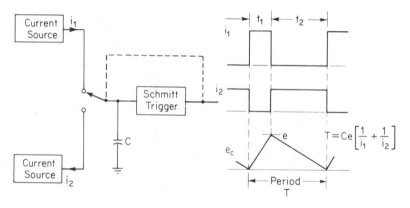

Figure 8-27 Simplified current source operation. (Courtesy of Hewlett-Packard Company.)

quency dial controls the sum of the two currents from the current sources by applying appropriate control voltages to the bases of the current control transistors in the current generators. The size of the ramp capacitor is selected by the *multiplier* switch. These last two controls provide decade switching and vernier control of the frequency of the output.

The *upper* current source, supplying a constant current to the ramp capacitor charges this capacitor at a constant rate, and the ramp voltage increases linearly. When the positive slope of the ramp voltage reaches the upper limit set by internal circuit components, the Schmitt trigger (a bistable multivibrator) changes state. The trigger circuit output goes negative, reversing the condition of the current control switch, and the capacitor starts discharging. The discharge rate is linear, controlled by the *lower* current source. When the negative ramp reaches a predetermined lower level, the Schmitt trigger switches back to its original state. This now provides a positive trigger circuit output that reverses the condition of the current switch again, cutting off the lower current source and switching on the upper current source. One cycle of operation has now been completed. The entire process, of course, is repetitive and the Schmitt trigger circuit provides negative pulses at a continuous rate.

The output of the Schmitt circuit is passed to the trigger output circuit and to the 50-Ω and 600-Ω amplifiers. The trigger output circuit differentiates the square-wave output from the Schmitt trigger, inverts the resulting pulse, and provides a positive triggering pulse. The 50-Ω amplifier is provided with an output attenuator to allow a vernier control of the signal output voltage. In addition to its free-running mode of operation, the generator can be synchronized or locked in to an external signal. This is accomplished by triggering the Schmitt circuit by an external synchronization pulse.

The unit is powered by an internal supply that provides regulated voltages for all stages of the instrument.

8-8 FUNCTION GENERATOR

A function generator is a versatile instrument that delivers a choice of different waveforms whose frequencies are adjustable over a wide range. The most common output waveforms are the sine, triangular, square, and sawtooth waves. The frequencies of these waveforms may be adjusted from a fraction of a hertz to several hundred kilohertz.

The various outputs of the generator may be available at the same time. For instance, by providing a square wave for linearity measurements in an audio system, a simultaneous sawtooth output may be used to drive the horizontal deflection amplifier of an oscilloscope, providing a visual display of the measurement results. The capability of the function generator to *phase lock* to an external signal source is another useful feature. One function generator may be used to phase lock a second function generator, and the two output signals can be displaced in phase by an adjustable amount. In addition, one generator may be phase locked to a harmonic of the sine wave of another generator. By adjusting the phase and the amplitude of the harmonics, almost any waveform may be generated by the summation of the fundamental frequency generated by the one function generator and the harmonic generated by the other function generator. The function generator can also be phase locked to a frequency standard, and all its output waveforms are then generated with the frequency accuracy and stability of the standard source.

The function generator can supply output waveforms at very low frequencies. Since the low frequency of a simple *RC* oscillator is limited, a different approach is used in the function generator of Fig. 8-28. This instrument delivers sine, triangular, and square waves with a frequency range of 0.01 Hz to 100 kHz. The frequency control network is governed by the frequency dial on the front panel of the instrument or by an externally applied control voltage. The frequency control voltage regulates two current sources.

The *upper* current source supplies a constant current to the triangle integrator whose output voltage increases linearly with time. The output voltage is given by the well-known relationship

$$e_{\text{out}} = -\frac{1}{C} \int i \, dt$$

An increase or a decrease in the current supplied by the upper current source increases or decreases the slope of the output voltage. The voltage comparator multivibrator changes state at a predetermined level on the positive slope of the integrator's output voltage. This change of state cuts off the upper current supply to the integrator and switches on the lower current supply.

The *lower* current source supplies a reverse current to the integrator so that its output decreases linearly with time. When the output voltage reaches a predetermined level on the negative slope of the output waveform, the voltage comparator again switches and cuts off the lower current source while at the same time switching on the upper source again.

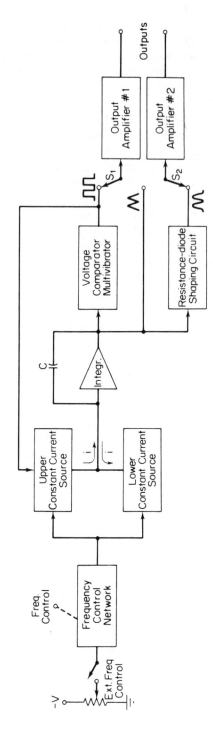

Figure 8-28 Basic elements of a function generator.

The voltage at the output of the integrator has a triangular waveform whose frequency is determined by the magnitude of the current supplied by the constant current sources. The comparator delivers a square-wave output voltage of the same frequency. The third output waveform is derived from the triangular waveform, which is synthesized into a sine wave by a diode resistance network. In this circuit the slope of the triangular wave is altered as its amplitude changes, resulting in a sine wave with less than 1 per cent distortion.

The output circuitry of the function generator consists of two output amplifiers that provide two simultaneous, individually selected outputs of any of the waveform functions.

8-9 AUDIOFREQUENCY SIGNAL GENERATION

Audiofrequency signal generators share many of the characteristics of their higher-frequency counterparts with a few notable differences. First, and probably most significant, the audiofrequency signal generator is not likely to use an oscillator controlled by LC-tuned circuits, but rather by a controlled phase shift through a resistor and capacitor, RC, network.

The requirements for an RC audio oscillator are identical to those required for the LC oscillator and are shown in Fig. 8-2. The Wien bridge oscillator produces clean sine waves using an RC network for feedback. Figure 8-29 shows a Wien bridge feedback network and an amplifier connected as an oscillator to determine at what frequency the Wien bridge will provide the required criterion for oscillation. Because the amplifier shown has a theoretical gain of infinity, and the loop gain for oscillations should be unity, the voltage from point A to B should be zero. Of course, it is not possible to have infinite gain, and likewise the voltage will not be zero but some small voltage such that the voltage from A to B times the actual, hopefully large, gain of the amplifier is equal to 1. Relative to ground, the voltage at A can be represented by

$$V_a = \frac{Z_1}{Z_1 + Z_2} V_i \qquad (8\text{-}24)$$

On the other hand,

$$V_b = \frac{R_2}{R_1 + R_2} V_i \qquad (8\text{-}25)$$

Since V_a and V_b are essentially the same,

$$V_a = V_b = \frac{Z_1}{Z_1 + Z_2} V_i = \frac{R_2}{R_1 + R_2} V_i \qquad (8\text{-}26)$$

$$\frac{Z_1}{Z_1 + Z_2} = \frac{R_2}{R_1 + R_2}$$

It can also be shown that the phase angle between V_a and the output is zero degrees at a frequency of $f_0 = 1/2\pi RC$.

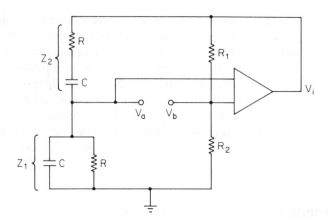

Figure 8-29 Wien bridge feedback network with amplifier.

The Wien bridge oscillator can be tuned by varying either the resistance or capacitance, or both. In practice, the Wien bridge oscillator is tuned with a variable capacitance while the oscillator is bandswitched using the resistance. To cover the lower end of the audiofrequency band with an *RC*-tuned oscillator requires that the resistances be rather large in order to use a conventional variable capacitor.

The Wien bridge oscillator usually is the heart of a general-purpose audio generator capable of reasonable stability and a dial accuracy of a few per cent. Harmonic distortion can be usually held to less than a few tenths of a per cent.

REFERENCES

8-1. Hayward, W. H., *Introduction to Radio Frequency Design*, chap. 7. Englewood Cliffs, N.J.: Prentice-Hall, Inc., 1982.

8-2. Krauss, Herbert L., Bostian, Charles W., and Raab, Frederick H., *Solid State Radio Engineering*, chaps. 5 and 6. New York: John Wiley & Sons, Inc., 1980.

8-3. Lenk, John D., *Handbook of Practical Electronic Circuits*, chap. 4. Englewood Cliffs, N.J.: Prentice-Hall, Inc., 1982.

8-4. Manassewitsch, Vadim, *Frequency Synthesizers, Theory and Design*. New York: John Wiley & Sons, Inc., 1980.

8-5. Prensky, Sol D., and Castellucis, Richard L., *Electronic Instrumentation*, 3rd. ed., chap. 11. Englewood Cliffs, N.J.: Prentice-Hall, Inc., 1982.

PROBLEMS

8-1. What will be the ratio of the highest to the lowest frequency of an oscillator if a 50- to 350-pF variable capacitor is used in the tuned circuit?

8-2. How many inductors and what values should be used with the oscillator described in Problem 8-1 to cover the frequency range from 1 to 30 MHz? Suggest the tuning ranges to allow for some overlap.

8-3. Convert the following: +5 dBw to dBm; −60 dBw to dBm; +56 dBm to dBw; +13 dBm to volts; 2 W to dBw; 1 V to dBw; −120 dBm to volts.

8-4. How much power is dissipated in a 50-Ω, 6-dB attenuator if it is fed from a 50-Ω generator with 10 W and is terminated with 50 Ω? How much power is delivered to the load?

8-5. What are the resistor values required for a 10-dB, 50-Ω attenuator?

8-6. Using the techniques presented in this chapter, determine the formula for determining the resistor values for the T pad shown in Fig. P8-6.

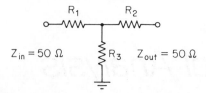

Figure P8-6

8-7. How much attenuation would be obtained from a piston-type attenuator if the cylinder diameter is 2 cm and the distance between the loops is 5 cm?

8-8. Would a *PIN* diode be a viable attenuator at audiofrequencies? Why?

8-9. What is the need for inserting isolation between the signal generator output and the oscillator in a simple signal generator? What are the ways this can be accomplished?

8-10. Why is a fixed attenuator inserted between a piston attenuator and the oscillator of a signal generator?

8-11. What would be the maximum reference frequency for a phase-locked loop that is to cover the frequency range from 20 to 40 MHz in 10-kHz steps?

8-12. What are some of the advantages of using direct synthesis rather than indirect synthesis?

8-13. Without requiring the swept oscillator to cover more than an octave, what is the minimum frequency the swept oscillator can be operated in a sweep generator covering 0 to 30 MHz?

8-14. What would the resonant frequency of a Wien bridge circuit be if the resistors are 100 kΩ and the capacitors are 0.1 μF?

9

Signal Analysis

9-1 INTRODUCTION

In the previous chapter the subjects of spectral purity, sidebands, and distortions were discussed referring to the outputs of signal generators. In this chapter the tools that can evaluate the amount of such distortions will be discussed under the topic of signal or spectrum analysis.

The first instrument to measure any sort of spectral content of signals was the harmonic distortion analyzer as applied to audiofrequency signals. In the early days of electronics, one of the most important tasks facing electronics engineers, typically called ''radio'' engineers, was the development of entertainment and communications radio systems. Audiofrequency harmonic distortion was an important part of this early development primarily because it could be easily heard and was objectionable to the listener. These early distortion analyzers measured total harmonic distortion without any indication of which harmonic was responsible. More sophisticated analyzers, called wave analyzers, could separate the harmonics and nonharmonic distortions and evaluate each one. These instruments were the first true spectrum analyzers.

9-2 WAVE ANALYZERS

9-2.1 Frequency-Selective Wave Analyzer

A *wave analyzer* is an instrument designed to measure the relative amplitudes of single-frequency components in a complex or distorted waveform. Basically, the instrument acts as a *frequency-selective voltmeter* which is tuned to the frequency of one signal component while rejecting all the other signal components. Two basic circuit configurations are generally used. For measurements in the audiofrequency range (from 20 Hz to 20 kHz), the analyzer has a filter section with a very narrow passband that can be tuned to the frequency component of interest. An instrument of this type is given in the functional block diagram of Fig. 9-1(a).

The waveform to be analyzed in terms of its separate frequency components is applied to an input attenuator that is set by the *meter range* switch on the front panel. A driver amplifier feeds the attenuated waveform to a high-*Q* active filter. The filter consists of a cascaded arrangement of *RC* resonant sections and filter amplifiers. The passband of the total filter section is covered in decade steps over the entire audio range by switching capacitors in the *RC* sections. Close-tolerance polystyrene capacitors are generally used for selecting the frequency ranges. Precision potentiometers are used to tune the filter to any desired frequency within the selected passband.

A final amplifier stage supplies the selected signal to the meter circuit and to an untuned buffer amplifier. The buffer amplifier can be used to drive a recorder or an electronic counter. The meter is driven by an average-type detector and usually has several voltage ranges as well as a decibel scale.

The bandwidth of the instrument is very narrow, typically about 1 per cent of the selected frequency. Figure 9-1(b) shows a typical attenuation curve of a wave analyzer (GenRad Type 1568-A Wave Analyzer). The initial attenuation rate is approximately 600 dB per octave; the attenuation at one-half and twice the selected frequency is about 75 dB. The filter characteristic also shows that the attenuation is still increasing far from the center frequency, well into the noise level of the instrument itself. The analyzer must have extremely low input distortion, so low, in fact, that it cannot be detected by the analyzer itself.

9-2.2 Heterodyne Wave Analyzer

Measurements in the megahertz range are usually done with another wave analyzer that is particularly suited to the higher frequencies. The input signal to be analyzed is *heterodyned* to a higher intermediate frequency (IF) by an internal local oscillator. Tuning the local oscillator shifts the various signal frequency components into the passband of the IF amplifier. The output of the IF amplifier is rectified and applied to the metering circuit. An instrument that uses the heterodyning principle is often called a *heterodyning tuned voltmeter*.

A wave analyzer using the heterodyning principle is shown in the block diagram of Fig. 9-2. The operating frequency range of this instrument is from 10 kHz to 18 MHz in 18 overlapping bands selected by the frequency range control of

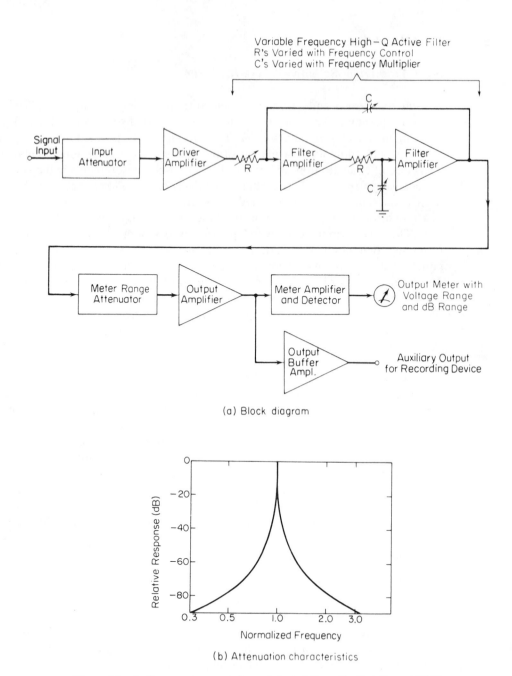

(a) Block diagram

(b) Attenuation characteristics

Figure 9-1 Audio-range wave analyzer (adapted from GenRad Type 1568A). Characteristics of the active filter show the extremely sharp attenuation at the selected frequency. (Courtesy of GenRad, Inc.)

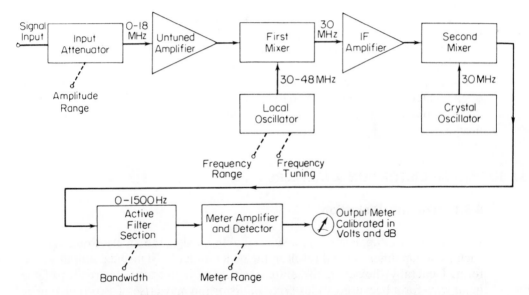

Figure 9-2 Functional block diagram of the heterodyning wave analyzer (adapted from HP model 312-A).

the local oscillator. The bandwidth is controlled by an active filter and can be selected at 200, 1,000, and 3,000 Hz.

The input signal enters the instrument through a probe connector that contains a unity-gain isolation amplifier. After appropriate attenuation, the input signal is heterodyned in the mixer stage with the signal from a local oscillator. The output of the mixer forms an intermediate frequency that is uniformly amplified by the 30-MHz IF amplifier. This amplified IF signal is then mixed again with a 30-MHz crystal oscillator signal, which results in information centered on a zero frequency. An active filter with controlled bandwidth and symmetrical slopes of 72 dB per octave then passes the selected component to the meter amplifier and detector circuit. The output from the meter detector can be read off a decibel-calibrated scale or may be applied to a recording device.

9-2.3 Applications

Applications of the wave analyzer are found in the fields of electrical measurements and sound and vibration analysis. For example, *harmonic distortion* of an amplifier can readily be measured, and the contribution of each harmonic to the total distortion figure can be determined. When the passband of the analyzer of Fig. 9-1(a) is tuned to the second harmonic, the fundamental frequency is sufficiently attenuated to reduce its level to substantially less than that of the harmonic. The curve of Fig. 9-1(b) shows that half-frequency attenuation is at least 75 dB. When the third harmonic is selected, the fundamental frequency is attenuated by more than 85 dB. A complete harmonic analysis can be carried out by resolving the individual components of a periodic signal and measuring or displaying these components. It is not unusual to be able to separate and display about 50 harmonics.

The wave analyzer is applied industrially in the field of reduction of sound and vibration generated by machines and appliances. The source of the noise or vibration generated by a machine must first be identified before it can be reduced or eliminated. A fine-spectrum analysis with the wave analyzer will show various discrete frequencies and resonances that can then be related to motions within the machine.

9-3 HARMONIC DISTORTION ANALYZERS

9-3.1 Harmonic Distortion

In the ideal case, application of a sinusoidal input signal to an electronic device, such as an amplifier, should result in the generation of a sinusoidal output waveform. Generally, however, the output waveform is not an exact replica of the input waveform because various types of distortion may arise. Distortion may be a result of the inherent nonlinear characteristics of the transistors in the circuit or of the circuit components themselves. Nonlinear behavior of circuit elements introduces harmonics of the fundamental frequency in the output waveform, and the resultant distortion is often referred to as *harmonic distortion* (HD).

A measure of the distortion represented by a particular harmonic is simply the ratio of the amplitude of the harmonic to that of the fundamental frequency, expressed as a percentage. Harmonic distortion is then represented by

$$D_2 = \frac{B_2}{B_1}, \qquad D_3 = \frac{B_3}{B_1}, \qquad D_4 = \frac{B_4}{B_1} \qquad (9\text{-}1)$$

where $D_n(n = 2, 3, 4, \ldots)$ represents the distortion of the nth harmonic, B_n represents the amplitude of the nth harmonic, and B_1 is the amplitude of the fundamental.

The *total harmonic distortion*, or *distortion factor*, is defined as

$$D = \sqrt{D_2^2 = D_3^2 + D_4^2 + \cdots} \qquad (9\text{-}2)$$

Several methods have been devised to measure the harmonic distortion caused either by a single harmonic or by the sum of all the harmonics. Some of the better-known methods are described in the following sections.

9-3.2 Tuned-Circuit Harmonic Analyzer

One of the oldest methods of determining the harmonic content of a waveform uses a tuned circuit, as in Fig. 9-3. A series-resonant circuit, consisting of inductor L and capacitor C, is tuned to a specific harmonic frequency. This harmonic component is transformer-coupled to the input of an amplifier. The output of the amplifier is rectified and applied to a meter circuit. After a reading is obtained on the meter, the resonant circuit is returned to another harmonic frequency and the

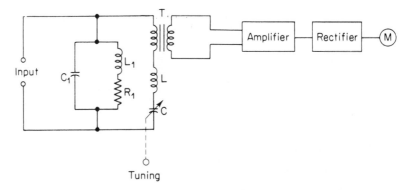

Figure 9-3 Functional block diagram of the tuned-circuit harmonic analyzer.

next reading is taken, and so on. The parallel-resonant circuit consisting of L_1, R_1, and C_1 provides compensation for the variation in the ac resistance of the series-resonant circuit and also for the variations in the amplifier gain over the frequency range of the instrument.

Although numerous modifications of this basic circuit have been developed, tuned-circuit analyzers generally have two major drawbacks: (1) At low frequencies, very large values for L and C are required and their physical size becomes rather impractical. (2) Harmonics of the signal frequency are often very close in frequency so that it becomes extremely difficult to distinguish between them. Some circuit refinements can lessen this problem and the analyzer does find useful application whenever it is important to measure each harmonic component individually rather than to take a single reading for the total harmonic distortion.

9-3.3 Heterodyne Harmonic Analyzer or Wavemeter

The difficulties of the tuned circuit are overcome in the heterodyne analyzer by using a highly selective, fixed-frequency filter. The simplified block diagram of Fig. 9-4 shows the basic functional sections of the heterodyne harmonic analyzer.

The output of a variable-frequency oscillator is mixed (*heterodyned*) successfully with each harmonic of the input signal, and either the sum or the difference frequency is made equal to the frequency of the filter. Since now each harmonic frequency is converted to a constant frequency, it is possible to use highly selective filters of the quartz-crystal type. With this technique, only the constant-frequency signal, corresponding to the particular harmonic being measured, is passed and delivered to a metering circuit. The mixer usually consists of a balanced modulator since it offers a simple means of eliminating the original frequency of the harmonic. The low harmonic distortion generated by the balanced modulator is another advantage over different types of mixers. Excellent selectivity is obtained by using quartz-crystal filters or inverse feedback filters.

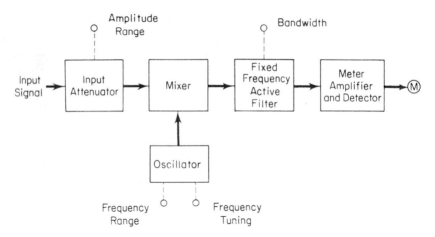

Figure 9-4 Block diagram of a heterodyne-type harmonic analyzer or wavemeter.

On some heterodyne analyzers the meter reading is calibrated directly in terms of voltage; other analyzers compare the harmonics of the impressed signal with a reference voltage, usually by making the reference voltage equal to the amplitude of the fundamental. Direct-reading instruments of the heterodyne type are sometimes known as *frequency-selective voltmeters*. In these instruments the frequency of the input signal is read off a calibrated dial. A low-pass filter in the input circuit excludes the sum of the mixed frequencies and passes only the difference frequency. This voltage is compared to the input signal and read off on a calibrated voltmeter in dBm and volts. The level range for most of these meters is from -90 dBm to $+32$ dBm.

9-3.4 Fundamental-Suppression Harmonic Distortion Analyzer

The fundamental-suppression method of measuring distortion is used when it is important to measure total harmonic distortion (THD) rather than the distortion caused by each component. In this method the input waveform is applied to a network that *suppresses* or *rejects* the fundamental frequency but passes all the harmonic-frequency components for subsequent measurement. This instrument has two major advantages: (1) The harmonic distortion generated within the instrument itself is very small and can be neglected. (2) The selectivity requirements are not severe because only the fundamental frequency component must be suppressed.

The block diagram of the fundamental suppression HD analyzer is shown in Fig. 9-5. The instrument consists of four major sections: (1) the input circuit with impedance converter, (2) the rejection amplifier, (3) the metering circuit, (4) the power supply. The impedance converter provides a low-noise, high-impedance input circuit, independent of the signal source impedance placed at the input

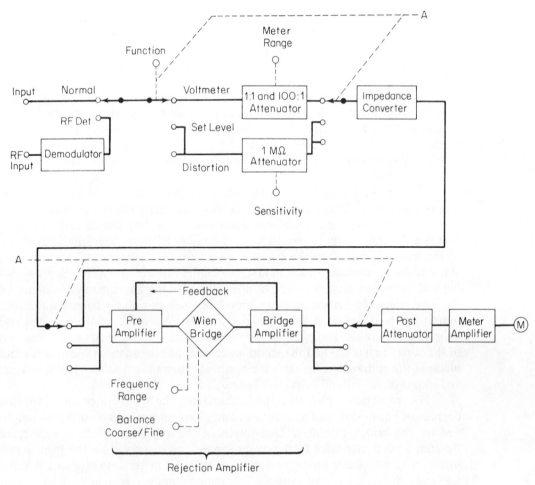

Figure 9-5 Block diagram of a fundamental-suppression distortion analyzer. (Courtesy of Hewlett-Packard Company.)

terminals to the instrument. The rejection amplifier rejects the fundamental frequency of the input signal and passes the remaining frequency components on to the metering circuit where the HD is measured. The metering circuit provides a visual indication of total HD in terms of a percentage of total input voltage.

Two modes of operation are possible: When the function switch is in the "voltmeter" position, the instrument operates as a conventional ac voltmeter, a very convenient feature. In this mode the input signal is applied to the impedance converter circuit through the 1/1 and 100/1 attenuator, which selects the appropriate meter range. The output of the impedance converter then bypasses the rejection amplifier and the signal is applied directly to the metering circuit. The voltmeter section can be used separately for general-purpose voltage and gain measurements.

When the function switch is in the "distortion" position, the rejection amplifier becomes part of the circuit and distortion measurements are made. In this mode the input signal is applied to a 1-MΩ input attenuator that provides 50-dB attenuation in 10-dB steps, controlled by a front panel switch marked *sensitivity*. When the desired attenuation is selected, the signal is fed to the impedance converter, which is a low-distortion, high input-impedance amplifier circuit whose gain is independent of the source impedance placed at the input terminals. The overall negative feedback in this amplifier results in unity gain and low distortion. Signals having a high source impedance can be measured accurately and the sensitivity selector can be used in the high-impedance positions without distorting the input signal.

The rejection amplifier circuit consists of a preamplifier, a Wien bridge, and a bridge amplifier. The preamplifier receives the signal from the impedance converter and provides additional amplification at extremely low distortion levels. The Wien bridge circuit is used as a rejection filter for the fundamental frequency of the input signal. With the function switch in the "distortion" position, the Wien bridge is connected as an interstage coupling element between the preamplifier and the bridge amplifier. The bridge is tuned to the fundamental frequency of the input signal by setting the *frequency range* selector and is balanced for zero output by the coarse and fine *balance* controls. When the bridge is tuned and balanced, the voltage and phase of the fundamental, which appears at the junction of the series reactance and the shunt reactance, are the same as the voltage and phase at the midpoint of the resistive branch. When these two voltages are equal and in phase, no output signal will appear.

For frequencies other than the fundamental, the Wien bridge offers varying degrees of phase shift and attenuation, and the resulting output voltage is amplified by the bridge amplifier. The output of the bridge amplifier is connected through a post-attenuator to the meter circuit and displayed on the front panel meter. The attenuator limits the signal level to the meter amplifier to 1 mV for full-scale deflection on all ranges. The meter amplifier is a multistage circuit designed for low drift and low noise, and with flat response characteristics. The meter is connected in a bridge-type rectifier and reads the average value of the signal impressed on the circuit. The meter scale is calibrated to the rms value of a sine wave.

The AM detector circuit allows measurement of envelope distortion in AM carriers. The input signal is applied to the demodulator, where the modulating signal is recovered from the RF carrier. The signal is then applied to the impedance converter through the 1-MΩ attenuator and is processed in the same manner as the normal distortion measurements.

The response characteristic of the Wien bridge rejection filter in Fig. 9-5 is modified by negative feedback from the bridge amplifier to the preamplifier. The result of this feedback is shown in the very sharp response curve of Fig. 9-6, causing the rejection of practically all frequency components except the fundamental.

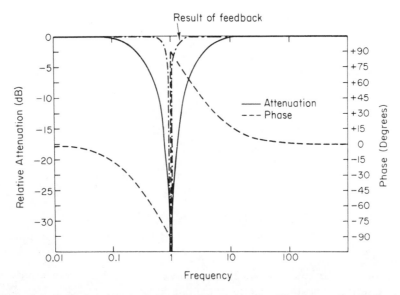

Result of feedback

Figure 9-6 Rejection characteristic of the Wien bridge, modified by negative feedback.

9-4 SPECTRUM ANALYSIS

The previously discussed wave analyzer is a simple example of a spectrum analyzer. If the wave analyzer could be swept in frequency, electronically, while an oscilloscope is used in lieu of the output meter, and if the sweep could be done at a rapid rate so that the display appeared constant, a real-time picture of the spectrum of the input signal could be observed. The wave analyzer shown in Fig. 9-1 does not lend itself to being electronically swept and, along with a host of other reasons, is not suited for this application.

Practical spectrum analyzers use the same principles as a superheterodyne receiver and can be represented by the block diagram in Fig. 9-7. There are so many variations in spectrum analyzers that it would be difficult in a text of this sort to present all the requirements of spectrum analyzer design. Therefore, a specific example will be given and described in detail. The spectrum analyzer shown in the block diagram of Fig. 9-7 is typical of a VHF spectrum analyzer covering the range from 10 kHz to 300 MHz.

The spectrum analyzer is similar to an up-converting superheterodyne receiver. The input of the spectrum analyzer is first converted to an IF higher than the highest input frequency, which in the case of the example is 400 MHz. As with every superheterodyne, the input image must be removed, which in this case represents the band of frequencies from 800 to 1,100 MHz, which can be easily removed with a low-pass filter. In addition to removing the image, the input low-pass filter should also attenuate any signals at the first IF of 400 MHz.

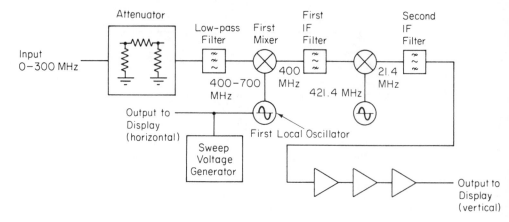

Figure 9-7 Block diagram of a general-purpose spectrum analyzer.

The example spectrum analyzer has a 1-kHz selectivity at its narrowest setting, and this selectivity cannot be achieved at 400 MHz. Therefore, the 400-MHz first IF must be heterodyned to a lower frequency. In the example spectrum analyzer, the second IF is 21.4 MHz, which allows the use of crystal filters to achieve the desired selectivity. Like the first frequency conversion, the second frequency conversion has an image and must be removed. The second local oscillator is 21.4 MHz above the first IF at 421.4 MHz, which sets the image frequency at 442.8 MHz, which must be removed by the first IF filter.

The frequency of the first local oscillator is swept electronically usually using varactor diodes in a manner similar to the sweep generator described in Chapter 8. The span of frequency that is swept is called the *dispersion* of the analyzer and represents the amount of frequency that can be displayed on the spectrum analyzer screen. The first local oscillator, which usually covers a frequency range of less than an octave, is easily turned with a varactor diode. Also, as was the case with the sweep generator, the voltage applied to the varactors must pass through a correcting circuit to cancel the nonlinearities. Unlike the sweep generator, the spectrum analyzer usually is required to sweep narrow frequency ranges, where the frequency instabilities of the first local oscillator will destroy the spectrum analyzer display.

Two types of frequency instabilities will cause difficulties when very narrow frequency ranges are displayed. The first type, called *long-term instability,* is the drift of the frequency of the first local oscillator. This will appear as the movement of the spectrum across the spectrum analyzer screen. This could be compensated by returning the spectrum analyzer to center the display. This is an annoyance, and if the frequency drift is too fast, the operator may not be able to keep the display centered.

A second type of frequency stability, called *phase noise,* is a rapid variation in frequency due to noise voltages in the tuned circuit or noise voltages picked up by the varactor circuit. Because the first local oscillator has a tuning range of several hundred megahertz, even microvolts of noise on the varactor tuning voltage can cause significant frequency modulation. It is not possible to correct for the frequency modulation due to phase noise, and an electronic means must be applied to the first local oscillator.

Phase noise is, in effect, frequency modulation, which, as is the case with any type of modulation, creates sidebands around the modulated carrier. The phase noise of a local oscillator is transferred to the input signal to be analyzed, so the resulting sidebands are evident on the signal after it is converted to the first IF, by a process called *reciprocal mixing*. Thus, if the first local oscillator of a spectrum analyzer had phase noise present, it would not be suitable for narrow spectrum analysis. To remove the phase noise from the first local oscillator, the frequency of the oscillator is phase locked to a harmonic of a crystal oscillator, as shown in Fig. 9-8. In this example a 1-MHz crystal oscillator is fed to a harmonic generator, which creates harmonics at each megahertz through the entire frequency range of the VCO. A double balanced mixer is used as a phase detector, and a loop amplifier closes the loop. Because the reference frequency, in this case literally hundreds of megahertz, is very high, the noise of the first local oscillator can be practically eliminated. Because there are several harmonics of the 1-MHz crystal that can be locked on near the unlocked frequency of the VCO, when the phase-locked circuits are switched on it may not be known which harmonic the system has locked. Systems are included in the phase-locked circuitry to locate the correct harmonic. These circuits tend to be more complex than the phase-locked loop and will not be discussed in this text.

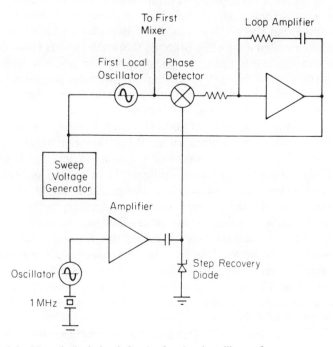

Figure 9-8 Phase-locked circuit for the first local oscillator of a spectrum analyzer.

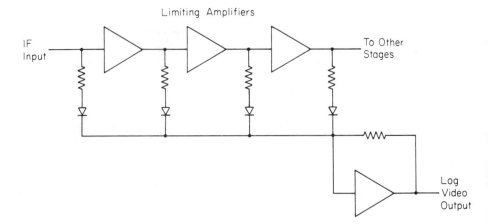

Figure 9-9 Successive-limiting type of log IF amplifier.

When the first local oscillator has been phase locked, the second local oscillator is swept to provide the necessary frequency scan. Although the second local oscillator is not stabilized and operates at a relatively high frequency, the fact that the tuning range of this oscillator is much narrower than the first local oscillator means that the noise level will be much reduced. In spectrum analyzers where even this noise level is excessive, the second local oscillator is derived from stabilized sources, and frequency scanning is added.

The majority of the spectrum analyzer gain is obtained at the second IF after the ultimate selectivity. The spectrum analyzer display is typically logarithmic, meaning that the display is in decibels, usually dBm. This requires a special IF amplifier called a log IF. Although there are various types of logarithmic amplifiers, the successive-limiting type, shown in Fig. 9-9, is most often used in spectrum analyzers. This form of logarithmic IF amplifier does not produce a perfect logarithmic relationship between the input and output, but a close piecewise-linear approximation. Each amplifier in the limiting log amplifier is a limiter with specific limiting threshold. In addition, each limiter stage has a rectifier diode and sums current to the output node. When no input signal is present, only noise is present, and none of the amplifier stages are in limiting. If a low-level signal were present, all the limiter stages would not be in limiting and each amplifier, primarily the last stage where the signal is the greatest, would add current to the output node. When the input signal is increased further, the last stage will enter limiting first. When an amplifier stage enters limiting, its contribution to the output current remains constant, and therefore the stage in limiting no longer has gain. Therefore, the slope of the input/output plot changes and becomes less every time an amplifier enters limiting. The result is shown in Fig. 9-10. The closeness of the fit to a true log transfer function depends on the number of decibels between the successive-limiting and typical integrated-circuit log amplifiers, which have about 9 or 10 dB of range.

The log display of a spectrum analyzer is typically 60 to 90 dB, which implies that the log IF amplifier would require between six and nine log amplifier inte-

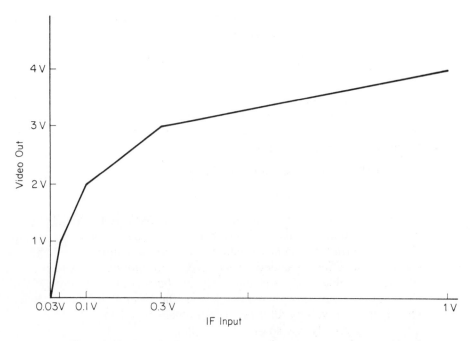

Figure 9-10 Transfer function for the log IF amplifier shown in Fig. 9-9.

grated circuits. Not only does the IF amplifier provide the required logarithmic conversion, it supplies the bulk of the spectrum analyzer gain.

The usefulness of the spectrum analyzer depends on its dynamic range. The dynamic range can be described as the range of signals between the smallest that can be detected above the system noise and the largest signal that does not cause any spurious signals greater than the smallest signal that can be seen.

It should be clear just what the smallest signal that can be seen above the system noise is, but the larger signals are limited by the generation of other signals. Typically, spurious signals are due to intermodulation. If a device is absolutely linear, any two signals, regardless of signal level (that is, assuming that they are not so huge that they destroy the device), applied to the device will result in an output of only the same two signals. If there are any nonlinearities, the two signals will interact and produce other signals at various frequencies not represented by the two original input frequencies. The fact that the device is not linear means that the two signals are not simply added, but that there is some form of interaction. To represent this mathematically, the output voltage is written as a sum of terms, where the output is dependent on all powers of the input voltage:

$$V_{\text{out}} = K_0 + K_1 V_{\text{in}} + K_2 V_{\text{in}}^2 + K_3 V_{\text{in}}^3 + \cdots \qquad (9\text{-}3)$$

In a linear device, the output is described only by the first two terms. For a linear system any input signal, regardless of how complex, appears at the output undistorted. When the transfer function of a device includes the higher-order terms, input signals are distorted and produce spurious outputs. As an example, consider a sine function applied to a device with distortion products and consider the effect of the second-order term:

$$K_2 V_{in}^2 = K_2(A \sin \omega t)^2 = \frac{K_2 A}{2} (1 - \cos 2\omega t) \qquad (9\text{-}4)$$

The distortion produced by this term on a sine-wave input is the generation of a second harmonic. In some systems this could cause a problem, but the real trouble occurs when more than one sine wave appears at the input at the same time. In this case, the second-order product yields

$$K_2 V_{in}^2 = K_2 (A \sin \omega_1 t + B \sin \omega_2 t)^2$$

$$= K_2 A^2 \sin^2 \omega_1 t + K_2 B^2 \sin^2 \omega_2 t + 2 K_2 AB \sin \omega_1 t \sin \omega_2 t \qquad (9\text{-}5)$$

There are now three terms; the first two are sine-squared, which have a frequency of twice the input signals and represent the second harmonic of each input signal. The third term's frequency contains the sum and difference of the two input signals. This effect is called *second-order intermodulation* and is not usually a significant problem of spectrum analyzers because the frequency of the cross-modulation is far enough removed from the desired frequency that it can be effectively removed by filtering. The problem becomes significant when one frequency is relatively low so that the resultant cross-modulation is near the desired frequency.

Investigating the effects of the third-order term, if a single input signal were present, the third-order term would introduce the third harmonic of the input signal. However, when two signals are present the result is

$$V_{out} = K_3(A \sin \omega_1 t + B \sin \omega_2 t)^3 = K_3(A^3 \sin^3 \omega_1 t + B^3 \sin^3 \omega_2 t$$

$$+ 3A^2 B \sin^2 \omega_1 t \sin \omega_2 t + 3AB^2 \sin \omega_1 t \sin^2 \omega_2 t) \qquad (9\text{-}6)$$

The contribution of the third-order term is not only the third harmonic of each input frequency but two new frequencies, twice one input plus or minus the other. This distortion, called *third-order intermodulation distortion,* is troublesome, because when two signals are close in frequency the distortion products are also close, which means that in many applications it will not be possible to filter the spurious signals.

There are higher orders of intermodulation, such as fifth order, that occur at frequencies of twice one frequency plus or minus three times the other. Third-order intermodulation is usually much stronger than any other order of intermodulation and is the prime limiting factor in the dynamic range of a spectrum analyzer.

The dynamic range of the spectrum analyzer can be determined in the following manner. An input is applied to the spectrum analyzer, and the level of the signal is adjusted so that it is only 3 dB above the internal noise generated by the analyzer. A second signal is applied along with the original signal, and the levels of the two input signals are kept the same while the input level of both is increased until the third-order intermodulation, as shown in Fig. 9-11, is 3 dB above the internally generated noise. The difference between the original signal level 3 dB

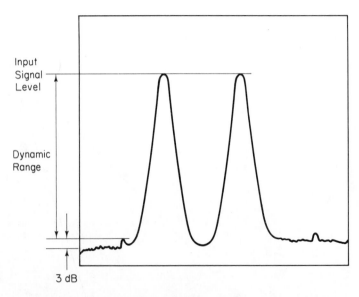

Figure 9-11 Third-order intermodulation products as they appear on a spectrum analyzer display with two input signals.

from the noise and the level of the two large signals when they generated the same level of spurious signal is the dynamic range of the spectrum analyzer.

Another method of stating the intermodulation performance of the spectrum analyzer is the concept of third-order intermodulation intercept point. If two signals at the same level are applied to a spectrum analyzer, or any other electronic device, a third-order intermodulation will be generated. The intermodulation could be generated anywhere in the device; however, the level of the intermodulation product is referenced to the input. That is, the intermodulation product at the output of the device is equivalent to an input signal of a certain level. If the levels of the two input signals are increased, such that the signals remain at the same level, the third-order intermodulation product will increase three times the decibel increase of the two input signals, indicating that the relationship involves the third order. This means that the third-order intermodulation increases at a much more rapid rate than the increase in the two input signals, which implies that eventually the intermodulation product will have the same level as the two input signals, as shown in Fig. 9-12. This level, where the input signals and the spurious third-order intermodulation are the same, is called the *third-order intercept point* and is an indication of the upper limit of the dynamic range of the spectrum analyzer. The third-order intercept point is strictly a theoretical point. Seldom does an electronic device have the capability of ever achieving operation at the third-order intercept point, and certainly the unit would be practically useless at that level of distortion.

The third-order intercept point can be used to calculate the level of the third-order products. The level of any third-order product can be determined from the

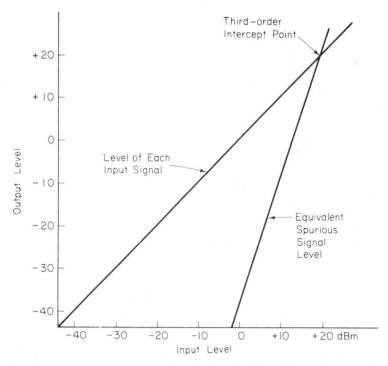

Figure 9-12 Third-order products as a function of the level of two input signals.

relationship

$$P_3 = 3P_{in} - 2I_p \qquad (9\text{-}7)$$

where P_3 = level of the third-order product (dBm)

I_p = power level of the third-order intercept (dBm)

P_{in} = power of the two input signals.

To determine the dynamic range of the spectrum analyzer, the third-order inter-modulation products must be the same as the minimum signal that can be seen by the analyzer, that is, the signal that is just visible above the noise level. For the sake of simplicity, assume that the intermodulation is equal to the noise level. Therefore, the equation can set the third-order intermodulation products equal to the minimum detectable signal:

$$P_3 = 3P_{in} - 2I_p = MDS \qquad (9\text{-}8)$$

where MDS is the minimum detectable signal, essentially the spectrum analyzer noise level, in dBm.

Rewriting Eq. (9-8),

$$3(P_{in} - MDS) = 2(I_p - MDS) \qquad (9\text{-}9)$$

Signal Analysis   Chap. 9

The dynamic range of the spectrum analyzer is the difference in level between the minimum detectable signal and the input that produces a spurious signal equal to the MDS, or

$$P_{\text{in}} - \text{MDS} = \tfrac{2}{3}(I_p - \text{MDS}) \tag{9-10}$$

EXAMPLE 9-1

What is the dynamic range of a spectrum analyzer with a third-order intercept point of $+25$ dBm and a noise level of -85 dBm?

SOLUTION Using the formula

$$\text{dynamic range} = \tfrac{2}{3}(I_p - \text{MDS}) = \tfrac{2}{3}[25 - (-85)] = 73$$

and substituting the given data, the dynamic range is found to be 73 dB.

The minimum detectable signal or the noise level of the spectrum analyzer is determined by two characteristics, the bandwidth of the IF filter in use and the noise figure of the analyzer. The noise figure of the analyzer is set by the design of the front end of the unit, while the IF filter bandwidth is a parameter from a later stage of the analyzer. The noise level of the spectrum analyzer can be related to the noise figure and the IF bandwidth by the following:

$$\text{MDS} = -114 \text{ dBm} + 10 \log (\text{BW}/1 \text{ MHz}) + \text{NF} \tag{9-11}$$

where BW is the 3-dB bandwidth in megahertz of the IF filter, and NF is the noise figure in decibels.

EXAMPLE 9-2

What is the minimum detectable signal of a spectrum analyzer with a noise figure of 20 dB and using a 1-kHz, 3-dB filter?

SOLUTION

$$-114 \text{ dBm} + 10 \log 1 \text{ kHz}/1 \text{ MHz} + 20 = -124 \text{ dBm}$$

The ability of the spectrum analyzer to separate signals is a function of the second IF bandwidth. To resolve two signals that are close in frequency, a narrow IF filter is required. In addition, signals that are close in frequency and are at two different amplitudes are even more difficult to resolve. Consider, as an example, two signals that are the same amplitude but are separated by 10 kHz. These signals could be resolved using an IF filter with a 3-dB bandwidth of 10 kHz, as shown in Fig. 9-13. The dip in the spectrum display is only 3 dB but it is clearly visible. On the other hand, if the two signals were separated not only by 10 kHz but by 10 dB, they would not be resolvable with the 3-dB, 10-kHz filter. The resolution of a spectrum analyzer is defined as the 6-dB bandwidth of the second IF filter.

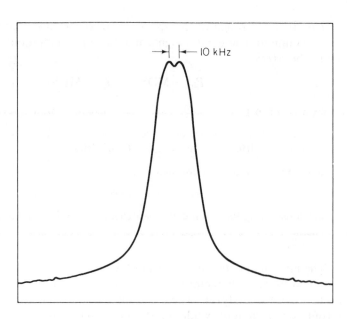

Figure 9-13 Two signals 10 kHz apart as displayed with an IF filter bandwidth of 10 kHz.

It may appear that a filter with sharper skirts would solve the problem of resolving closely situated signals. The ratio of the 6-dB point of a filter to the 60-dB point is an indication of the steepness of the skirts of the filter. Thus, it would appear that a filter with a lower shape factor would resolve signals close in frequency, and to a certain extent it would. However, there are significant disadvantages to sharp-skirted filters in spectrum analyzers.

The reader should be familiar with the nature of distortions introduced when a modulated signal is passed through a filter with a bandwidth too narrow to pass the entire modulation bandwidth. Not only will the high-frequency components be reduced, but the high-Q circuits of the filter will introduce ringing. Even though the signal being analyzed with the spectrum analyzer may not be modulated, the fact that the signal sweeps through the filter center frequency causes ringing in the filter. Relative to the second IF filter, the CW input signal of the spectrum analyzer is modulated as a function of the sweep speed of the first local oscillator. If the sweep speed, that is, the megahertz per second rate of the first local oscillator, is too great, the amplitude of the signal out of the second IF filter will be reduced and possibly distorted. Sharp-skirted filters cause the most distortion, and a special filter shape called Gaussian causes the least amount of distortion. The maximum permissible sweep rate as a function of the Gaussian filter bandwidth is given by the following equation:

$$\text{maximum sweep rate} = 2.3 \ (\text{bandwidth})^2 \quad \text{Hz/s} \qquad (9\text{-}12)$$

As can be seen, the maximum sweep rate for a narrow-bandwidth filter can be quite slow, and usually a spectrum analyzer is equipped with a storage display.

9-4.1 Spectrum Analyzers for Higher Frequencies

Spectrum analysis at frequencies higher than about 100 MHz is a very important tool for the development of circuits and systems at these higher frequencies. With the exception of a few higher-frequency oscilloscopes, there are no tools for the analysis of signals at frequencies above a few hundred megahertz. Most signal analysis is done with the oscilloscope for lower frequencies, such as the determination of amplitude, phase, and distortion. The spectrum analyzer provides a sensitive instrument to investigate these parameters at higher frequencies.

The frequency of the VCO for a spectrum analyzer is required to extend from a frequency higher than the highest input frequency to a frequency at least twice the highest input frequency. For spectrum analyzers operating above 1,000 MHz, this implies an oscillator from at least 1,000 to 2,000 MHz and, in practical designs, more on the order of 2,500 to 3,500 MHz. This frequency range usually requires an oscillator with a tuned circuit other than the typical coil and capacitor found in lower-frequency oscillators.

An oscillator circuit suitable for this frequency range is the *YIG-tuned oscillator*. YIG, yttrium iron garnet, is a ferromagnetic material that has some very useful properties at microwave frequencies. YIG, like many ferromagnetics, has the property that the molecules of the garnet have magnetic moments that normally are randomly aligned. The magnetic moments can be aligned in one direction by the application of a static magnetic field. The application of an alternating magnetic field will cause the magnetic moments to precess much like a toy top. The precession frequency is a function of the type and size of the YIG material and the strength of the applied static magnetic field. The greatest amplitude of the precession occurs when the applied alternating field is equal to the precession frequency of the YIG crystal. Therefore, this resonance can be used to create oscillators and filters. The resonance frequency is typically in the gigahertz region, and the Q of a YIG resonator can be quite high.

YIG resonators are typically made from highly polished spheres of YIG about 0.25 mm in diameter. The sphere is placed in a static magnetic field of field intensity H, as shown in Fig. 9-14. A pickup coil is arranged at right angles to the static magnetic field and is used as the method of coupling energy into and out of the YIG sphere. In some application a second coupling coil would be added, which is orthogonal to both the static field and the other coupling coil.

The equivalent circuit of the YIG resonator is essentially a parallel-tuned circuit with a small fixed series inductance. The resonant frequency of the parallel circuit can be electronically tuned by varying the current through the magnetic field coils. Unlike the typical electronically tuned oscillator using a varactor, where the resonant frequency is varied by changing only the capacitor of the tuned circuit, the YIG resonator tunes both the equivalent capacitance and inductance. This allows for a more constant impedance of the resonant circuit in an oscillator and also allows for a tuning range of several octaves, rather than just the two that are typical of a varactor-tuned oscillator.

The YIG resonant circuit can be used in an oscillator as the frequency-determining element, as shown in Fig. 9-15. In this example the resonant circuit is

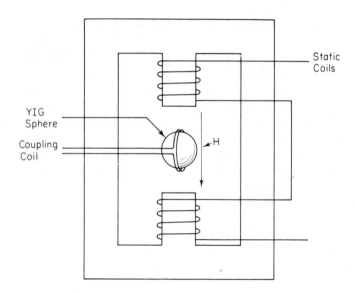

Figure 9-14 YIG sphere and the associated coupling coil and static magnetic field.

placed in the emitter, while positive feedback is introduced by the choke in the base lead.

The frequency of this circuit is controlled electronically by varying the current through the static magnetic field coils much in the same fashion as the voltage across a varactor diode would be used to tune a conventional oscillator. There are some significant differences between the YIG-tuned oscillator and the varactor-tuned oscillator. First, the ratio of the maximum-to-minimum frequency can be much greater than 2, which is the recommended limit for varactor-tuned oscillators. Second, the high Q of the YIG oscillator brings improved phase noise performance for spectrum analyzers and sweep generators.

The frequency range of the spectrum analyzer can be extended without resorting to a higher-frequency local oscillator by a technique called *harmonic mixing*. A mixer will convert an input signal to an IF by taking the sum or

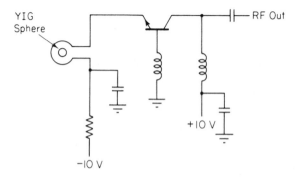

Figure 9-15 Oscillator circuit using a YIG resonator.

Signal Analysis   Chap. 9

difference between the local oscillator and the input signal. Many mixers will also convert an input signal with harmonics of the local oscillator.

An example of a simple harmonic mixer is shown in Fig. 9-16. In this example a single diode is used to mix an input RF signal with the third harmonic of the local oscillator. If the level of the local oscillator is sufficiently high, the diode can be thought of as a switch being switched at the rate of the local oscillator. Mixing is essentially multiplying two signals together, and the switch action of the diode can be thought of as multiplying a square wave with an amplitude of 1 with the input waveform. Because a square wave is made of the summation of the fundamental and all the odd harmonics of the base frequency, it would be expected that the simple diode mixer would not only mix the RF input with the local oscillator but with all the odd harmonics as well. In a practical circuit, because the duty cycle is not an exact 50 per cent, more than just the odd harmonics are present, and the example diode mixer will mix the input RF signal with all harmonics of the local oscillator.

In the spectrum analyzer described previously, the possibility of generating any spurious inputs with the harmonics of the local oscillator was eliminated by the input low-pass filter. If this low-pass filter were eliminated, or if a band-pass filter were placed at the input of the spectrum analyzer, certain harmonics of the local oscillator could be used to extend the range of the spectrum analyzer. As an example of how this might work, the previous spectrum analyzer example will be used.

The local oscillator covers from 400 to 700 MHz and the first IF is 400 MHz. If the second harmonic of the local oscillator were used, 800 to 1,400 MHz, the second harmonic minus the first IF would give an input range from 400 to 1,000 MHz, while the sum would yield 1,200 to 1,800 MHz.

The third harmonic of the local oscillator, 1,200 to 2,100 MHz, would allow the conversion of 800 to 1,700 MHz and 1,600 to 2,500 MHz for difference and sum, respectively.

Other harmonics can be used for extending the frequency range further.

One other method of gaining yet another frequency range from the same mixer and local oscillator is to simply use the sum of the first IF and the local oscillator frequency that covers the frequency range of 800 to 1,100 MHz.

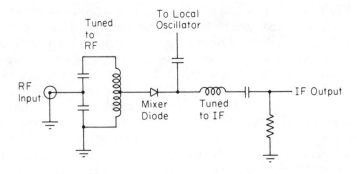

Figure 9-16   Simple series diode mixer capable of mixing by harmonics.

It should be noticed that, although the range of the spectrum analyzer can be extended by this technique, there is a range of frequencies in this example that is not covered and that is from 300 to 400 MHz. When complete frequency coverage is required, the second IF is often used in lieu of the first IF to provide the required frequency coverage.

When harmonic mixing is used, several corrections are required to the spectrum analyzer display. First, when the harmonic mix is used, the center frequency dial of the analyzer must have the correct frequency calibrations. This is usually handled by having a mechanical dial arrangement that simply displays the correct frequencies. Electronic dials can manipulate the numbers electronically. Second, because a harmonic of the local oscillator is used, the rate of change of frequency, relative to the Nth harmonic, per volt is N times that at the fundamental, so therefore the spectrum analyzer display must be corrected for this. This is corrected in a simple fashion; if the Nth harmonic is being used, the local oscillator tuning voltage is simply divided by N. Finally, the mixing efficiency of the mixer at harmonics, especially the higher-order harmonics, is less than at the fundamental. Therefore, the display will have to be corrected for this loss of signal. This is accomplished by simply offsetting the display by the number of decibel difference between the fundamental mixer loss and the harmonic mixer loss. A block diagram of a spectrum analyzer with harmonic mixing is shown in Fig. 9-17 with all the required correcting circuits.

The chief problem in using the harmonic mixing spectrum analyzer is that the input low-pass filter is removed and all the possible harmonic mixing ranges are present at the input of the spectrum analyzer. Therefore, there is considerable ambiguity in the display as some signals can appear at more than one point on the display. Various signal-identifying techniques can discern between the correct and incorrect signals, but the best technique is to place an external band-pass filter between the system being tested and the spectrum analyzer, which will eliminate many of the spurious signals. An example of a spectrum analyzer is shown in Fig. 9-18.

9-4.2 Fourier Transform Spectrum Analyzers

The discussion of spectral analysis to this point involved manipulation of the signal to be analyzed by bandpass filtering, notch filtering, frequency translation, and various combinations of these techniques. There are mathematical methods of calculating the spectrum of a signal if the signal were reduced to a mathematical equation or a set of data points. The most direct mathematical method is called the Fourier transform. A signal that can be represented as an equation, a graph, or a set of data points where the independent variable is time can be transformed into another equation, graph, or set of data points where the variable is frequency. The transformation produces the spectrum of the waveform. If a signal is transformed into a mathematical set of data points by digitizing an analog signal, a digital computer could be programmed with a form of Fourier transform that would calculate the spectrum of the waveform. The method used to calculate the

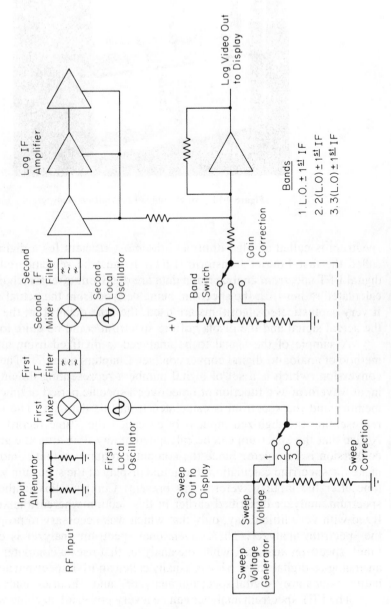

Figure 9-17 Harmonic mixing spectrum analyzer showing all the corresponding circuits.

Figure 9-18   An example of a spectrum analyzer.

spectrum is called an algorithm, and the most efficient for a digital computer is called the fast Fourier transform (FFT). It must be remembered that with the digital FFT spectrum analyzer, the data are digitized, after which the spectrum is calculated rather than the spectrum being derived from the actual signal present. If very sophisticated algorithms are used, the time delay from the occurrence of the actual signal and the display of the spectrum could be quite long.

A sample of the signal to be analyzed is digitized using any appropriate method of analog-to-digital conversion (see Chapters 6 and 7). The results of this conversion, which is a set of digital numbers representing the amplitude of the input waveform as a function of time, over a specific period of time, is stored in a memory and the spectrum is calculated from this data set. The set of numbers representing the digitized input is often called the "time record" of the input. Notice that the spectrum can be calculated at any time after the analog-to-digital conversion is complete. Since the computer requires a finite amount of time to make the spectrum calculation, the actual display of the spectrum will occur some time after the input waveform was present. Compare this to the conventional spectrum analyzer described earlier in this section wherein the signal was analyzed with very little delay, only that which was necessary to propagate through the spectrum analyzer. The conventional spectrum analyzer is called a "real-time" spectrum analyzer, while the analyzer that uses a computer algorithm and an analog-to-digital conversion is usually called an FFT spectrum analyzer. Alternative names are "digital spectrum analyzer" and "Fourier analyzer."

The FFT spectrum analyzer can be a very powerful machine without becoming a very expensive unit, as the power of the analyzer comes from the computer algorithms, which can be enhanced without adding large amounts of hardware to the analyzer. Of course, the analyzer is no better than its analog-to-digital converters or the size of the memory. Another advantage that is exploited to enhance the power of the FFT analyzer is the fact that the input signal is captured and "frozen" in time. Thus long and complex mathematical operations may be performed on the input signal.

Despite the fact that the FFT analyzer is potentially a very powerful unit, there are some inherent limitations to the technique. First, the fast Fourier transform is not a true continuous transform but produces a transform with a finite resolution. This means that the spectrum can only be found at specific intervals. The nature of the spectrum can only be inferred between the intervals. Generally, the nature of the waveform being analyzed allows a simple interpolation between the discrete spectral lines. However, for some waveforms this assumption will produce an erroneous result.

The FFT analyzer samples the input signal for a specific period of time and this is called the window. The signal to be analyzed is considered to be a periodic signal where the digitized signal within the window is repeated indefinitely.

To gain an insight into how this affects the FFT spectrum analyzer, assume that a signal is sampled and digitized for a period of 1 s. The spectrum analyzer has only 1 s of data to arrive at a spectrum calculation. If the signal were a very slowly changing one, the 1 s of data would not contain as much information about the signal changes as necessary for an accurate spectrum calculation. However, if the signal were a rapidly changing signal, the 1-s sample would provide lots of data, covering many cycles of a periodic waveform, some possibly redundant, to calculate an accurate spectrum. The more data that are available, the more than can be calculated about the spectrum. A slow-changing signal is one where the signal can be described with only low-frequency components. Rapidly changing signals require high-frequency spectral components to describe them. If 1 s of data were obtained, a spectral calculation could be generated with spectral information or resolution of 1 Hz. The FFT spectrum analyzer calculates the spectrum as if the sampled data within the window repeated indefinitely. This is because nothing is known about the input signal beyond the sample window, and this assumption is necessary. If the window time is chosen carefully so that sufficient data are available, an accurate spectral calculation may be made.

The narrowest possible resolution of a sampled signal is

$$f_r = \frac{1}{T}$$

where f_r is the resolution frequency and T is the sampling window time.

The number of data points within the window has an effect on the quality of the calculated spectrum. The Nyquist sampling theorem states that the highest-frequency component of a complex signal that can be accurately sampled is one-half the sampling rate. Relating this to our example, if 1 s of data were obtained at a 1-kHz sampling rate, the calculated spectrum would have a range of 1 to 500 Hz with a spectral display point every 1 Hz. Therefore, 500 points would be displayed.

The frequency range of the input to the FFT spectrum analyzer must be restricted to no more than one-half of the sampling frequency to prevent the generation of spurious spectral components called aliases. This requirement is similar to any sampled system except that the aliases are quite visible in the spectrum display.

The resolution of the analog-to-digital conversion will affect the quality of the spectrum calculation. Clearly, the finer the resolution of the digital conversion of the data, the more accurate the calculated spectrum display will be. Roughly, the ratio of the largest increment to the smallest increment that can be resolved by the analog-to-digital conversion is called the dynamic range and is usually expressed as a decibel number. This can be represented as

$$R_d = 20 \log 2^N$$

where N is the number of bits in the digitization.

The dynamic range represents the difference in level between the greatest signal that can be measured without overload and the smallest signal that can be displayed together with the larger signal. This is essentially the same as the definition of dynamic range applied to the real-time spectrum analyzer. Remember that in the real-time spectrum analyzer, the overload condition was represented by the generation of intermodulation products that appeared on the spectrum analyzer display. The lower level of the dynamic range was limited by the noise level of the analyzer. There is a type of noise associated with analog-to-digital conversion called quantizing noise, which was explained in previous chapters, and this is the limiting factor for small signals in the FFT spectrum analyzer. Therefore, both types of analyzers are limited on the high end by overload and on the lower end by noise.

EXAMPLE 9-3

What resolution, total frequency display, and dynamic range would be available from an input signal that was sampled for 4 s at a sampling rate of 20 kHz using a 10-bit conversion?

SOLUTION The resolution of the spectral calculation is the reciprocal of the sample window and is

$$f_r = \frac{1}{T} = \frac{1}{4} = 0.25 \text{ Hz}$$

The maximum calculated spectral frequency is one-half of the sample frequency and is

$$f_h = \frac{f_s}{2} = \frac{20 \text{ kHz}}{2} = 10 \text{ kHz}$$

The dynamic range is

$$R_d = 20 \log 2^N = 20 \log 1{,}024 = 60 \text{ dB}$$

To gain an insight into the number of samples and the amount of computer data involved in an FFT spectrum calculation, determine the number of samples and bits required for Example 9-3. Four seconds of data sampled at a 20-kHz rate would result in 80,000 data words. Since each data word is a 10-bit analog-to-digital conversion, 800,000 bits of computer data is involved in the input data set.

Because the FFT spectrum analyzer samples a fixed amount of time of the desired signal to be analyzed, the resultant spectrum determination represents a spectrum of a periodic function, where the sample is repeated infinitely. The sample represents a window and the data are considered as a periodic function where the data in the window are repeated. Therefore, the spectrum display is made up of lines that are separated by $1/T$ hertz, where T is the window duration. The shape of the window will affect the spectrum to a degree depending on the type of waveform that is being analyzed. The simplest type of window is represented by an on/off switch. The switch is activated, the signal is digitized, and the switch is closed. As a repeated waveform this sampling can contain sudden discontinuities when the switch is turned on and off. This type of window produces the most distortion. In many cases, however, this simple window does not produce any significant problems. There are some waveforms that the on/off window, sometimes called a uniform window or rectangular window, produces unacceptable degradation of the calculated spectrum.

The solution to the windowing problem is not to open the window suddenly, but gradually. Instead of using an on/off switch, this is accomplished by using a variable attenuator, which is more like opening a valve and admitting the signal to be digitized. This will reduce the sharp transitions obtained with a simple on/off gating of the input signal.

Even how the valve is opened has an effect on the distortion produced by sampling. Various mathematical functions can be used to control the opening of the valve and minimize the distortions of certain waveforms.

Figure 9-19 shows some popular windows and their mathematical equations.

The rectangular window or the uniform window is generally used for transients. For sine waves and periodic functions without a lot of harmonics, the Hamming window may be used. There are, however, several subtle problems of the Hamming window, and a "flattop" or Hann window is used when inaccuracies introduced by the Hamming window are unacceptable. Most FFT spectrum analyzers have several selectable windows.

FFT spectrum analysis is restricted primarily to low-frequency analyzers because of the limitations of the speed of analog-to-digital converters. Chapter 7 discussed fast A/D conversion for use in digitizing oscilloscopes, but these converters were limited to about 8 bits. To obtain a reasonable dynamic range, 10 or more bits are desired for the digital conversion. This hampers the speed of the analog-to-digital conversion and rules out the use of some of the more rapid converters such as the flash converter. Typically, FFT analyzers are limited to frequencies below 500 kHz.

Figure 9-20 shows the block diagram of an FFT type of spectrum analyzer. At the input of the spectrum analyzer is a low-pass filter which prevents aliasing. In many cases this is an automatically selected filter with a cutoff frequency determined by the spectrum analyzer parameter settings. An attenuator follows the low-pass filter, which sets the level of the signal fed to the analog-to-digital

$$\text{Rectangular Window} = \begin{cases} 1 & \text{for } |n| \le \dfrac{N-1}{2} \\ 0 & \text{elsewhere} \end{cases}$$

$$\text{Hann Window} = \begin{cases} 0.5 + 0.5 \cos \dfrac{2\pi n}{N-1}; & |n| \le \dfrac{N-1}{2} \\ 0 & \text{elsewhere} \end{cases}$$

$$\text{Hamming Window} = \begin{cases} 0.54 + 0.46 \cos \dfrac{2\pi n}{N-1}; & |n| \le \dfrac{N-1}{2} \\ 0 & \text{elsewhere} \end{cases}$$

n = Sample Number

N = Total Samples

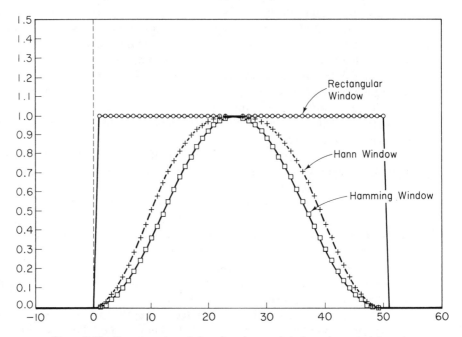

Figure 9-19 Some popular window functions and their mathematical formulae.

converter to prevent overload of the converter. Maximizing the dynamic range of a spectrum analyzer is so important that many instruments automatically set the attenuator to the optimum value. This is done by monitoring the A/D converter output with a computer and adjusting the attenuator to allow the greatest input signal without overload. The analog-to-digital converter immediately follows the low-pass filter. The converted data words are stored in the computer memory for calculation. The sample rate, the window time, and the starting time are determined by the setting of the front panel controls and the microprocessor.

Once all the samples have been digitized, the FFT calculation will begin.

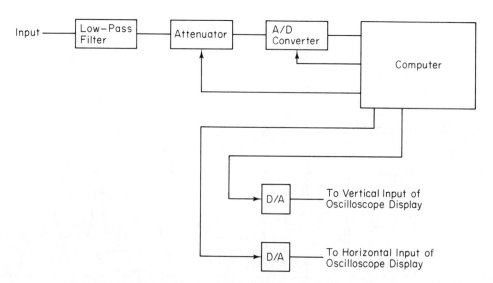

Figure 9-20 Block diagram of a fast Fourier Transform spectrum analyzer.

The spectral components are calculated and the values are stored in the computer memory. The spectral samples are retrieved from the computer memory, fed to a digital-to-analog converter, and displayed on a CRT.

The nature of the display, the frequency range, the resolution, the input amplitude levels, and so on, are set by the algorithm used by the computer. The sample windows can be modified for improved spectral display. Averaging techniques can be used to increase the signal-to-noise performance of the analyzer. Unlike the real-time spectrum analyzer, all this is accomplished without additional filters, phase-locked circuits, or complicated electrical switching requirements.

Since the analyzer contains a computer to perform the FFT algorithm, this computer can be used to perform other mathematical operations on the spectral display. As an example, the results of several spectral displays can be averaged to improve noisy displays. More sophisticated "averages" such as root mean square can be employed to reduce the effects of a noisy signal. All the averaging routines require additional time to obtain data and to make the FFT calculation and the average calculation. The result, however, is a marked improvement in the spectral determination. The highest levels of performance of an FFT spectrum analyzer are achieved using these statistical methods.

9-4.3 Applications of the Spectrum Analyzer

The spectrum analyzer such as the unit shown in Fig. 9-18 is a powerful tool and has many applications. To those who have never used the instrument, these applications may not be readily apparent. To illustrate some of the applications, the following signals, their descriptions, and the spectrum analyzer display, as shown in Fig. 9-21, will be presented.

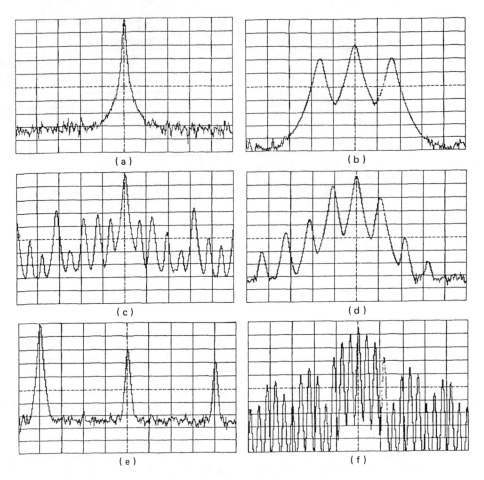

Figure 9-21 Applications of the spectrum analyzer.

(a) *Pure sinusoid with no modulation or harmonic distortion.* This signal is characterized by a single spectral line regardless of what the dispersion of the spectrum analyzer is or the IF filter bandwidth.

(b) *Amplitude modulation.* When a carrier is modulated with amplitude modulation, two sidebands are generated, one above the carrier frequency and a second below the carrier frequency. The separation in frequency between the carrier and the sidebands is equal to the modulation frequency. The power contained in the sidebands is dependent on the percentage of modulation. One hundred per cent modulation produces sidebands that are 6 dB below the carrier. The amplitude of the carrier, on the other hand, does not change, regardless of the percentage of modulation.

(c) *Frequency modulation.* Frequency modulating a carrier produces sidebands that are centered around the carrier as in the case of amplitude modulation, except that more than one sideband is generated. The number of sidebands and the amplitude of those sidebands is described by complex formulas based on the Bessel functions. The sidebands are all multiples of the modulating frequency, and the amplitude of the carrier is affected by the amount of modulation supplied. The precise amount of frequency modulation can be determined if the modulation is adjusted so that the amplitude of the carrier or other sidebands goes to zero.

(d) *Asymmetrical spectra.* The generation of a spectrum that is not symmetrical about the carrier is usually an indication that both frequency and amplitude modulation are occurring simultaneously. This could occur in an FM system where the passband of an amplifier is not flat and the frequency modulation is introducing amplitude modulation. Likewise, amplitude modulation applied to a carrier that also causes frequency instabilities, which is a common problem with phase-locked loops, will cause a similar spectrum.

(e) *Harmonic distortion.* Harmonics appear as additional signals in the spectrum analyzer display at multiples of the carrier frequency. It is often required that the harmonic content of a signal be kept low, on the order of 60 or more dB below the carrier. As an example, this may be required so that a transmitter operating at an assigned frequency will not interfere with other radio services at twice the assigned frequency that may be located near the transmitter.

(f) *Pulse modulation.* Examining pulse modulation was the first application of the spectrum analyzer. Determining the pulse modulation of radar transmitters was a difficult task in the early development of radar, and the spectrum analyzer was used to evaluate the quality of the pulse modulation. The spectrum of a rectangular amplitude pulse is shown in Fig. 9-21(f). The structure of the sidebands shows the rise- and fall-times of the pulse modulation, and the symmetry indicates the presence or absence of frequency modulation, which is a problem with modulated oscillators such as those used with high-power radar transmitters.

REFERENCES

9-1. Engleson, Morris, and Tewlewski, Fred, *Spectrum Analyzer Theory and Applications.* Dedham, Mass.: Artech House, Inc., 1974.

9-2. *The Fundamentals of Signal Analysis,* Application note AN-243, Hewlett-Packard Company, Palo Alto, Calif.

9-3. Hayward, W. H., *Introduction to Radio Frequency Design,* chap. 6. Englewood Cliffs, N.J.: Prentice-Hall, Inc., 1982.

9-4. Krauss, Herbert L., Bostian, Charles W., and Raab, Frederick H., *Solid State Radio Engineering,* chaps. 2 and 7. New York: John Wiley & Sons, Inc., 1980.

PROBLEMS

9-1. What is the dynamic range of a spectrum analyzer if the noise level of the display is equal to −80 dBm and two −10-dBm signals produce third-order intermodulation products that just appear above the noise?

9-2. What is the resolution of a spectrum analyzer using an IF filter with a 3-dB bandwidth of 30 kHz?

9-3. What is the maximum sweep rate in kilohertz per second that could be used with a spectrum analyzer without introducing distortion with a 3-kHz Gaussian filter?

9-4. Single sideband is amplitude modulation with only one sideband and no carrier. What would this modulation look like displayed on a spectrum analyzer?

9-5. What would be the third-order intermodulation products relative to the input of a device if two input signals of −10 dBm were applied to a device with a third-order intercept of +15 dBm?

9-6. What is the dynamic range of a spectrum analyzer with a 30-kHz, 3-dB bandwidth, a noise figure of 15 dB, and a third-order intercept of +25 dBm?

9-7. How does placing a fixed attenuator ahead of a spectrum analyzer affect **(a)** the third-order intercept; **(b)** the dynamic range; **(c)** the noise figure?

9-8. What frequency ranges could be covered with a spectrum analyzer having a first IF of 2,050 MHz and an input of 0 to 1,000 MHz using harmonic mixing up to the third harmonic?

9-9. What is the maximum frequency and resolution for an analyzer using a 1.5-s window and a 100-kHz sample rate?

9-10. Compare the time to sweep from 0 Hz to 100 kHz without scan loss using a 100-Hz filter and the minimum time required to sample the same frequency range using an FFT analyzer.

10

Frequency Counters and Time-Interval Measurements

10-1 SIMPLE FREQUENCY COUNTER

Standards of time and frequency (time and frequency being essentially the same standard) are unique in that they may be transmitted by radio from one location to another without the actual movement of the standard. Therefore, it is possible to have traceability to the primary standard without difficulty. Additionally, the primary standard is related to the structure of matter, and primary standards can be easily duplicated throughout the world to allow high-accuracy measurements anywhere. Because of the relative ease with which frequency and time can be measured to great accuracies, electronics systems have developed around this capability. Consider, as an example, the tolerance expected of radio-transmission equipment. The spectrum required by a voice-modulated two-way radio transmitter using frequency modulation is on the order of 15 kHz. This implies that if the frequency of the transmitter carrier could be held to absolute precision a communications channel could be assigned every 15 kHz and make the most efficient use of the radio spectrum. Because accurate measurement techniques are available and standards can be made available, the communications channels are assigned every 20 kHz in the UHF (450 MHz) band. This requires a carrier frequency accuracy and stability of only 5 kHz, which is approximately 0.001 per cent, which is easily achieved with modern frequency control and measurement techniques.

Although relatively stable frequency standards have been available for many years, precise frequency measurement has not always been an easy measurement task. Early frequency measurement required precision standards, frequency comparators and interpolation oscillators, as well as a lot of operator skill. This came to an abrupt end with the introduction of digital logic and the development of the frequency counter.

Figure 10-1 shows the block diagram of a simple frequency counter. Although referred to as "simple," this basic counter is capable of great precision if the parts are constructed properly. The frequency counter operates on the principle of gating the input frequency into the counter for a predetermined time. As an example, if an unknown frequency were gated into the counter for an exact 1 second (s), the number of counts allowed into the counter would be precisely the frequency of the input. The term *gated* stems from the fact that an AND or an OR gate is used to allow the unknown input into the counter to be accumulated. Figure 10-2 shows the waveforms associated with this action. This example shows an AND gate; however, an OR gate could be used in a similar circuit. A positive-going pulse having a period of exactly 1 s is applied to one input of the AND gate. As long as the 1-s pulse is a logic 1, the output of the AND gate is the same as the unknown input. When the 1-s pulse returns to logic 0, the output of the AND gate is zero. Thus, exactly 1 s of unknown input pulses is allowed at the output of the AND gate. It is necessary to count these pulses and display the result.

If the gate is open for exactly 1 s, the count accumulated is equal to the average frequency of the unknown input in hertz (Hz). If, as an example, the gate was open for 10 s, the accumulated count would be the average frequency in 0.1 Hz. Likewise, if the gate were open for 0.1 s, the count would be the average

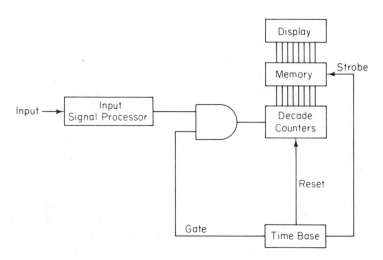

Figure 10-1 Basic block diagram of a frequency counter.

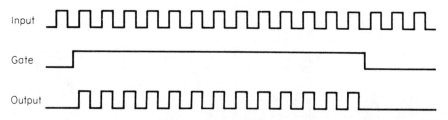

Input

Gate

Output

Figure 10-2 Waveforms associated with the gating function of a frequency counter.

frequency in tens of hertz. When a frequency counter has more than one gate time interval available, the decimal point of the display is switched with the gate time selector switch to correct the frequency display.

10-1.1 Display Counters

The actual counting circuits are, in practice, constructed from integrated circuit counters, but it is constructive to understand the internal operation of a digital counter.

The heart of a frequency counter is the decade counter, which can be constructed from four flip-flops and an AND gate, as shown in Fig. 10-3. This form of decade counter is called a ripple counter owing to the fact that the clock of one flip-flop is derived from the output of the previous flip-flop, which requires that the clock pulses ripple through the counter from the first stage to the last stage. The last stage, however, derives its clock from the first stage, which reduces the propagation delay to a certain degree.

A superior method of constructing a counter is to use a synchronous counter. This circuit, shown in Fig. 10-4, requires that all the flip-flop clocks be connected together, which greatly reduces the propagation delay and allows higher counting speeds.

The output of the decade counter follows the sequence shown in Fig. 10-5 and is called binary coded decimal (BCD), which implies that the normal binary

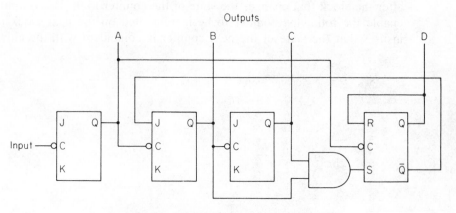

Figure 10-3 Ripple binary-coded-decimal counter.

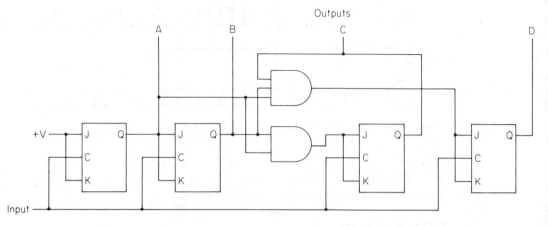

Figure 10-4 Logic diagram of a binary synchronous counter.

code is used except that each digit is defined only for values between 0 and 9. As an example, the decimal number 138 is 0001 0111 1000 in BCD.

Each BCD counter allows one decade of counting and thus the BCD counters must be cascaded. For example, three cascaded BCD counters are required to count from 0 to 999. There are two methods of cascading BCD counters, ripple cascading and synchronous. Ripple cascading is usually reserved for ripple counters and, unfortunately, makes the slow ripple counter even slower. With the exception of low-frequency counters, the ripple counter is not used in serious frequency-measuring equipment. The ripple connection requires the last output of the least significant counter to drive the clock input of the next more significant counter, as shown in Fig. 10-6. The clock input to the next stage must respond to the negative edge of the clock as the last bit, which has a binary weight of 8, goes low at the transition from 9 to 0.

The synchronous counter has a *terminal count* or *carry* output for the purpose of cascading counters, as shown in Fig. 10-7. This output goes to a logic 1 after the clock that changes the state of the counter to 9. This output is used to enable the following counter to be incremented on the next clock pulse. This insures that the state of the next counter is coincident with the clock and pre-

Clock	Counter State			
	D	C	B	A
1	0	0	0	1
2	0	0	1	0
3	0	0	1	1
4	0	1	0	0
5	0	1	0	1
6	0	1	1	0
7	0	1	1	1
8	1	0	0	0
9	1	0	0	1
10	0	0	0	0

Figure 10-5 Binary-coded-decimal counting sequence.

Frequency Counters and Time-Interval Measurements   Chap. 10

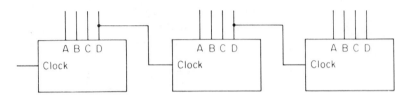

Figure 10-6 Cascading ripple counters.

serves the synchronous counter operation when the counters are cascaded. When more than two counters are cascaded, the requirement for any one counter to change state is that all of the less significant counters must be at 9. Some integrated-circuit counters have internal cascading logic that propagates the "nine" state from the least significant digit through all the intervening counters to the most significant digit. When there is a large number of cascaded counters, the delay can limit the count frequency of the counter. Therefore, other techniques called *look ahead* or *carry forward* are used to reduce the amount of propagation delay.

The BCD information available at the output of the counter must be converted to some form of visible display. The conversion depends on the type of display desired. For example, conversion from BCD to the very popular seven-segment display requires a single, inexpensive integrated circuit. Figure 10-8 shows a 4-bit counter including the seven-segment code conversion.

It is desirable in a frequency counter to display the count continuously. Since the counter is reset to zero and allowed to count during the gate period, during this time the output of the counter is constantly changing. The output of the counter cannot be displayed during this period as it would appear as a meaningless blur. Therefore, the count at the end of the measurement period is stored in a simple memory and displayed during the next counting period, after which the next count is stored in the memory and displayed. This memory is required to store only 4 bits, the entire BCD word, for each decade of the counter and is typically a simple 4-bit latch, which consists of four D-type flip-flops all clocked together, with each flip-flop storing 1 bit of data.

Digital logic usually cannot supply the required current for driving a display. Even those displays that require minimal amounts of current, such as liquid crystals, require special signals, which are not readily available from the decoder output. Therefore, a display driver is included between the decade counter and the displays.

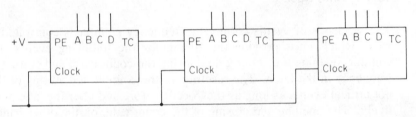

Figure 10-7 Cascaded synchronous counters.

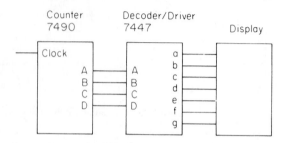

Figure 10-8 Block diagram of a decade counter interfaced with a seven-segment display.

For counters requiring a large number of digits, typically 10 or more, there are various techniques to reduce the required hardware, one of which is shown in Fig. 10-9. This technique is called *display multiplexing* and reduces the number of drivers and decoders required to implement large counters. In this example a common decoder and driver are shared between all the display digits. A multiplexer selects the BCD data from one of the latches and routes these data to the input of the seven-segment decoder. The decoded seven-segment information is applied to the proper display. The entire process is driven by an oscillator and a counter called the *scan oscillator and counter*. When this process is done at a rapid rate, the display appears constant to the eye. It would seem that the inclusion of the multiplexer, scan oscillator, and multiplexed display drivers is hardly worth the aggravation to save a few simple decoders. However, this technique has significant advantages when the frequency counter circuits are integrated into a single silicon chip.

Consider, as an example, a 10-digit frequency counter. This scale of frequency counter could be integrated onto a single silicon chip except that 70 outputs would be required for the readouts alone if they are of the seven-segment type. Add to this the power and ground, a time-base input, and other inputs required for the frequency counter and the net result is 80 or more pins, which does not allow for inexpensive packaging. The readout output could be multiplexed with seven outputs for the segments and a 4-bit binary output for selecting each digit, which results in only 11 output pins for the display interface. Adding the other required pins results in a package size that can be handled with conventional packaging technology.

10-1.2 Time Base

The sequence of events within the frequency counter is controlled by the time base, which must provide the timing for the following events: resetting the counter, opening the count gate, closing the count gate, and storing the counted frequency in the latch. The resetting of the counter and storing of the count are not critical events as long as they occur before and after the gate period, respectively. The opening and closing of the count gate, on the other hand, determine the accuracy of the frequency counter and are very critical in its timing.

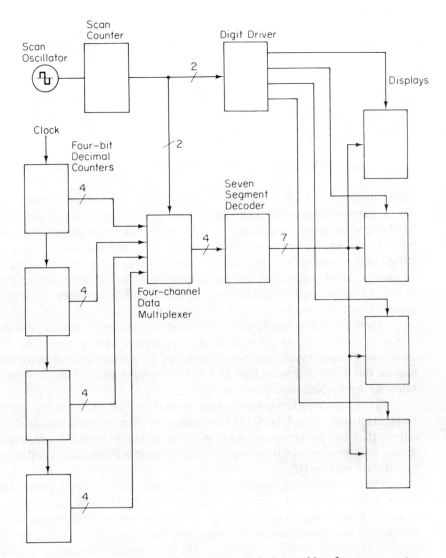

Figure 10-9 Block diagram of a multiplexed display used in a frequency counter.

Since the accuracy of the frequency counter depends directly on the accuracy of the time-base signals, the time base is driven from an accurate crystal-controlled oscillator. This element of the time base is typically a temperature-compensated crystal oscillator operating at several megahertz. A crystal oven could be used to supply a similar accuracy, except that the oven requires a relatively long period after the initial application of power, up to 24 hours, to stabilize. The temperature-compensated oscillator does not require the application of power to provide the correct frequency and is available for use immediately after power-on. Figure 10-10 shows a simplified diagram of a temperature-compensated crystal oscillator. A conventional crystal oscillator is used as the

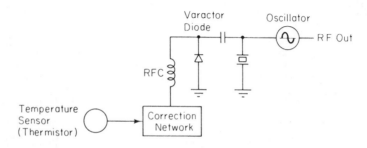

Figure 10-10 Block diagram of a temperature-compensated crystal oscillator.

basic building block of the compensated oscillator, except that a varactor diode is placed across the crystal. The varactor allows the frequency of the oscillator to be changed by minute amounts. The crystal oscillator frequency error is characterized over the desired operating temperature, and the error characteristic is stored in the correction network. This can either be a digital storage technique or an analog circuit with nonlinear characteristics. The ambient temperature is fed to the correction network, which adjusts the oscillator frequency by varying the varactor voltage as a function of temperature.

Aside from the temperature variation of frequency of a crystal oscillator, quartz crystals tend to age and change frequency over a period of time. This undesired frequency change can be reduced by special crystal fabrication techniques, but it still can be as high as 5×10^{-7} part per year. This must be compensated for by periodic recalibration.

Many temperature-compensated crystal oscillators have the capability of being electronically adjusted. If the frequency counter has a standard frequency output that can be compared to one of the available broadcast frequency standards, the frequency of the time-base oscillator in a frequency counter can be set to within 1 part in 10^9.

Three outputs are required from the time base: a reset pulse, the gating pulse, and a strobe pulse, in that order. Figure 10-11 shows a simple circuit for generating the three required pulses without overlap. The crystal oscillator is divided by powers of ten, as the period of the frequency of the crystal is much shorter than the desired gate time. The final digital divider is a 4-bit binary counter that has 16 states. The zero state of the counter is decoded to provide the reset pulse for the frequency counter. The 2 state is decoded to provide the gate open pulse. The 1 state was not used so as to provide a delay after the reset pulse to allow the counters to be fully recovered from the reset. The gate remains open for exactly 10 clock pulses, and thus the 12 state of the counter is decoded to provide the gate close pulse. The 13 state of the counter is not decoded so as to provide a delay period before the counter is stored in the latch during the 14 state. The 15 state is not decoded and provides the necessary nonoverlap between the store and reset pulses, which occur immediately after the 15 state of the counter.

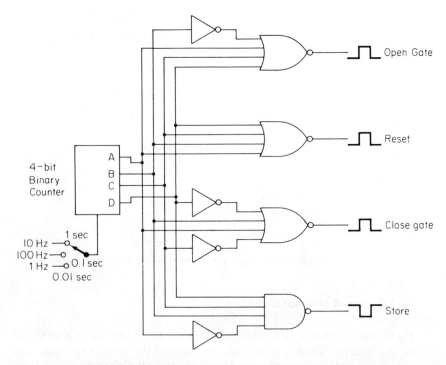

Figure 10-11 Logic diagram of a time base for a frequency counter.

It is important that the propagation delay from the input clock to the edges of the open and close pulses be the same for each so that the gate is exactly equal to the correct number of clock pulses. This requires fast logic and careful design.

Most frequency counters have several available gate time intervals that can be selected by a switch. As shown in Fig. 10-11, the input of the binary counter can be selected from a choice of 1 Hz, 10 Hz, 100 Hz, and 1 kHz. These frequencies provide gate times of 10, 1, 0.1, and 0.01 s, respectively.

10-1.3 Input Signal Processing

The unknown frequency input is not guaranteed to be of the correct logic level to drive the frequency counter, and a processing circuit is required. Typically, this is an amplifier to increase the signal level, an attenuator to adjust for variations in input amplitudes, and a comparator so that the slow risetime of the input waveforms can be reduced to provide reliable operation of the internal logic circuits. A schematic diagram of a typical frequency counter input circuit is shown in Fig. 10-12. Amplitudes of a few millivolts can be used to trigger the frequency counter using this circuit.

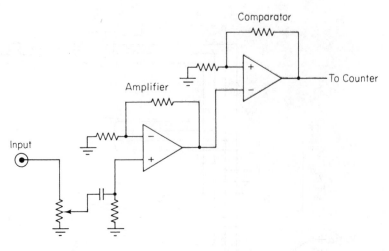

Figure 10-12 Input circuits for a simple frequency counter.

10-1.4 Period Measurement

If two input signals were substituted for the open and close gate signals, and one of the internal clock signals, that is, one of the available frequencies that are powers of 10 Hz, is supplied to the count gate, the time interval between the two input signals could be measured. The arrangement of this period measuring is shown in Fig. 10-13. The input signals must be processed in the same fashion as the count input signal, and the same circuit can be used for period measurement. A second identical circuit will have to be supplied for the period measurement.

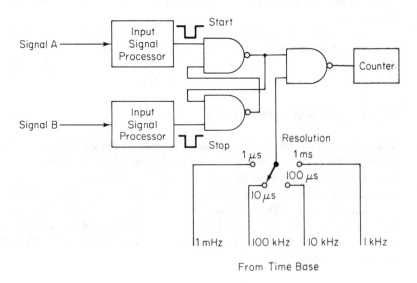

Figure 10-13 Circuit arrangement for making period measurements.

Another period measurement can be made using a single input. This would be useful for determining the period of pulses and other signals. In this mode of operation, the gating signal is the input, and the internal frequency clocks are used as timing sources. To measure the period of a pulse waveform, it is necessary to open the count gate at the rising edge of the pulse and to close the gate at the falling edge of the pulse. In the case of a negative-going pulse, this procedure would be reversed, that is, opening the gate on the negative edge and closing the gate at the positive edge. If the risetimes and falltimes of the input pulse are short, compared to the resolution of the period measurement, the actual trigger point is not critical. A sophisticated frequency counter will have independent control over the voltage level of both the rising and falling edges, as shown in Fig. 10-14. Although this results in the most flexible and accurate measurements, this type of frequency counter requires operator skill and a method of viewing the trigger points, such as an oscilloscope. Because most period measurements involve pulses with fast rise- and falltimes, a simple alternative is to ac couple the input signal and open and close the count gate at the zero crossings of the ac-coupled signal. Figure 10-15 shows a typical pulse waveform input and the resulting trigger points after ac coupling.

One very important period measurement is the period measurement to determine frequency. This measurement is not made from rising edge to falling edge but from a point in an input cycle to the same point in the next cycle, which is the period of the input signal. In this case, the gate is to be opened at a point of the input waveform and closed at precisely the same point in the next cycle. This is accomplished in the following fashion. The input signal is ac coupled, and a zero crossing detector triggers a flip-flop. The following zero crossing is of the opposite slope and does not trigger the flip-flop. The next zero crossing, however, occurs after a time period equal to the period of the input waveform and toggles the flip-flop, which provides a gate time exactly equal to the period of the input waveform, as shown in Fig. 10-16.

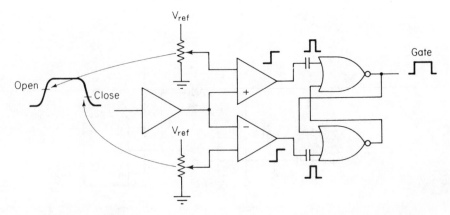

Figure 10-14 Frequency counter input circuits showing the ability to set rising and falling edges individually.

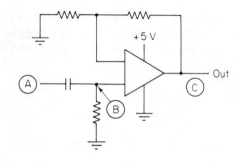

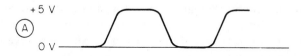

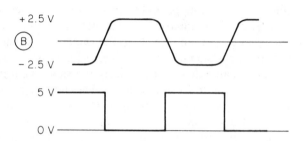

Figure 10-15 Zero-crossing detector for a frequency counter and the associated waveforms.

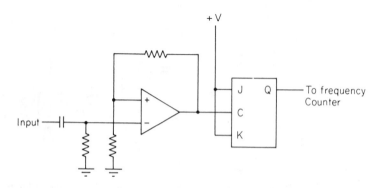

Figure 10-16 Input circuit configuration for measuring the period of a waveform.

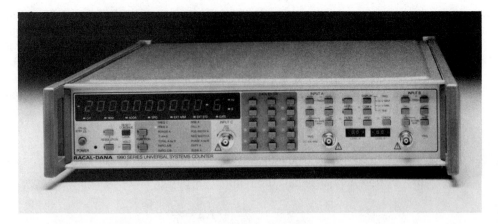

Figure 10-17 Microprocessor-controlled computing counter. (Courtesy of Racal-Dana Instruments, Inc.)

The typical laboratory counter, such as that shown in Fig. 10-17, has both input period measurement and independent control of risetime and falltime triggering selectable from a front-panel switch.

10-2 MEASUREMENT ERRORS

10-2.1 Gating Error

Frequency and time measurements made by an electronic counter are subject to several inaccuracies inherent in the instrument itself. One very common instrumental error is the *gating error,* which occurs whenever frequency and period measurements are made. For frequency measurement the main gate is opened and closed by the oscillator output pulse. This allows the input signal to pass through the gate and be counted by the decade counters. The gating pulse is not synchronized with the input signal; they are, in fact, two totally unrelated signals.

In Fig. 10-18 the gating interval is indicated by waveform (c). Waveforms (a) and (b) represent the input signal in different phase relationships with respect to the gating signal. Clearly, in one case, six pulses will be counted; in the other case, only five pulses are allowed to pass through the gate. We have therefore a ±1 count ambiguity in the measurement. In measuring low frequencies, the gat-

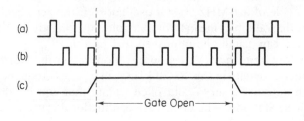

Figure 10-18 Gating error.

ing error may have an appreciable effect on the results. Take, for example, the case where a frequency of 10 Hz is to be measured and the gating time equals 1 s (a reasonable assumption). The decade counters would indicate a count of 10 ± 1 count, an inaccuracy of 10 per cent. *Period* measurements are therefore to be preferred over *frequency* measurements at the *lower* frequencies.

The dividing line between frequency and period measurements may be determined as follows: Let

$$f_c = \text{crystal (or clock) frequency of the instrument}$$

$$f_x = \text{frequency of the unknown input signal}$$

In a *period* measurement the number of pulses counted equals

$$N_p = \frac{f_c}{f_x} \qquad (10\text{-}1)$$

In a *frequency* measurement with a 1-s gate time the number of pulses counted is

$$N_f = f_x \qquad (10\text{-}2)$$

The *crossover* frequency (f_o) at which $N_p = N_f$ is

$$\frac{f_c}{f_o} = f_o \quad \text{or} \quad f_o = \sqrt{f_c} \qquad (10\text{-}3)$$

Signals with a frequency *lower* than f_o should therefore be measured in the "period" mode; signals of frequencies *above* f_o should be measured in the "frequency" mode in order to minimize the effect of the ±1 count gating error. The accuracy degradation at f_o caused by the ±1 count gating error is $100/\sqrt{f_c}$ per cent.

10-2.2 Time-Base Error

Inaccuracies in the time base also cause errors in the measurement. In frequency measurements the time base determines the opening and closing of the signal gate, and it provides the pulses to be counted. Time-base errors consist of oscillator calibration errors, short-term crystal stability errors, and long-term crystal stability errors.

Several methods of *crystal calibration* are in common use. One of the simplest calibration techniques is to zero-beat the crystal oscillator against the standard frequency transmitted by a standards radio station, such as WWV. This method gives reliable results with accuracy on the order of 1 part in 10^6, which corresponds to 1 cycle of a 1-MHz crystal oscillator. If the zero-beating is done with visual (rather than audible) means, for example, by using an oscilloscope, the calibration accuracy can usually be improved to 1 part in 10^7.

Several very low frequency (VLF) radio stations cover the North American continent with precise signals in the 16–20-kHz range. Low-frequency receivers are available with automatic servo-controlled tuning that can be slaved to the signal of one of these stations. The error between the local crystal oscillator and the incoming signal can then be recorded on a strip-chart recorder. A simplified diagram of this procedure is given in Fig. 10-19. Improved calibration accuracy can be obtained by using VLF stations rather than HF stations because the transmission paths for very low frequencies is shorter than for high-frequency transmissions.

Short-term crystal stability errors are caused by momentary frequency variations due to voltage transients, shock and vibration, cycling of the crystal oven, electrical interference, etc. These errors can be *minimized* by taking frequency measurements over *long* gate times (10 s to 100 s) and multiple-period-average measurements. A reasonable figure for short-term stability of a standard crystal-oven combination is on the order of 1 or 2 parts in 10^7.

Long-term stability errors are the more subtle contributors to the inaccuracy of a frequency or time measurement. Long-term stability is a function of aging and deterioration of the crystal. As the crystal is temperature-cycled and kept in continuous oscillation, internal stresses induced during manufacture are relieved, and minute particles adhering to the surface are shed reducing its thickness. Generally, these phenomena will cause an *increase* in the oscillator frequency.

A typical curve of frequency change versus time is shown in Fig. 10-20. The *initial* rate of change of crystal frequency may be on the order of 1 part in 10^6 per day. This rate will decrease, provided that the crystal is maintained at its operating temperature, normally about 50°C to 60°C, with ultimate stabilities of 1 part in 10^9. If, however, the instrument containing the crystal is unplugged from the power source for a period of time sufficient to allow the crystal to cool appreciably, a new slope of aging will ensue when the instrument is put back into operation. It is possible that the actual frequency of oscillation after cool off will vary by several cycles and that the original frequency will not again be reached unless calibration is done.

To show the effect of long-term stability on the absolute accuracy of the measurement, assume that the oscillator was calibrated to within 1 part in 10^9 and that a long-term stability of 1 part in 10^8 per day was reached. Assume further that calibration was done 60 days ago. The guaranteed accuracy at this time is then $1 \times 10^{-9} + 60 \times 10^{-8} = 6.01 \times 10^{-7}$, or 6 parts in 10^7. It can be seen therefore that maximum absolute accuracy can be achieved only if an exact calibration is performed a relatively short time *before* the measurement is taken.

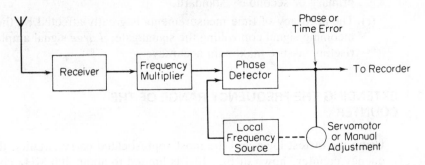

Figure 10-19 Calibration of a local frequency source.

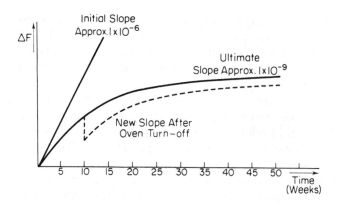

Figure 10-20 Frequency changes versus time for an oven-controlled crystal.

10-2.3 Trigger Level Error

In time-interval and period measurements the signal gate is opened and closed by the input signal. The accuracy with which the gate is opened and closed is a function of the *trigger level error*. In the usual application the input signal is amplified and shaped, and then it is applied to a Schmitt trigger circuit that supplies the gate with its control pulses. Usually the input signal contains a certain amount of unwanted components or noise, which is amplified along with the signal. The time at which triggering of the Schmitt circuit occurs is a function of the input signal amplification and of its signal-to-noise ratio. In general, we can say that trigger time errors are reduced with large signal amplitudes and fast risetimes.

Maximum accuracy can be obtained if the following suggestions are followed:

(a) The effect of the one-count gating error can be minimized by making frequency measurements above $\sqrt{f_c}$ and period measurements below $\sqrt{f_c}$, where f_c is the clock frequency of the counter.

(b) Since long-term stability has a cumulative effect, the accuracy of measurement is mostly a function of the time since the last calibration against a primary or secondary standard.

(c) The accuracy of time measurements is greatly affected by the *slope* of the incoming signal controlling the signal gate. *Large* signal amplitude and *fast* risetime assure maximum accuracy.

10-3 EXTENDING THE FREQUENCY RANGE OF THE COUNTER

Using the fastest logic and the most sophisticated carry circuits, the simple frequency counter shown in Fig. 10-1 is limited to about 100-MHz counting speed. To increase the frequency range of the counter, several techniques can be used.

Figure 10-21 Using a prescaler to extend the range of a frequency counter.

One technique is to use a prescaler as shown in Fig. 10-21. A *prescaler* is a fast digital counter that, typically, divides the input frequency by 10. The prescaler does not drive a display, is not gated, nor are the output data strobed into the storage latch. Therefore, the propagation delay of the prescaler is not important as long as the prescaler can operate at the desired frequency. If a divide-by-10 prescaler were used ahead of a 10-MHz counter, the counter frequency would be increased by a factor of 10 and the system would be capable of counting to 100 MHz. Prescalers are available for frequencies up to 1 GHz with divisions of 10 or 100, which can extend the range of the example 10-MHz counter to 1 GHz.

There is a penalty to be paid for the use of the prescaler. The resolution of the frequency counter is reduced by the same factor as the prescaler. As an example, if a 10-MHz counter were used with a prescaler, the frequency displayed would be multiplied by 10, all the digits including the least significant. This implies that if the counter had a resolution of 1 Hz, which is the value of the last digit, when multiplied by 10 the resolution would be reduced to 10 Hz. This can be overcome by simply using a longer time base and restoring the resolution. This can become a practical problem if the prescaler has a large division and very accurate frequency measurements are to be made. For example, if the divide-by-100 prescaler were used to extend the frequency range of the 10-MHz counter to 1 GHz, and a measurement of 1-Hz resolution were desired, the gate time would be 100 s, which could be a significant problem. Typically, frequency measurements with resolution of better than 1 kHz at 1 GHz are rare.

The prescaler, as effective as it can be, is limited to frequencies below about 1.5 GHz with the current state of technology. For making frequency counter measurements at higher frequencies, heterodyning techniques are used. Figure 10-22 shows a heterodyning converter for a frequency counter. This converter is

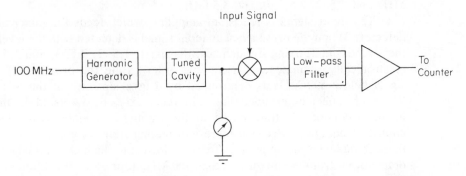

Figure 10-22 Manually tuned heterodyning frequency converters for extending the frequency range of frequency counters.

Sec. 10-3   Extending the Frequency Range of the Counter

331

used with a 50-MHz counter, which requires that the converter reduce the input frequency to 50 MHz or less, which it does with mixing frequencies every 100 MHz. Because both the sum and the differences are used, the converter frequency never exceeds 50 MHz. A 100-MHz source, which is derived from the frequency counter's time base, feeds a harmonic generator using a step recovery diode. The step recovery diode has a unique reverse recovery characteristic in that it stops conducting very abruptly, which generates harmonics of the driving waveform to several gigahertz. The harmonic content of the diode generator extends well into the 5-GHz region. Harmonics from the fundamental at 100 MHz to 5 GHz are selected by a tuned cavity that tunes one of the harmonics. It is necessary to know which of the 50 harmonics is being tuned, and a calibrated dial is provided as a tuning meter to peak the desired signal. The setting of the harmonic tuner dial does not affect the accuracy of the measurement unless the incorrect harmonic is tuned. The 50 harmonics represent a 2 per cent resolution, which can be easily achieved with a mechanical assembly.

The selected harmonic is mixed with the input and the difference is filtered, amplified, and fed to the counter. Because there is a harmonic available every 100 MHz, the input signal is never more than 50 MHz from one of the harmonics. To select the correct harmonic, the input frequency must be known to within 10 MHz or so, which can be done with another measurement technique such as a wavemeter or spectrum analyzer.

Since either the sum or difference between the selected harmonic and the input signal may be counted, the operator is required to make the necessary calculations to determine the actual frequency. This involves adding or subtracting, depending on whether the sum or difference is counted, the harmonic frequency that is read from the harmonic tuner dial.

Modern frequency counters are capable of tuning the harmonic and making the necessary calculation automatically. Figure 10-23 shows a block diagram of an automatic heterodyning unit for converting frequencies up to 4 GHz to extend the range of a 500-MHz counter. A 100-MHz signal from the frequency counter is multiplied using a bipolar transistor frequency multiplier to 500 MHz. This signal is amplified and used to drive a step recovery diode frequency multiplier. The output of the step recovery diode multiplier is filtered to recover signals at 1,000 MHz and 1.5, 2, 2.5, 3.0, and 3.5 GHz.

The input signal is fed to an amplifier, which feeds the mixer and a level detector. When the presence of an input signal is detected with the level detector, the six possible mixing frequencies, that is, 1, 1.5, 2, 2.5, 3, and 3.5 GHz, are electronically sequenced in ascending order, while the presence of an output signal below 500 MHz is determined by a level detector at the mixer output. When it has been determined that a difference exists below 500 MHz, the selected mixing frequency is transmitted to the frequency counter and added to the counted frequency. Because there is a mixing frequency every 500 MHz, and these frequencies are sequenced from the lowest to the highest, the first detection of an output from the mixer less than 500 MHz represents the difference between the input frequency and the selected mixing frequency. It is possible to obtain an

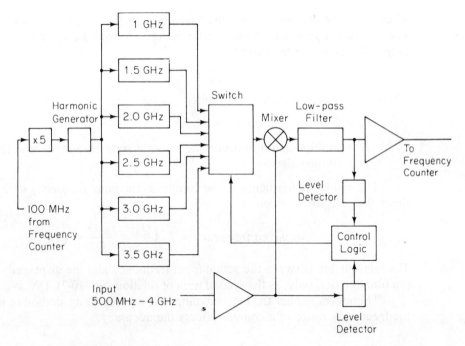

Figure 10-23 Automatic heterodyning unit for extending the frequency range of frequency counters to 4 GHz.

output using the next-higher mixing frequency, but this is avoided by selecting the first mixing frequency to supply an output below 500 MHz.

It is informative to calculate the effects on the accuracy of both the prescaler and the heterodyning methods of frequency extension.

For the case of the prescaler, assuming that the prescaler does not miss counts, and this is generally true, the output frequency is simply the input frequency divided by N, the prescaler ratio. The displayed frequency is the input frequency to the counter times the gate time, which is

$$\text{displayed frequency} = \frac{f_{in}}{N} t \tag{10-4}$$

Because N is a constant, the accuracy of the display is simply a function of t, the gate time. Thus, the accuracy of the counter with a prescaler is exactly the same as the accuracy of the counter without a prescaler.

Consider the case when using the heterodyning frequency converter where the mixing signal is derived from the same clock as is used to derive the time base within the counter. The gate time is an integer number of cycles of the time-base clock, or

$$\text{gate time} = \frac{Q}{f_c} \tag{10-5}$$

where Q is the division of the time base and f_c is the time-base clock frequency. The mixing frequency in the converter is derived from the same source, and the output frequency of the converter is

$$f_{in} = f'_{in} \pm N f_c \qquad (10\text{-}6)$$

where f_{in} = frequency into the counter

f'_{in} = frequency into the converter

N = multiplication between the internal time-base clock and the hetero-dyning signal

The displayed frequency of the counter is the input frequency of the counter times the gate time, which is

$$\text{displayed frequency} = f_{in}\left(\frac{Q}{f_c}\right) = \frac{f'_{in} Q}{f_c} + QN \qquad (10\text{-}7)$$

The relationship between the actual input frequency and the displayed frequency is a function of f_c only, as the second term of relationship (10-7), QN, is a constant.

Therefore, neither the heterodyning nor the prescaling method of increasing the frequency range of a counter affects the accuracy.

10-4 AUTOMATIC AND COMPUTING COUNTERS

The frequency counter, being an intensely digital machine, is an excellent candidate for automating and computerizing. One excellent measurement that can be handled by a calculating counter is the measurement of low frequencies with accuracy. One significant problem with the frequency counter is the measurement of low frequencies. If a signal of less than 1 Hz was to be measured with a resolution of 0.01 Hz, the time required would be 100 s if the conventional gate-controlled counter were used. An alternative measurement technique is to measure the period of the input waveform and calculate the frequency from the relationship:

$$\text{frequency} = \frac{1}{\text{period}} \qquad (10\text{-}8)$$

The time required to display the frequency is the period of the unknown input plus the computation time. For the example of a frequency on the order of 1 Hz, the period is 1 s, while the computation time is on the order of 1 ms or less. Essentially, the frequency of any waveform can be measured within the time of one period plus a small increase for the computation. However, the determination of frequency from a single period measurement has a statistical probability of error that is very great. A second frequency calculation made from a second period measurement would improve the probability of error, while a third calculation would further improve the error. The calculating frequency counter would continue to make frequency calculations from the period of the input as long as the input were present and display the arithmetic mean of the calculations.

Not only are low-frequency measurements improved from the calculating ability of a frequency counter; the measurement of pulsed carriers can be improved by the calculating counter. It is often necessary to determine the frequency of bursts of energy that do not last for long periods of time. As an example, consider a 1-μs burst of a 1-GHz carrier. To measure the frequency of the burst, only 1,000 complete cycles are available that can be counted. The frequency counter has an ambiguity of + or − one count which in this case represents an error of 1 part in 1,000, or 0.1 per cent. If the accuracy of the measurement is to be better than this, more than one burst has to be counted and ultimately used for the frequency calculation. The calculating frequency counter can make several measurements, average the result of each measurement, and display a statistically determined frequency.

In the section on frequency counter accuracy, it was discussed that there is a point where measuring the period of an input with a certain clock frequency produces improved accuracy over the measurement of the input frequency for a fixed gate time. A block diagram of an automated frequency counter with the capability of automatically making period or input frequency measurement and then performing the necessary mathematics to display the correct frequency is shown in Fig. 10-24. In this counter, rather than a conventional gate, there are two gated counters. One counter is used to accumulate the input frequency, while the second counter accumulates a precision clock. Both counters are gated simultaneously, such that the number of input cycles has been accumulated in counter

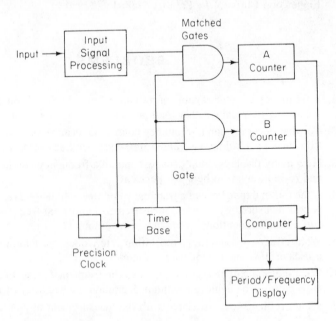

Figure 10-24 Precision computing counter using dual counters.

A while a precision clock, or the elapsed time is accumulated in counter B. The frequency of the input can be determined from the following relationship:

$$\text{input frequency} = \frac{\text{count in A}}{\text{count in B}} \tag{10-9}$$

The opening and closing of the gate are controlled from either the input signal or the internal precision clock. Essentially, if the gate is controlled by the internal clock a conventional frequency measurement will be made, and if the gate is controlled by the input signal, a period measurement is being made. As previously explained, the frequency where the accuracy changes from a period measurement to a frequency measurement is $(f_c)^{1/2}$, where f_c is the clock frequency from which the time base is derived and the clock used for the period measurement. In this example the precision clock used for the period measurement is 500 MHz, which places the changeover point at 22 MHz. From the setting of the input switches, which can select the number of significant digits and the resolution, the automatic frequency counter will select the method of measurement.

REFERENCES

10-1. Prensky, Sol D., and Castellucis, Richard L., *Electronic Instrumentation,* 3rd ed. Englewood Cliffs, N.J.: Prentice-Hall, Inc., 1982.

10-2. Tocci, Ronald J., *Digital Systems: Principles and Applications,* chaps. 4, 5, and 7. Englewood Cliffs, N.J.: Prentice-Hall, Inc., 1980.

PROBLEMS

10-1. A frequency counter capable of measuring an unknown frequency to within 1 Hz by measuring frequency rather than period would require what minimum gate time?

10-2. To what accuracy can a frequency counter determine an unknown frequency of 450 kHz, using a 1-s time base and a time-base accuracy of 0.01 per cent?

10-3. How many displays (total decades) should a frequency counter have if its accuracy and resolution are to be 0.001 per cent?

10-4. If the internal time base of a frequency counter is 10.000 MHz, what frequency range is best measured by a period measurement, and what frequency range is best measured by a conventional frequency measurement?

10-5. What effects on accuracy, resolution, etc., does the addition of a fixed modulus prescaler have on a frequency counter?

10-6. What method can be used to increase the frequency range of a frequency counter? How can this be achieved without degrading the accuracy of the counter?

10-7. What problems are associated with the measurement of pulsed signals?

11

Transducers as Input Elements to Instrumentation Systems

11-1 CLASSIFICATION OF TRANSDUCERS

An electronic instrumentation system consists of a number of components which together are used to perform a measurement and record the result. An instrumentation system generally consists of three major elements: an *input* device, a signal-conditioning or *processing* device, and an *output* device. The input device receives the quantity under measurement and delivers a *proportional* electrical signal to the signal-conditioning device. Here the signal is amplified, filtered, or otherwise modified to a format acceptable to the output device. The output device may be a simple indicating meter, an oscilloscope, or a chart recorder for visual display. It may be a magnetic tape recorder for temporary or permanent storage of the input data, or it may be a digital computer for data manipulation or process control. The kind of system depends on what is to be measured and how the measurement result is to be presented.

The input quantity for most instrumentation systems is *nonelectrical*. In order to use electrical methods and techniques for measurement, manipulation, or control, the nonelectrical quantity is converted into an electrical signal by a device called a *transducer*. One definition states "a transducer is a device which, when actuated by energy in one transmission system, supplies energy in the same form or in another form to a second transmission system." This energy transmission may be electrical, mechanical, chemical, optical (radiant), or thermal.

This broad definition of a transducer includes, for example, devices that convert *mechanical* force or displacement into an electrical signal. These devices form a very large and important group of transducers commonly found in the industrial instrumentation area, and the instrumentation engineer is primarily concerned with this type of energy conversion. Many other physical parameters (such as heat, light intensity, humidity) may also be converted into electrical energy by means of transducers. These transducers provide an output signal when stimulated by a *nonmechanical* input: a thermistor reacts to temperature variations, a photocell to changes in light intensity, an electron beam to magnetic effects, and so on. In all cases, however, the electrical output is measured by standard methods, yielding the magnitude of the input quantity in terms of an *analog* electrical measure.

Transducers may be classified according to their application, method of energy conversion, nature of the output signal, and so on. All these classifications usually result in overlapping areas. A sharp distinction between, and classification of, types of transducers is difficult. Table 11-1 shows a classification of transducers according to the *electrical principles* involved. The first part of the table lists transducers that require external power. These are the *passive* transducers, producing a variation in some electrical parameter, such as resistance, capacitance, and so on, which can be measured as a voltage or current variation.

TABLE 11-1 TYPES OF TRANSDUCERS

Electrical parameter and class of transducer	Principle of operation and nature of device	Typical application
Passive Transducers (Externally Powered)		
Resistance		
Potentiometric device	Positioning of the slider by an external force varies the resistance in a potentiometer or a bridge circuit.	Pressure, displacement
Resistance strain gage	Resistance of a wire or semiconductor is changed by elongation or compression due to externally applied stress.	Force, torque, displacement
Pirani gage or hot-wire meter	Resistance of a heating element is varied by convection cooling of a stream of gas.	Gas flow, gas pressure
Resistance thermometer	Resistance of pure metal wire with a large positive temperature coefficient of resistance varies with temperature.	Temperature, radiant heat
Thermistor	Resistance of certain metal oxides with negative temperature coefficient of resistance varies with temperature.	Temperature
Resistance hygrometer	Resistance of a conductive strip changes with moisture content.	Relative humidity
Photoconductive cell	Resistance of the cell as a circuit element varies with incident light.	Photosensitive relay

TABLE 11-1   TYPES OF TRANSDUCERS (*continued*)

Electrical parameter and class of transducer	Principle of operation and nature of device	Typical application
Capacitance		
Variable capacitance pressure gage	Distance between two parallel plates is varied by an externally applied force.	Displacement, pressure
Capacitor microphone	Sound pressure varies the capacitance between a fixed plate and a movable diaphragm.	Speech, music, noise
Dielectric gage	Variation in capacitance by changes in the dielectric.	Liquid level, thickness
Inductance		
Magnetic circuit transducer	Self inductance or mutual inductance of ac-excited coil is varied by changes in the magnetic circuit.	Pressure, displacement
Reluctance pickup	Reluctance of the magnetic circuit is varied by changing the position of the iron core of a coil.	Pressure, displacement, vibration, position
Differential transformer	The differential voltage of two secondary windings of a transformer is varied by positioning the magnetic core through an externally applied force.	Pressure, force, displacement, position
Eddy current gage	Inductance of a coil is varied by the proximity of an eddy current plate.	Displacement, thickness
Magnetostriction gage	Magnetic properties are varied by pressure and stress.	Force, pressure, sound
Voltage and current		
Hall effect pickup	A potential difference is generated across a semiconductor plate (germanium) when magnetic flux interacts with an applied current.	Magnetic flux, current
Ionization chamber	Electron flow induced by ionization of gas due to radioactive radiation.	Particle counting, radiation
Photoemissive cell	Electron emission due to incident radiation on photoemissive surface.	Light and radiation
Photomultiplier tube	Secondary electron emission due to incident radiation on photosensitive cathode.	Light and radiation, photosensitive relays
Self-Generating Transducers (No External Power)		
Thermocouple and thermopile	An emf is generated across the junction of two dissimilar metals or semiconductors when that junction is heated.	Temperature, heat flow, radiation
Moving-coil generator	Motion of a coil in a magnetic field generates a voltage.	Velocity, vibration
Piezoelectric pickup	An emf is generated when an external force is applied to certain crystalline materials, such as quartz.	Sound, vibration, acceleration, pressure changes
Photovoltaic cell	A voltage is generated in a semiconductor junction device when radiant energy stimulates the cell.	Light meter, solar cell

The second category are transducers of the *self-generating* type, producing an analog voltage or current when stimulated by some physical form of energy. The self-generating transducers do *not* require external power. Although it would be almost impossible to classify all sensors and measurements, the devices listed in Table 11-1 represent a good cross section of commercially available transducers for application in instrumentation engineering. Some of the more common transducers and their application are discussed in the following sections.

11-2 SELECTING A TRANSDUCER

In a measurement system the transducer is the input element with the critical function of transforming some physical quantity to a proportional electrical signal. Selection of the appropriate transducer is therefore the first and perhaps most important step in obtaining accurate results. A number of elementary questions should be asked before a transducer can be selected, for example,

(a) What is the physical quantity to be measured?

(b) Which transducer principle can best be used to measure this quantity?

(c) What accuracy is required for this measurement?

The first question can be answered by determining the *type* and *range* of the measurand. An appropriate answer to the second question requires that the *input* and *output* characteristic of the transducer be compatible with the recording or measurement system. In most cases, these two questions can be answered readily, implying that the proper transducer is selected simply by the addition of an *accuracy* tolerance. In practice, this is rarely possible due to the complexity of the various transducer parameters that affect the accuracy. The accuracy requirements of the total *system* determine the degree to which individual factors contributing to accuracy must be considered. Some of these factors are:

(a) *Fundamental transducer parameters:* type and range of measurand, sensitivity, excitation

(b) *Physical conditions:* mechanical and electrical connections, mounting provisions, corrosion resistance

(c) *Ambient conditions:* nonlinearity effects, hysteresis effects, frequency response, resolution

(d) *Environmental conditions:* temperature effects, acceleration, shock and vibration

(e) *Compatibility of the associated equipment:* zero balance provisions, sensitivity tolerance, impedance matching, insulation resistance

Categories (a) and (b) are basic electrical and mechanical characteristics of the transducer. Transducer accuracy, as an independent component, is contained in categories (c) and (d). Category (e) considers the transducer's compatibility with its associated system equipment.

The total measurement error in a transducer-activated system may be reduced to fall within the required accuracy range by the following techniques:

(a) Using in-place system *calibration* with corrections performed in the data reduction
(b) Simultaneously *monitoring* the environment and correcting the data accordingly
(c) Artificially *controlling* the environment to minimize possible errors

Some *individual* errors are predictable and can be calibrated out of the system. When the entire system is calibrated, these calibration data may then be used to correct the recorded data. *Environmental* errors can be corrected by data reduction if the environmental effects are recorded simultaneously with the actual data. Then the data are corrected by using the known environmental characteristics of the transducers. These two techniques can provide a significant increase in system accuracy.

Another method to improve overall system accuracy is to control *artificially* the environment of the transducer. If the environment of the transducer can be kept unchanged, these errors are reduced to zero. This type of control may require either physically moving the transducer to a more favorable position or providing the required isolation from the environment by a heater enclosure, vibration isolation, or similar means.

11-3 STRAIN GAGES

11-3.1 Gage Factor

The strain gage is an example of a passive transducer (Table 11-1) that converts a *mechanical* displacement into a change of *resistance*. A strain gage is a thin, wafer-like device that can be attached (bonded) to a variety of materials to measure applied strain. *Metallic* strain gages are manufactured from small-diameter resistance wire, such as Constantan,* or etched from thin foil sheets. The resistance of the wire or metal foil changes with length as the material to which the gage is attached undergoes tension or compression. This change in resistance is proportional to the applied strain and is measured with a specially adapted Wheatstone bridge.

The sensitivity of a strain gage is described in terms of a characteristic called the *gage factor, K,* defined as the unit change in resistance per unit change in length, or

$$\text{gage factor } K = \frac{\Delta R/R}{\Delta l/l} \tag{11-1}$$

---

* A trade name. Constantan is a copper-nickel alloy consisting of 60 per cent copper and 40 per cent nickel.

where K = gage factor

R = *nominal gage resistance*

ΔR = change in gage resistance

l = normal specimen length (unstressed condition)

Δl = change in specimen length

The term $\Delta l/l$ in the denominator of Eq. (11-1) is the strain σ, so that Eq. (11-1) can be written as

$$K = \frac{\Delta R/R}{\sigma} \tag{11-2}$$

where σ is the strain in the lateral direction.

The resistance change ΔR of a conductor with length l can be calculated by using the expression for the resistance of a conductor of uniform cross section:

$$R = \rho \frac{\text{length}}{\text{area}} = \frac{\rho \times l}{(\pi/4)d^2} \tag{11-3}$$

where ρ = specific resistance of the conductor material

l = length of the conductor

d = diameter of the conductor

Tension on the conductor causes an increase Δl in its length and a simultaneous decrease Δd in its diameter. The resistance of the conductor then changes to

$$R_s = \rho \frac{(l + \Delta l)}{(\pi/4)(d - \Delta d)^2} = \rho \frac{l(1 + \Delta l/l)}{(\pi/4)d^2 (1 - 2 \Delta d/d)} \tag{11-4}$$

Equation (11-4) may be simplified by using Poisson's ratio, μ, defined as the ratio of strain in the lateral direction to strain in the axial direction. Therefore

$$\mu = \frac{\Delta d/d}{\Delta l/l} \tag{11-5}$$

Substitution of Eq. (11-5) into Eq. (11-4) yields

$$R_s = \rho \frac{l}{(\pi/4)d^2} \left(\frac{1 + \Delta l/l}{1 - 2\mu \, \Delta l/l} \right) \tag{11-6}$$

which can be simplified to

$$R_s = R + \Delta R = R \left[1 + (1 + 2\mu) \frac{\Delta l}{l} \right] \tag{11-7}$$

The increment of resistance ΔR as compared to the increment of length Δl can then be expressed in terms of the gage factor K where

$$K = \frac{\Delta R/R}{\Delta l/l} = 1 + 2\mu \tag{11-8}$$

Poisson's ratio for most metals lies in the range of 0.25 to 0.35, and the gage factor would then be on the order of 1.5 to 1.7.

For strain-gage applications, a *high sensitivity* is very desirable. A large gage factor means a relatively large resistance change which can be more easily measured than a small resistance change. For Constantan wire, K is about 2, whereas Alloy 479 gives a K value of about 4.

It is interesting to carry out a simple calculation to find out what effect an applied stress has on the resistance change of a strain gage. Hooke's law gives the relationship between stress and strain for a linear stress-strain curve, in terms of the modulus of elasticity of the material under tension. Defining *stress* as the applied force per unit area and *strain* as the elongation of the stressed member per unit length, Hooke's law is written as

$$\sigma = \frac{s}{E} \qquad (11\text{-}9)$$

where σ = strain, $\Delta l / l$ (no units)

s = stress (kg/cm^2)

E = Young's modulus (kg/cm^2)

EXAMPLE 11-1

A resistance strain gage with a gage factor of 2 is fastened to a steel member subjected to a stress of 1,050 kg/cm^2. The modulus of elasticity of steel is approximately 2.1×10^6 kg/cm^2. Calculate the change in resistance, ΔR, of the strain-gage element due to the applied stress.

SOLUTION Hooke's law, Eq. (11-9), yields

$$\sigma = \frac{\Delta l}{l} = \frac{s}{E} = \frac{1{,}050}{2.1 \times 10^6} = 5 \times 10^{-4}$$

The sensitivity of the strain gage $K = 2$. Therefore, from Eq. (11-2),

$$\frac{\Delta R}{R} = K\sigma = 2 \times 5 \times 10^{-4} = 10^{-3} \quad \text{or} \quad 0.1\%$$

Example 11-1 illustrates that the relatively high stress of 1,050 kg/cm^2 results in a resistance change of only 0.1 per cent, a very small change indeed. Actual measurements generally involve resistance changes of very much lower values, and the bridge measuring circuit must be very carefully designed to be able to detect these small changes in resistance.

11-3.2 Metallic Sensing Elements

Metallic strain gages are formed from thin resistance wire or etched from thin sheets of metal foil. Wire gages are generally small in size, are subject to minimal leakage, and can be used in high temperature applications. Foil elements are somewhat larger in size and are more stable than wire gages. They can be used under extreme temperature conditions and under prolonged loading, and they dissipate self-induced heat easily.

Various resistance materials have been developed for use in wire and foil gages. Some of these are described in the following paragraphs.

Constantan is a copper-nickel alloy with a low temperature coefficient. Constantan is commonly found in gages that are used in dynamic strain measurements, where alternating strain levels do not exceed $\pm$ 1,500 μcm/cm. Operating temperature limits are from 10°C to 200°C.

Nichrome V is a nickel-chrome alloy used for static strain measurements to 375°C. With temperature compensation, the alloy may be used for static measurements to 650°C and dynamic measurements to 1,000°C.

Dynaloy is a nickel-iron alloy with a high gage factor and a high resistance to fatigue. This material is used in dynamic strain applications when high temperature sensitivity can be tolerated. The temperature range of dynaloy gages is generally limited by the carrier materials and the bonding cement.

Stabiloy is a modified nickel-chrome alloy with a wide temperature compensation range. These gages have excellent stability from cryogenic temperatures to approximately 350°C and good fatigue life.

Platinum-tungsten alloys offer excellent stability and high resistance to fatigue at elevated temperatures. These gages are recommended for static tests to 700°C and dynamic measurements to 850°C. Because the material has a relatively large temperature coefficient, some form of temperature compensation must be used to correct this error.

Semiconductor strain gages are often used in high-output transducers such as load cells. These gages have very high sensitivities, with gage factors from 50 to 200. They are, however, sensitive to temperature fluctuations and often behave in a nonlinear manner.

The size of the finished gage, and the manner in which the wire or foil pattern is arranged, varies with the application. Some bonded gages can be as small as $\frac{1}{8}$ in. by $\frac{1}{8}$ in., although they are generally somewhat larger, and are manufactured to a maximum size of 1 in. long by $\frac{1}{2}$ in. wide. In the usual application, the strain gage is cemented to the structure whose strain is to be measured. The problem of providing a good bonding between the gage and the structure is very difficult. The adhesive material must hold the gage firmly to the structure, yet it must have sufficient elasticity to give under strain without losing its adhesive properties. The adhesive should also be resistant to temperature, humidity, and other environmental conditions.

11-3.3 Gage Configuration

The shape of the sensing element is selected according to the strain to be measured: uniaxial, biaxial, or multidirectional. Uniaxial applications most often use long, narrow sensing elements, as in Fig. 11-1, to maximize the strain sensing material in the direction of interest. End loops are few and short, so that sensitivity to transverse strains is low. Gage length is selected according to the strain field

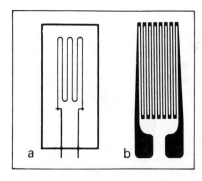

Figure 11-1 Uniaxial strain gages: (a) wire;
(b) foil. (Courtesy of Statham Division,
Schlumberger Industries.)

to be investigated. For most strain measurements, the 6 mm gage length offers good performance and easy installation.

Simultaneous measurement of strains in more than one direction can be accomplished by placing single-element gages at the proper locations. However, to simplify this task and provide greater accuracy, multielement, or *rosette*, gages are available.

Two-element rosettes, shown in Fig. 11-2, are often used in force transducers. The gages are wired in a Wheatstone bridge circuit to provide maximum output. For stress analysis, the axial and transverse elements may have different resistances that can be so selected that the combined output is proportional to stress while the output of the axial element alone is proportional to strain. Three-element rosettes are often used to determine the direction and magnitude of principal strains resulting from complex structural loading. The most popular types have 45°- or 60°-angular displacements between the sensing elements, as shown in Fig. 11-3. The 60°-rosettes are used when the direction of the principal strains is unknown. The 45°-rosettes provide greater angular resolutions and are normally used when the directions of the principal strains are known.

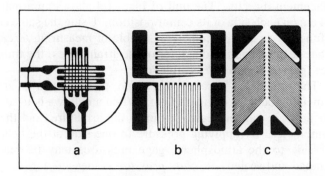

Figure 11-2 Two-element rosettes: (a) 90° stacked foil;
(b) 90° planar foil; (c) 90° shear planar foil. (Courtesy
of Statham Division, Schlumberger Industries.)

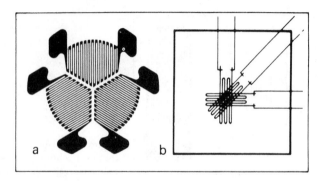

Figure 11-3   Three-element rosettes: (a) 60° planar foil; (b) 45° stacked wire. (Courtesy of Statham Division, Schlumberger Industries.)

11-3.4 Unbonded Strain Gage

The unbonded strain gage consists of a stationary frame and an armature that is supported in the center of the frame. The armature can move only in one direction. Its travel in that direction is limited by four filaments of strain-sensitive wire, wound between rigid insulators that are mounted on the frame and on the armature. The filaments are of equal length and arranged as shown in Fig. 11-4(a).

When an external force is applied to the strain gage, the armature moves in the direction indicated. Elements A and D increase in length, whereas elements B and C decrease in length. The resistance change of the four filaments is proportional to their change in length, and this change can be measured with a Wheatstone bridge, as shown in Fig. 11-4(b). The unbalance current, indicated by the current meter, is calibrated to read the magnitude of the displacement of the armature.

The unbonded strain-gage transducer can be constructed in a variety of configurations, depending on the required use. Its principal use is a *displacement* transducer: A linkage pin can be attached to the armature in order to measure displacement directly. The unit of Fig. 11-4 allows an armature displacement of 0.004 cm on each side of its center position. Using the same construction, the unit will function as a dynamometer, capable of measuring *force*. Depending on the number of turns and the diameter of the strain wires, the transducer will measure forces from ±40 g to ±2 kg, full-scale.

The transducer becomes a *pressure* pickup when its armature is connected to a metallic bellows or diaphragm. When a bellows is used, force on the end of the bellows is transmitted by a pin to the armature, and the unit functions as a dynamometer. By applying pressure to one side of the bellows and venting the other side to the atmosphere, gage pressures may be read. If the bellows is evacuated and sealed, *absolute pressure* is measured.

Another modification is provided by two pressure connections, one to each side of the bellows or diaphragm, for the measurement of *differential* pressure.

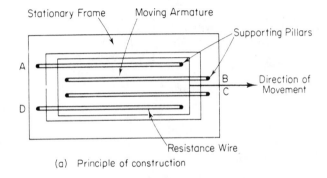

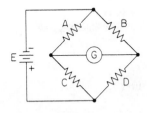

(a) Principle of construction

(b) Wheatstone bridge circuit

Figure 11-4 Unbonded strain gage: (a) principle of construction; (b) Wheatstone bridge circuit.

Finally, when a weight is fastened to the armature, the transducer becomes an *accelerometer*.

11-4 DISPLACEMENT TRANSDUCERS

The concept of converting an applied force into a displacement is basic to many types of transducers. The mechanical elements that are used to convert the applied force into a displacement are called *force-summing* devices. The force-summing members generally used are the following:

(a) Diaphragm, flat or corrugated
(b) Bellows
(c) Bourdon tube, circular or twisted
(d) Straight tube
(e) Mass cantilever, single or double suspension
(f) Pivot torque

Examples of these force-summing devices are shown in Fig. 11-5. *Pressure* transducers generally use one of the first four types of force-summing members; categories (e) and (f) will be found in *accelerometer* and *vibration* pickups.

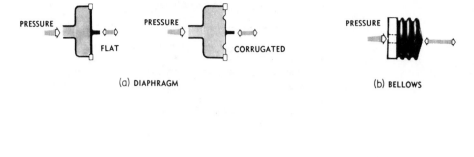

(a) DIAPHRAGM

(b) BELLOWS

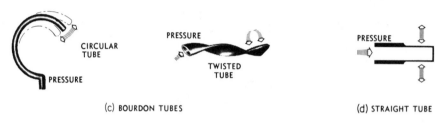

(c) BOURDON TUBES

(d) STRAIGHT TUBE

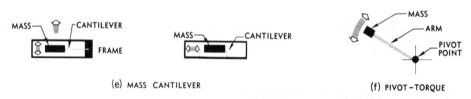

(e) MASS CANTILEVER

(f) PIVOT–TORQUE

Figure 11-5 Force-summing devices. (Courtesy of Statham Division, Schlumberger Industries.)

The displacement created by the action of the force-summing device is converted into a change of some *electrical* parameter. The electrical principles most commonly used in the measurement of displacement are

(a) Capacitive
(b) Inductive
(c) Differential transformer
(d) Ionization
(e) Oscillation
(f) Photoelectric
(g) Piezoelectric
(h) Potentiometric
(i) Velocity

These principles are discussed and illustrated in the following sections.

11-4.1 Capacitive Transducer

The capacitance of a parallel-plate capacitor is given by

$$C = \frac{kA\varepsilon_0}{d} \text{ (farads)} \qquad (11\text{-}10)$$

where A = area of each plate (m²)

$\quad d$ = plate spacing (m)

$\quad \varepsilon_0 = 9.85 \times 10^{12}$ (F/m)

$\quad k$ = dielectric constant

Since the capacitance is inversely proportional to the spacing of the parallel plates, any variation in d causes a corresponding variation in the capacitance. This principle is applied in the capacitive transducer of Fig. 11-6. A force, applied to a diaphragm that functions as one plate of a simple capacitor, changes the distance between the diaphragm and the static plate. The resultant change in capacitance could be measured with an ac bridge, but it is usually measured with an oscillator circuit. The transducer, as part of the oscillatory circuit, causes a change in the frequency of the oscillator. This change in frequency is a measure of the magnitude of the applied force.

The capacitive transducer has excellent frequency response and can measure both static and dynamic phenomena. Its disadvantages are sensitivity to temperature variations and possibility of erratic or distorted signals due to long lead length. Also, the receiving instrumentation may be large and complex, and it often includes a second fixed-frequency oscillator for heterodyning purposes. The difference frequency, thus produced, can be read by an appropriate output device such as an electronic counter.

11-4.2 Inductive Transducer

In the inductive transducer the measurement of force is accomplished by the change in the inductance ratio of a pair of coils or by the change of inductance in a single coil. In each case, the ferromagnetic armature is displaced by the force

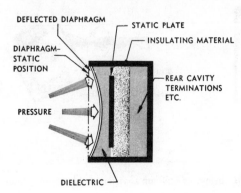

DEFLECTED DIAPHRAGM
STATIC PLATE
INSULATING MATERIAL
DIAPHRAGM-STATIC POSITION
REAR CAVITY TERMINATIONS ETC.
PRESSURE
DIELECTRIC

Figure 11-6 Capacitive transducer. (Courtesy of Statham Division, Schlumberger Industries.)

being measured, varying the *reluctance* of the magnetic circuit. Figure 11-7 shows how the air gap is varied by a change in position of the armature. The resultant change in inductance is a measure of the magnitude of the applied force.

The coil can be used as a component of an LC oscillator whose frequency then varies with applied force. This type of transducer is used extensively in telemetry systems, with a single coil that modulates the frequency of a local oscillator.

Hysteresis errors of the transducer are almost entirely limited to the mechanical components. When a diaphragm is used as the force-summing member, as shown in Fig. 11-7(a), it may form part of the magnetic circuit. In this arrangement the overall performance of the transducer is somewhat degraded because the desired mechanical characteristics of the diaphragm must be compromised to improve the magnetic performance.

The inductive transducer responds to static and dynamic measurements, and it has continuous resolution and a fairly high output. Its disadvantages are that the frequency response (variation of the applied force) is limited by the construction of the force-summing member. In addition, external magnetic fields may cause erratic performance.

11-4.3 Variable Differential Transformer Transducer

The differential transformer transducer measures force in terms of the displacement of the ferromagnetic core of a transformer. The basic construction of the linear variable differential transformer (LVDT) is given in Fig. 11-8. The transformer consists of a single primary winding and two secondary windings which are placed on either side of the primary. The secondaries have an equal number of turns but they are connected in series opposition so that the emfs induced in the coils oppose each other. The position of the movable core determines the flux linkage between the ac-excited primary winding and each of the two secondary windings.

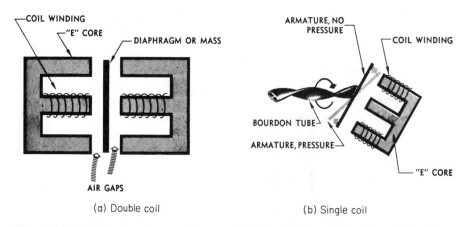

(a) Double coil   (b) Single coil

Figure 11-7 Inductive transducers. (Courtesy of Statham Division, Schlumberger Industries.)

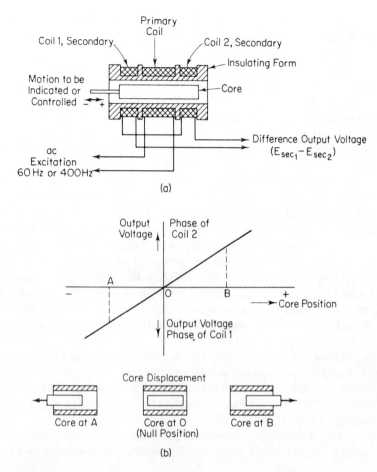

Figure 11-8 Linear variable differential transformer (LVDT): (a) essential components of the LVDT; (b) relative positions of the core generate the indicated output voltages. The linear characteristics impose limited core movements, which are typically up to 5 mm from the null position.

With the core in the center, or *reference,* position, the induced emfs in the secondaries are equal, and since they oppose each other, the output voltage will be 0 V. When an externally applied force moves the core to the left-hand position, more magnetic flux links the left-hand coil than the right-hand coil. The induced emf of the left-hand coil is therefore larger than the induced emf of the right-hand coil. The magnitude of the output voltage is then equal to the *difference* between the two secondary voltages, and it is *in phase* with the voltage of the left-hand coil. Similarly, when the core is forced to move to the right, more flux links the right-hand coil than the left-hand coil and the resultant output voltage is now in phase with the emf of the right-hand coil, while its magnitude again equals the difference between the two induced emfs. Figure 11-8(b) shows the LVDT output voltage as a function of the core position.

The output of the differential transformer may serve as a component in a force-balancing servo system. This is indicated schematically in Fig. 11-9. The output terminals of an *input* transformer and a *balancing* transformer are connected in series opposition. The algebraic sum of the two voltages is fed to an amplifier that drives a two-phase motor. When the two transformers are in their reference positions, the sum of their output voltages is zero and no voltage is delivered to the servo motor. When the core of the input transformer is moved away from its reference position by an externally applied displacement input, an output voltage is delivered to the amplifier and the motor rotates. The motor shaft is mechanically coupled to the core of the balancing transformer. Since the output of the balancing transformer opposes the output of the input transformer, the motor continues to rotate until the outputs of the two transformers are equal. The indicator on the motor shaft is calibrated to read the displacement of the balancing transformer and, indirectly, the displacement of the input transformer.

A variation of the moving-core differential transformer is given in Fig. 11-10. Here the primary winding is wound on the center leg of an E core, and the secondary windings are wound on the outer legs of the E core. The armature is rotated by the externally applied force about a pivot point above the center leg of the core. When the armature is displaced from its balance or reference position, the reluctance of the magnetic circuit through one secondary coil is decreased, while, simultaneously, the reluctance of the magnetic circuit through the other secondary coil is increased. The induced emfs in the secondary windings, which are equal in the reference position of the armature, will now be different in magnitude as a result of the applied displacement. The induced emfs again oppose each other, and the transformer operates in the same manner as the moving-core transformer of Fig. 11-8.

The differential transformer provides continuous resolution and shows low hysteresis. Relatively large displacements are required, and the instrument is

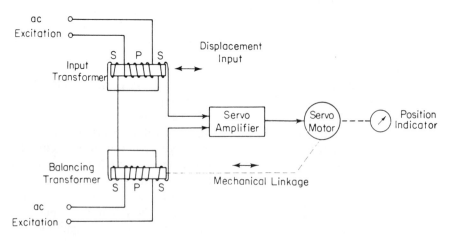

Figure 11-9 Displacement measurement using two differential transformers in a closed-loop servo system.

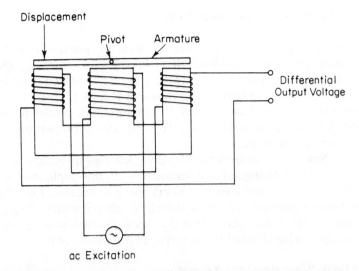

Figure 11-10   Differential transformers with an E core and pivoted armature.

sensitive to vibration. The receiving instrument must be selected to operate on ac signals, or a demodulator network must be used if a dc output is required.

11-4.4 Oscillation Transducer

This class of transducer uses the force-summing member to change the capacitance or inductance in an LC oscillator circuit. Figure 11-11 shows the basic elements of an LC oscillator whose frequency is affected by a change in the inductance of the coil. The stability of the oscillator must be excellent in order to detect changes in oscillator frequency caused by the externally applied force.

   This transducer measures both static and dynamic phenomena and is convenient for use in telemetry applications. Its limited temperature range, poor thermal stability, and low accuracy restrict its use to low-accuracy applications.

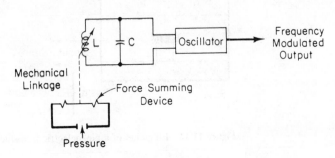

Figure 11-11   Basic elements of an oscillation transducer.

11-4.5 Piezoelectric Transducer

Asymmetrical crystalline materials, such as quartz, Rochelle salt, and barium titanite, produce an emf when they are placed under stress. This property is used in piezoelectric transducers, where a crystal is placed between a solid base and the force-summing member, as shown in Fig. 11-12. An externally applied force, entering the transducer through its pressure port, applies pressure to the top of a crystal. This produces an emf across the crystal, proportional to the magnitude of the applied pressure.

Since the transducer has a very good high-frequency response, its principal use is in high-frequency accelerometers. In this application its output voltage is typically on the order of 1 to 30 mV per g of acceleration. The device needs no external power source and is therefore self-generating. The principal disadvantage of this transducer is that it cannot measure static conditions. The output voltage is also affected by temperature variations of the crystal.

11-4.6 Potentiometric Transducer

A potentiometric transducer is an electromechanical device containing a resistance element that is contacted by a movable slider. Motion of the slider results in a resistance change that may be linear, logarithmic, exponential, and so on, depending on the manner in which the resistance wire is wound. In some cases, deposited carbon, platinum film, and other techniques are used to provide the resistance element. The basic elements of the potentiometric transducer are given in Fig. 11-13.

The potentiometric principle is widely used despite its many limitations. Its electric efficiency is very high and it provides a sufficient output to permit control operations without further amplification. The device may be ac- or dc-excited and can therefore serve a wide range of functions. Because of the mechanical friction of the slider against the resistance element, its life is limited and noise may

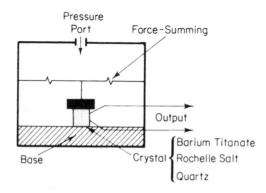

Figure 11-12 Elements of a piezoelectric transducer.

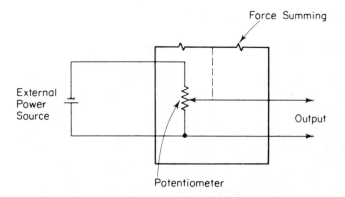

External
Power
Source

Output

Potentiometer

Figure 11-13 Principle of the potentiometric transducer.

develop as the element wears out. Large displacements are usually required to move the slider along the entire working surface of the potentiometer.

11-4.7 Velocity Transducer

The velocity transducer essentially consists of a moving coil suspended in the magnetic field of a permanent magnet, as shown in Fig. 11-14. A voltage is generated by the motion of the coil in the field. The output is proportional to the velocity of the coil, and this type of pickup is therefore generally used for the measurement of velocities developed in a linear, sinusoidal, or random manner. Damping is obtained electrically, thus assuring high stability under varying temperature conditions.

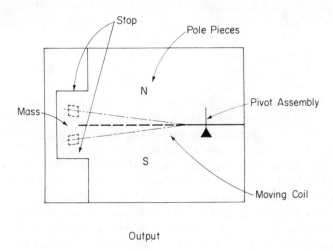

Output

Figure 11-14 Elements of a velocity transducer.

11-5 TEMPERATURE MEASUREMENTS

11-5.1 Resistance Thermometers

Resistance-temperature detectors, or resistance thermometers, employ a sensitive element of extremely pure platinum, copper, or nickel wire that provides a definite resistance value at each temperature within its range. The relationship between temperature and resistance of conductors in the temperature range near 0°C can be calculated from the equation

$$R_t = R_{ref} (1 + \alpha \, \Delta t) \tag{11-11}$$

where R_t = resistance of the conductor at temperature t (°C)

R_{ref} = resistance at the reference temperature, usually 0°C

α = temperature coefficient of resistance

Δt = difference between operating and reference temperature

Almost all metallic conductors have a positive temperature coefficient of resistance so that their resistance increase with an increase in temperature. Some materials, such as carbon and germanium, have a negative temperature coefficient of resistance that signifies that the resistance decreases with an increase in temperature. A high value of α is desirable in a temperature-sensing element so that a substantial change in resistance occurs for a relatively small change in temperature. This change in resistance (ΔR) can be measured with a Wheatstone bridge, which may be calibrated to indicate the temperature that caused the resistance change rather than the resistance change itself.

Figure 11-15 shows the variation of resistance with temperature for several commonly used materials. The graph indicates that the resistance of platinum and copper increases almost linearly with increasing temperature, while the characteristic for nickel is decidedly nonlinear.

The sensing element of a resistance thermometer is selected according to the intended application. Table 11-2 summarizes the characteristics of the three most commonly used resistance materials. Platinum wire is used for most laboratory work and for industrial measurements of high accuracy. Nickel wire and copper wire are less expensive and easier to manufacture than platinum wire elements, and they are often used in low-range industrial applications.

Resistance thermometers are generally of the probe type for immersion in the medium whose temperature is to be measured or controlled. A typical sensing element for a probe-type thermometer is constructed by coating a small platinum or silver tube with ceramic material, winding the resistance wire over the coated tube, and coating the finished winding again with ceramic. This small assembly is then fired at high temperature to assure annealing of the winding and then it is placed at the tip of the probe. The probe is protected by a sheath to produce the complete sensing element.

Practically all resistance thermometers for industrial applications are mounted in a tube or well to provide protection against mechanical damage and to

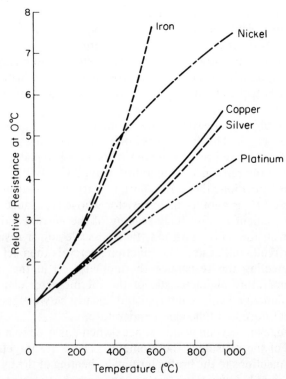

Figure 11-15 Relative resistance (R_t/R_{ref}) versus temperature for some pure metals.

guard against contamination and eventual failure. Protecting tubes are used at atmospheric pressure; when they are equipped with a pipe-thread bushing, they may be exposed to low or medium pressures. Metal tubes offer adequate protection to the sensing element at temperatures to 2,100°F, although they may become slightly porous at temperatures above 1,500°F and then fail to protect against contamination.

TABLE 11-2   RESISTANCE THERMOMETER ELEMENTS

Type	Temperature range	Accuracy	Advantages	Disadvantages
Platinum	−300°F to +1,500°F	±1°F	Low cost High stability Wide operating range	Relatively slow response time (15 s) Not as linear as copper thermometers
Copper	−325°F to +250°F	±0.5°F	High linearity High accuracy in ambient temperature range High stability	Limited temperature range (to 250°F)
Nickel	+32°F to +150°F	±0.5°F	Long life High sensitivity High temperature coeffi- cient	More nonlinear than copper Limited temperature range (to 150°F)

Protecting wells are designed for use in liquid or gases at high pressure such as in pipe lines, steam power plants, pressure tanks, pumping stations, and so on. The use of a protecting well becomes imperative at pressures above 3 at. Protective wells are drilled from solid bar stock, usually carbon steel or stainless steel, and the sensing element is mounted inside. A waterproof junction box with provision for conduit coupling is attached to the top of the well or tube, as shown in Fig. 11-16.

A typical bridge circuit with resistance thermometer R_t in the unknown position is shown in Fig. 11-17. The function switch connects three different resistors in the circuit. R_{ref} is a fixed resistor whose resistance is equal to that of the thermometer element at the reference temperature (say, 0°C). With the function switch in the "REF" position, the zero-adjust resistor is varied until the bridge indicator reads zero. R_{fs} is another fixed resistor whose resistance equals that of the thermometer element for full-scale reading of the current indicator. With the function switch in the "FS" position, the full-scale-adjust resistor is varied until the indicator reads full scale. The function switch is then set to the "MEAS" position, connecting the resistance thermometer R_t in the circuit. When the resistance-temperature characteristic of the thermometer element is linear, the galvanometer indication can be interpolated linearly between the set of values of reference temperature and full-scale temperature.

The Wheatstone bridge has certain disadvantages when it is used to measure the resistance variations of the resistance thermometer. These are the effect of contact resistances of connections to the bridge terminals, heating of the elements by the unbalance current, and heating of the wires connecting the thermometer to the bridge. Slight modifications of the Wheatstone bridge, such as the double slide-wire bridge, eliminate most of these problems. Despite these measurement difficulties, the resistance thermometer method is so accurate that it is one of the standard methods of temperature measurement within the range −183 to 630°C.

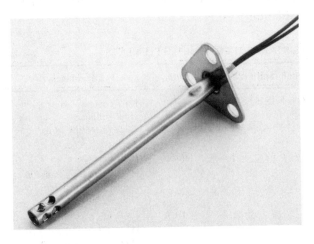

Figure 11-16(a) Thermistor housing for air/gas temperature measurements. (Courtesy of Fenwal Electronics.)

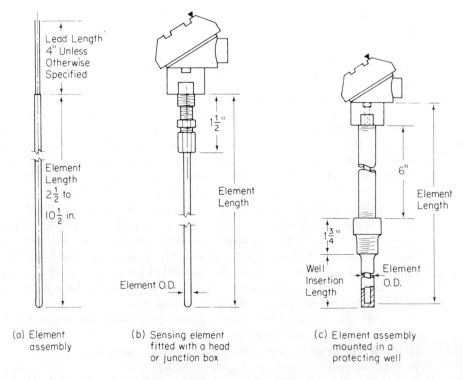

(a) Element assembly

(b) Sensing element fitted with a head or junction box

(c) Element assembly mounted in a protecting well

Figure 11-16(b)   Resistance thermometers.

11-5.2 Thermocouples

In 1821 Thomas Seebeck discovered that when two dissimilar metals were in contact, a voltage was generated where the voltage was a function of temperature. The device, consisting of two dissimilar metals joined together, is called a

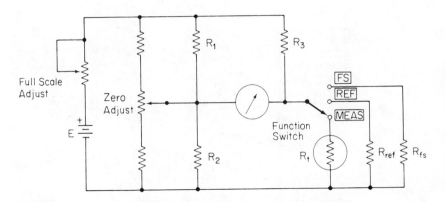

Figure 11-17   Bridge circuit with a resistance thermometer as one of the bridge elements.

thermocouple and the voltage is called the Seebeck voltage, in honor of the discoverer.

As an example, joining copper and Constantan (see the description of Constantan in Section 11-3.2) produces a voltage on the order of a few tens of millivolts (see Fig. 11-18) with the positive potential at the copper side. An increase in temperature causes an increase in voltage.

There are several methods of joining the two dissimilar metals. One is to weld the wires together. This produces a brittle joint, and if not protected from stresses, this type of thermocouple can fracture and break apart. During the welding process gases from the welding can diffuse into the metal and cause a change in the characteristic of the thermocouple. Another method of joining the two dissimilar metals is to solder the wires together. This has the disadvantage of introducing a third dissimilar metal. Fortunately, if both sides of the thermocouple are at the same temperature, the Seebeck voltage due to thermocouple action between the two metals of the thermocouple and the solder will have equal and opposite voltages and the effect will cancel. A more significant disadvantage is that the thermocouple is a desirable transducer for measuring high temperatures. In many cases the temperatures to be measured are higher than the melting point of the solder and the thermocouple will come apart.

It would appear to be a simple matter to measure the Seebeck voltage and create an electronic thermometer. To do this, wires could be connected as shown in Fig. 11-19 to make the measurement. This connection of the wires causes a problem of measurement, as shown in the figure.

Assume that the meter uses copper wires as shown. In this case, where the two copper wires come in contact there is no problem, but where the copper

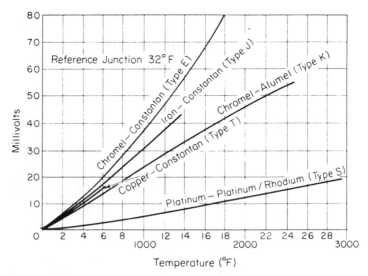

Figure 11-18 Thermocouple output voltage as a function of temperature for various thermocouple materials.

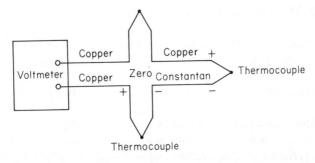

Figure 11-19 Effects of additional parasitic thermocouple.

comes in contact with another metal, such as the Constantan thermocouple wire, the two dissimilar metals create another thermocouple, which generates its own Seebeck voltage.

For this example, copper interconnecting wires were used and the thermocouple was copper and Constantan. The composition of the wires is immaterial, as any combination will produce these parasitic thermocouples with the problems of additional Seebeck voltages. It is an inescapable fact that there will be at least two thermocouple junctions in the system. To contend with this, it is necessary that the temperature of one of the junctions be known and constant. Therefore, there is a fixed offset voltage in the measuring system. It was customary a long time ago to place this junction in a mixture of ice and water, thus stabilizing the temperature to 0°C as shown in Fig. 11-20. More modern techniques use electronic reference junctions that are not necessarily at 0°C. This junction is called the reference or cold junction due to the fact that this junction was in the ice bath.

The classic method of measuring thermocouple voltages was the use of a potentiometer. This was a mechanical device and is no longer used. Completely electronic devices are used to measure thermocouple voltages and to convert from the Seebeck voltage to temperature, and to compensate for the reference junction.

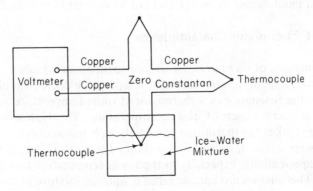

Figure 11-20 Application of a reference junction.

11-5.3 Thermocouple Error Sources

Open junction.   There are many sources of an open junction, some of which were outlined earlier. Usually, the error introduced by an open junction is of such an extreme magnitude that an open junction is easily spotted. By simply measuring the resistance of the thermocouple, the open junction can be identified.

Decalibration.   This error is a potentially serious fault, as it can cause more subtle errors that may escape detection. Decalibration is due to altering the characteristics of the thermocouple wire, thus changing the Seebeck voltage. This can be caused by subjecting the wire to excessively high temperatures, diffusion of particles from the atmosphere into the wire, or by cold working the wire. The last effect may be caused by straining the wire by drawing it through a long conduit.

Insulation degradation.   The thermocouple is often used at very high temperatures. In some cases the insulation can break down and cause a significant leakage resistance which will cause an error in the measurement of the Seebeck voltage. In addition, chemicals in the insulation can diffuse into the thermocouple wire and cause decalibration.

Galvanic action.   Chemicals coming in contact with the thermocouple wire can cause a galvanic action. This resultant voltage can be as much as 100 times the Seebeck effect, causing extreme errors.

Thermal conduction.   The thermocouple wire will shunt heat energy away from the source to be measured. For small masses to be measured, small-diameter thermocouple wire could be used. However, the smaller-diameter wire is more susceptible to the effects described previously. If a reasonable compromise between the degrading effects of small thermocouple wire and the loss of thermal energy and the resultant temperature error cannot be found, thermocouple extension wire can be used. This allows the thermocouple to be made of small-diameter wire, while the extension wire that covers the majority of the connecting distance is of a much larger diameter and not as susceptible to the degrading effects.

11-5.4 Thermistor Characteristics

Thermistors, or *thermal resistors*, are semiconductor devices that behave as resistors with a high, usually *negative*, temperature coefficient of resistance. In some cases, the resistance of a thermistor at room temperature may decrease as much as 6 per cent for each 1°C rise in temperature. This high sensitivity to temperature change makes the thermistor extremely well suited to precision temperature *measurement, control*, and *compensation*. Thermistors are therefore widely used in such applications, especially in the lower temperature range of −100°C to 300°C.

Thermistors are composed of a sintered mixture of metallic oxides, such as manganese, nickel, cobalt, copper, iron, and uranium. Their resistances range from 0.5 Ω to 75 MΩ and they are available in a wide variety of shapes and sizes. Smallest in size are the beads with a diameter of 0.15 mm to 1.25 mm. Beads may be sealed in the tips of solid glass rods to form probes that are somewhat easier to mount than beads. Disks and washers are made by pressing thermistor material under high pressure into flat cylindrical shapes with diameters from 2.5 mm to 25 mm. Washers can be stacked and placed in series or in parallel for increased power dissipation.

Three important characteristics of thermistors make them extremely useful in measurement and control applications: (a) the *resistance-temperature* characteristic, (b) the *voltage-current* characteristic, and (c) the *current-time* characteristic. Representative examples of these characteristic curves are shown in Fig. 11-21.

The resistance-temperature characteristic of Fig. 11-21(a) shows that the thermistor has a very high negative temperature coefficient of resistance, making it an ideal *temperature transducer*. The resistance-versus-temperature variations of the two industrial materials are compared to the characteristics for platinum (a widely used resistance thermometer material). Between the temperatures of $-100°C$ and $+400°C$, the resistance of type A thermistor material changes from 10^7 to 1 ohm-cm, while the resistance of platinum varies only by a factor of approximately 10 over the same temperature range.

The voltage-current characteristic of Fig. 11-21(b) shows that the voltage drop across a thermistor increases with increasing current until it reaches a peak value beyond which the voltage drop decreases as the current increases. In this portion of the curve, the thermistor exhibits a *negative resistance* characteristic. If a very small voltage is applied to the thermistor, the resulting small current does not produce sufficient heat to raise the temperature of the thermistor above ambient. Under this condition, Ohm's law is followed and the current is proportional to the applied voltage. Larger currents, at larger applied voltages, produce enough heat to raise the thermistor above the ambient temperature and its resistance then decreases. As a result, more current is then drawn and the resistance decreases further. The current continues to increase until the heat dissipation of the thermistor equals the power supplied to it. Therefore under any fixed ambient conditions, the resistance of a thermistor is largely a function of the power being dissipated within itself, provided that there is enough power available to raise its temperature above ambient. Under such operating conditions, the temperature of the thermistor may rise 100°C or 200°C, and its resistance may drop to one-thousandth of its value at low current.

This characteristic of *self-heat* provides an entirely new field of uses for the thermistor. In the self-heat state the thermistor is sensitive to anything that changes the *rate* at which heat is conducted away from it. It can so be used to measure *flow, pressure, liquid level, composition of gases,* etc. If, however, the rate of heat removal is fixed, the thermistor is sensitive to power input and can be used for *voltage* or *power-level* control.

The current-time characteristic curve of Fig. 11-21(c) indicates the time delay to reach maximum current as a function of the applied voltage. When the self-heating effect just described occurs in a thermistor network, a certain finite

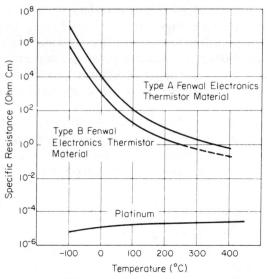

(a) Resistance-temperature characteristic

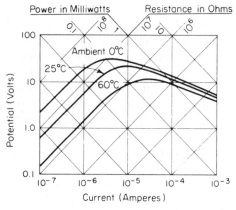

(b) Voltage-current characteristic

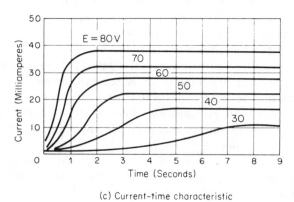

Time (Seconds)

(c) Current-time characteristic

Figure 11-21 Characteristic curves of thermistors. (Courtesy of Fenwal Electronics, Inc.)

time is required for the thermistor to heat and the current to build up to a maximum steady-state value. This time, although fixed for a given set of circuit parameters, may easily be varied by changing the applied voltage or the series resistance of the circuit. This time-current effect provides a simple and accurate means of achieving time delays from milliseconds to many minutes.

11-5.5 Interfacing Resistive Transducers to Electronic Circuits

Perhaps the major problem with resistive transducers, which includes strain gages and temperature sensors, is that for most operational ranges, the amount of resistance change is very small. As an example, consider measuring the current through a resistance transducer such as an RTD. A simple panel meter is used as an indicator to provide a remote reading of temperature. The change in the meter indication is very small for small temperature changes. As an example, the change in resistance of a platinum resistance thermometer is 0.385 per cent per degree Celsius. In this case, a 1-degree change in temperature will produce a 0.385 per cent change in the indicating meter, which will be hardly visible. A solution to this problem is to connect the resistance transducer in a bridge circuit as shown in Fig. 11-22. Bridge circuits, as discussed in Chapter 5, are no longer used for resistance measurements and are seldom used in the potentiometer configuration for precision voltage measurements. They are often used, however, for interfacing resistance-type transducers as shown in this chapter. First, the zero output voltage point can be set for a convenient point such as 0 degrees Celsius or 0 degrees Fahrenheit rather than absolute zero, which would be the situation if only the transducer current were measured. The setting of the zero output point can be achieved by adjusting the values of R_1, R_2, and R_3 to provide bridge

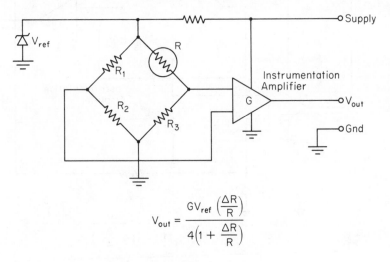

$$V_{out} = \frac{GV_{ref}\left(\frac{\Delta R}{R}\right)}{4\left(1 + \frac{\Delta R}{R}\right)}$$

Figure 11-22 Representative thermistors.

balance at the desired temperature. The change of output voltage in Fig. 11-22 can be given as

$$E_o = \frac{GV_{ref}\left(\dfrac{\Delta R}{R}\right)}{4\left(1 + \dfrac{\Delta R}{2R}\right)} \tag{11-12}$$

The output voltage is not a linear function of the change in resistance because of the ΔR term in the denominator of Eq. (11-12).

An improvement may be made by providing a constant-current source for the bridge as shown in Fig. 11-23. The equation for the output voltage as a function of resistance is

$$E_o = \frac{GV_{ref}\,\Delta R}{R_s 4\left(1 + \dfrac{\Delta R}{4R}\right)} \tag{11-13}$$

There is an improvement in the linearity of the output voltage as a function of the change in resistance, as the ΔR term in the denominator is divided by a factor of $4R$ rather than $2R$ as in Eq. (11-12), and thus the effects of the ΔR term in the denominator are reduced.

If the change in resistance is small, which is often the case with a resistance thermal device, the error due to the lack of linearity is small. Absolute linearity

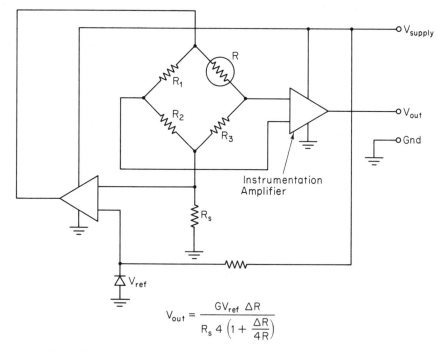

$$V_{out} = \frac{GV_{ref}\,\Delta R}{R_s\,4\left(1 + \dfrac{\Delta R}{4R}\right)}$$

Figure 11-23 Wheatstone bridge powered from a constant current source.

can be achieved by using two transducers as shown in Fig. 11-24. The output voltage as a function of R is

$$E_o = \frac{GV_{ref}}{2R_s} \Delta R \tag{11-14}$$

This technique requires the use of two matched transducers in the environment to be measured. Most resistive transducers are not expensive items, and providing two matched transducers on a common header is not a difficult task.

Often when a transducer is eventually connected to a digital system, a microprocessor can be employed to "linearize" the bridge output voltage.

11-5.6 Thermistor Applications

Although thermistors are probably best known for their function in the measurement and control of temperature, they can be used in a variety of other applications. A few common applications are described in the following paragraphs.

The thermistor's relatively large resistance change per degree change in temperature (called the *sensitivity*) makes it an obvious choice as a temperature transducer. A typical industrial-type thermistor with a 2,000-Ω resistance at 25°C and a temperature coefficient of 2.9 per cent/°C will exhibit a resistance change of 78 Ω/°C change in temperature. When this thermistor is connected in a simple series circuit consisting of a battery and a microammeter, any variation in temper-

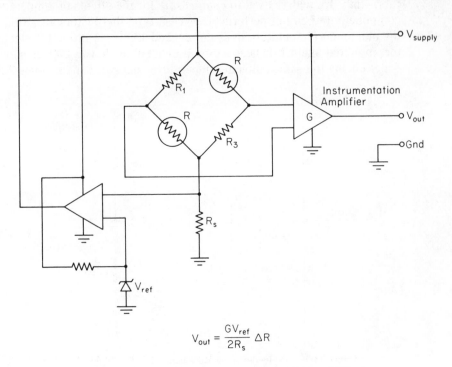

$$V_{out} = \frac{GV_{ref}}{2R_s} \Delta R$$

Figure 11-24 Wheatstone bridge employing two RTDs.

ature causes a change in thermistor resistance and a corresponding change in circuit current. The meter can be calibrated directly in terms of temperature and may be able to solve temperature variations of 0.1°C. Higher sensitivity is obtained by using the bridge circuit of Fig. 11-25. The 4-kΩ thermistor will readily indicate a temperature change of as little as 0.005°C.

This high sensitivity, together with the relatively high thermistor resistance that may be selected (for example, 100 kΩ), makes the thermistor ideal for *remote* measurement or control, since changes in contact or transmission line resistance due to ambient temperature effects are negligible.

A simple *temperature control* circuit may be constructed by replacing the microammeter in the bridge circuit of Fig. 11-25 with a relay. This is indicated in the typical thermistor temperature control circuit of Fig. 11-26 where a 4-kΩ thermistor is connected in an ac-excited bridge. The unbalance voltage is fed to an ac amplifier whose output drives a relay. The relay contacts are used to control the current in the circuit that generates the heat. These control circuits can be operated to a precision of 0.0001°F.

Thermistor control systems are inherently sensitive, stable, and fast acting, and they require relatively simple circuitry. The voltage output of the standard thermistor bridge circuit at 25°C will be approximately 18 mV/°C using a 4,000-Ω thermistor in the configuration of Fig. 11-25.

Because thermistors have a negative temperature coefficient of resistance—opposite to the positive coefficient of most electrical conductors and semiconductors—they are widely used to *compensate* for the effects of temperature on both component and circuit performance. Disk-type thermistors are often used when the maximum temperature does not exceed 125°C. A properly selected thermistor, mounted against or near a circuit element such as a copper meter coil, and experiencing the same ambient temperature changes, can be connected in such a

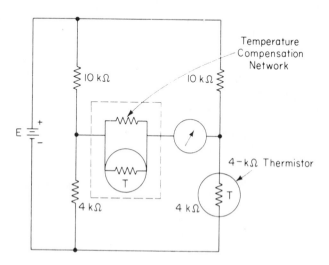

Figure 11-25 Temperature measurement with the thermistor in a bridge circuit with compensation to improve sensitivity.

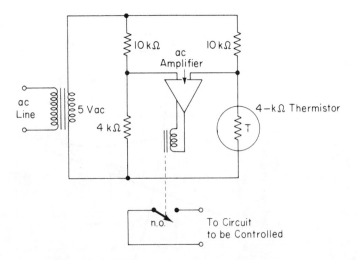

Figure 11-26 Thermistor temperature control circuit.

way that the total circuit resistance is constant over a wide range of temperatures. This is shown in the curves of Fig. 11-27, which illustrates the effect of a *compensation network.*

The compensator consists of a thermistor, shunted by a resistor. The negative temperature coefficient of this combination equals the positive coefficient of the copper meter coil. The coil resistance of 5,000 Ω at 25°C varies from approximately 4,500 Ω at 0°C to 5,700 Ω at 60°C, representing a change of about ±12 per cent. With a single thermistor compensation network, this variation can be reduced to approximately ±15 Ω, or ±$\frac{1}{4}$ per cent. With double or triple compensation networks, variations can be reduced even further.

In a *thermal conductivity measurement,* two thermistors are connected in adjacent legs of a Wheatstone bridge, as shown in Fig. 11-28. The bridge supply voltage is high enough to raise the thermistors above ambient temperature, typically to about 150°C. One thermistor is mounted in a static area to provide temperature compensation while the other is placed in the medium to be measured. Any change in the thermal conductivity of this medium will change the rate at which heat is dissipated from the sensing thermistor, thus changing its temperature. This results in bridge unbalance, which can be calibrated in appropriate units.

In another application, two thermistors are placed in separate cavities in a brass block. With air in both cavities, the bridge is balanced. When the air in one cavity is replaced by pure carbon dioxide, which has a lower conductivity than air, the bridge will be unbalanced because that thermistor becomes hotter and lower in resistance. This amount of unbalance represents 100 per cent CO_2 in the analyzer; 50 per cent CO_2 gives just half the meter reading, and the instrument may be calibrated with a linear scale to read per cent CO_2 in air. Similar calibration may be made for any other mixture of two gases.

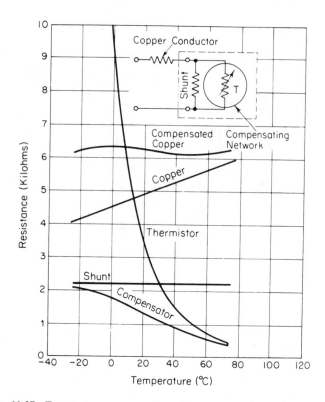

Figure 11-27 Temperature compensation of a copper conductor by means of a thermistor network.

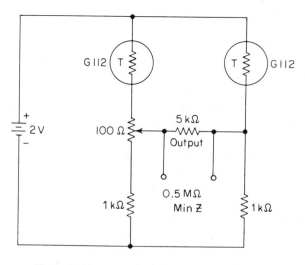

Figure 11-28 Thermal conductivity measurement.

If the same bridge uses one thermistor sealed in a cavity in a brass block and the other thermistor mounted in a small pipe, it may be used as a *flow meter*. When no air is flowing through the pipe, the bridge may be balanced. When air flows through the pipe, the thermistor is cooled, and its resistance increases which unbalances the bridge. The amount of cooling is proportional to the rate of flow of the air and the meter may be calibrated in terms of flow in the pipe. Such instruments have been made to measure flow rates as low as 0.001 cm^3 per minute.

11-6 PHOTOSENSITIVE DEVICES

Photosensitive elements are versatile tools for detecting radiant energy or light. They exceed the sensitivity of the human eye to all the colors of the spectrum and operate even into the ultraviolet and infrared regions.

The photosensitive device has found practical use in many engineering applications. This section deals with the following devices and their applications:

(a) *Vacuum-type phototubes,* used to best advantage in applications requiring the observation of light pulses of short duration, or light modulated at relatively high frequencies.

(b) *Gas-type phototubes,* used in the motion picture industry as sound-on-film sensors.

(c) *Multiplier phototubes,* with tremendous amplifying capability, used extensively in photoelectric measurement and control devices and also as scintillation counters.

(d) *Photoconductive cells,* also known as *photoresistors* or *light-dependent resistors,* find wide use in industrial and laboratory control applications.

(e) *Photovoltaic cells* are semiconductor junction devices used to convert radiation energy into electrical power. A fine example is the *solar cell* used in space engineering applications.

11-6.1 Vacuum Phototube

The photocathode emits electrons when stimulated by incident radiant energy. The most important photocathode now used in vacuum phototubes is the cesium-antimony surface, which is characterized by high sensitivity in the visible spectrum. The type of glass used in the glass envelope determines mainly the sensitivity of the device at other wavelengths. Usually the glass cuts off the transmitted radiation in the ultraviolet region.

Typical voltage-current characteristics are shown in Fig. 11-29(a). When sufficient voltage is applied between the photocathode and the anode, the collected current is almost entirely dependent on the amount of incident light. Vacuum phototubes are characterized by a photocurrent response that is linear over a wide range, so much so that these tubes are frequently used as standards in light-comparison measurements. Figure 11-29(b) shows the linear current-light relationship.

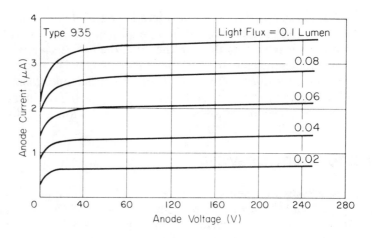

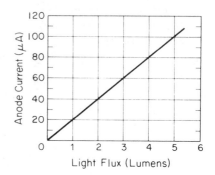

(a) Typical anode characteristics

(b) Output current as a function
of light intensity

Figure 11-29   Characteristics of a vacuum phototube.

11-6.2 Gas-Filled Phototube

The gas-filled phototube has the same general construction as the vacuum photo-tube, except that the envelope contains inert gas (usually argon) at a very low pressure. Electrons are emitted from the cathode by photoelectric action and accelerate through the gas by the applied voltage at the anode. If the energy of the electrons exceeds the ionization potential of the gas (15.7 V for argon), the collision of an electron and a gas molecule can result in ionization, i.e., the creation of a positive ion and a second electron. As the voltage is further increased above the ionization potential, the current collected by the anode increases because of the higher number of collisions between photoelectrons and gas molecules. If the anode voltage is raised to a very high value, the current becomes uncontrolled; all the gas molecules are then ionized and the tube exhibits a glow discharge. The condition should be avoided because it may permanently damage the phototube. Typical current-voltage characteristics for various light levels are given in Fig. 11-30.

11-6.3 Multiplier Phototubes

To detect very low light levels, special amplification of the photocurrent is necessary in most applications. The multiplier phototube, or photomultiplier, uses secondary emission to provide current amplification in excess of a factor 10^6 and then becomes a very useful detector for low light levels.

In a photomultiplier the electrons emitted by the photocathode are electrostatically directed toward a secondary emitting surface, called a *dynode*. When the proper operating voltage is applied to the dynode, three to six secondary electrons are emitted for every primary electron striking the dynode. These secondary electrons are focused to a second dynode, where the process is repeated. The original emission from the photocathode is therefore *multiplied* many times.

Figure 11-31 shows a photomultiplier with ten dynodes. The last dynode (10) is followed by the anode that collects the electrons and serves as the signal output electrode in most applications.

The linear photomultiplier of Fig. 11-32 (also known as the Matheson tube) has a specially designed focused cage structure with a large effective area for the collection of photoelectrons on the first dynode. The Matheson tube uses a curved cathode and annular rings for electrostatic focusing of the photoelectrons. This construction results in very effective collection of photoelectrons and also in very short transit times (high-frequency response).

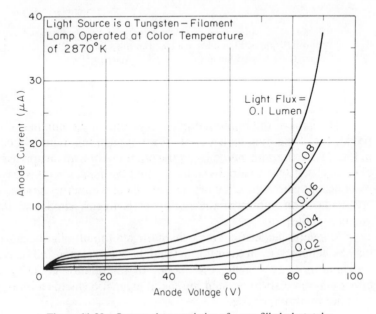

Figure 11-30 Output characteristics of a gas-filled phototube.

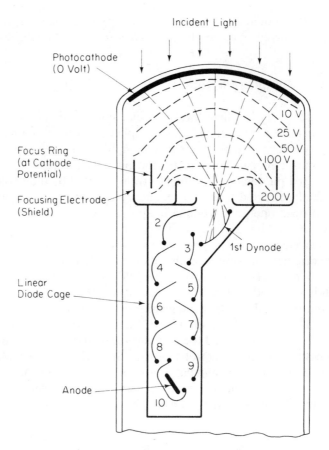

Figure 11-31 Linear photomultiplier with Matheson front-end configuration, showing equipotential lines and electron trajectories feeding into a linear diode cage. (Courtesy of Radio Corporation of America.)

The gain of the photomultiplier depends on the number of dynodes and the properties of the dynode material. For a typical ten-dynode tube, such as shown in Fig. 11-32, the gain would be on the order of 10^6 with an applied voltage of 100 V per stage (a 1,000-V supply source would be needed in this case). The spectral response may be controlled by the material of which the cathode and the dynodes are made. The output of the multiplier is linear, similar to the output of the vacuum phototube.

Magnetic fields affect the gain of the photomultiplier because some electrons may be deflected from their normal path between stages and therefore never reach a dynode nor, eventually, the anode. In scintillation-counting applications this effect may be disturbing, and mu-metal magnetic shields are often placed around the photomultiplier tube.

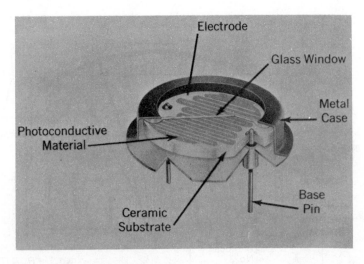

Electrode

Glass Window

Metal
Case

Photoconductive
Material

Base
Pin

Ceramic
Substrate

Figure 11-32   Cutaway view of a photoconductive cell.
(Courtesy of Radio Corporation of America.)

11-6.4 Photoconductive Cells

Photoconductive cells are elements whose conductivity is a function of the incident electromagnetic radiation. Many materials are photoconductive to some degree, but the commercially important ones are cadmium sulfide, germanium, and silicon. The spectral response of the cadmium sulfide cell closely matches that of the human eye, and the cell is therefore often used in applications where human vision is a factor, such as street light control or automatic iris control for cameras.

The essential elements of a photoconductive cell are the ceramic substrate, a layer of photoconductive material, metallic electrodes to connect the device into a circuit, and a moisture-resistant enclosure. A cutaway view of a typical photoconductive cell is shown in Fig. 11-32.

A typical application of a practical *on-off* photocell control circuit is given in Fig. 11-33. Resistors R_2, R_3, and R_4 are chosen so that the emitter-to-base bias of Q_2 is sufficiently positive to allow Q_2 to conduct. As a result, the relay in Q_2 collector circuit will be energized. When configuration A is used as the control circuit, the relay is energized when the light on the photocell is below a predetermined level. When the photocell is illuminated, the emitter-to-base bias of Q_1 becomes sufficiently positive to allow Q_1 to conduct. Its collector potential becomes considerably less positive, decreasing the bias on Q_2, and Q_2 cuts off, deenergizing the relay. When configuration B is used, the relay will be energized when the light incident on the photocell is above a predetermined level.

Semiconductor junction photocells are used in some applications. The volt-ampere characteristics of a *p-n* junction may appear as the solid line in Fig. 11-34, but when light is applied to the cell, the curve shifts downward, as shown by the broken line.

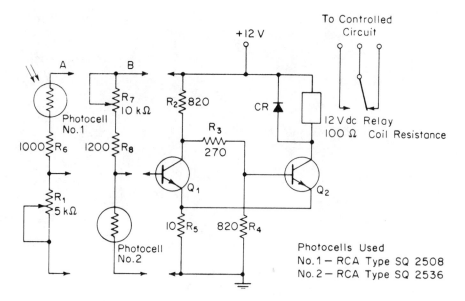

Figure 11-33 Twelve-volt photocell control circuit. (Courtesy of RCA, Electronic Components and Devices Division.)

In photoconductive applications the cell is biased in the reverse direction. When the cell is illuminated, the reverse current increases and an output voltage can be developed across an output resistor. This output voltage is then proportional to the amount of incident light. A typical order of magnitude for the increase in output current is approximately 0.7 μA for each 1 footcandle increase in illumination. This increase in photocurrent is linear with an increase in illumination. The time constant of *p-n* junction photocells is relatively fast, making these devices useful for optical excitation frequencies well above the audio range.

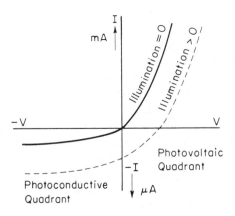

Figure 11-34 Current-voltage characteristics of a photojunction diode.

11-6.5 Photovoltaic Cells

Photovoltaic cells may be used in a number of applications. The silicon *solar cell* converts the radiant energy of the sun into electrical power. The solar cell consists of a thin slice of single crystal *p*-type silicon, up to 2 cm square, into which a very thin (0.5 micron) layer of *n*-type material is diffused. The conversion efficiency depends on the spectral content and the intensity of the illumination.

Multiple-unit silicon photovoltaic devices may be used for sensing light in applications such as reading punched cards in the data-processing industry.

Gold-doped germanium cells with controlled spectral response characteristics act as photovoltaic devices in the infrared region of the spectrum and may be used as infrared detectors.

REFERENCES

11-1. Bartholomew, Davis, *Electrical Measurements and Instrumentation,* chap. 11. Boston: Allyn and Bacon, Inc., 1963.

11-2. *Capsule Thermistor Course,* Fenwal Electronics, Inc., Framingham, Mass., n. d.

11-3. Fribance, Austin E., *Industrial Instrumentation Fundamentals,* chaps. 10, 12, 15, 16. New York: McGraw-Hill Book Company, 1962.

11-4. *Introduction to Transducers for Instrumentation,* Statham Instruments, Inc., Los Angeles, Calif., n. d.

11-5. Lion, Kurt S., *Instrumentation in Scientific Research.* New York: McGraw-Hill Book Company, 1959.

11-6. Minnar, E. J. (ed.), *Instrument Society of American Transducer Compendium.* New York: Plenum Press, 1963.

11-7. Partridge, G. R., *Principles of Electronic Measurements,* chap. 13. Englewood Cliffs, N.J.: Prentice-Hall, Inc., 1958.

11-8. Perry, C. C., and Lissner, H. R., *The Strain Gage Primer,* 2nd ed. New York: McGraw-Hill Book Company, 1962.

11-9. *Photomultiplier Handbook,* PMT-62, RCA Solid State Division, Optics and Devices, Lancaster, Pa., 1980.

11-10. *Phototubes and Photocells,* Technical Manual PT-60, Radio Corporation of America, Electronic Components and Devices, Lancaster, Pa., n. d.

11-11. Stout, Melville B., *Basic Electrical Measurements,* 2nd ed., chap. 16. Englewood Cliffs, N.J.: Prentice-Hall, Inc., 1960.

PROBLEMS

11-1. Name four types of electrical pressure transducer and describe one application of each type.

11-2. Under what conditions is a "dummy" strain gage used, and what is the function of that gage?

11-3. What is the difference between a photoemissive, a photoconductive, and a photovoltaic cell? Name one application for each cell.

11-4. A resistance strain gage with a gage factor of 2.4 is mounted on a steel beam whose modulus of elasticity is 2×10^6 kg/cm². The strain gage has an unstrained resistance of 120.0 Ω which increases to 120.1 Ω when the beam is subjected to a stress. Calculate the stress at the point where the strain gage is mounted.

11-5. The unstrained resistance of each of the four elements of the unbonded strain gage of Fig. 11-4 is 120 Ω. The strain gage has a gage factor of 3 and is subjected to a strain ($\Delta l/l$) of 0.0001. If the indicator is a high-impedance voltmeter, calculate the reading of this voltmeter for a battery voltage of 10 V.

11-6. The linear variable differential transformer (LVDT) of Fig. 13-8 produces an output of 2 V rms for a displacement of 50×10^{-6} cm. Calculate the sensitivity of the LVDT in μV/mm. The 2-V output of the LVDT is read on a 5-V voltmeter that has a scale with 100 divisions. The scale can be read to 0.2 division. Calculate the resolution of the instrument in terms of displacement in inches.

12

Analog and Digital Data Acquisition Systems

12-1 INSTRUMENTATION SYSTEMS

Data acquisition systems are used to measure and record signals obtained in basically two ways: (a) Signals originating from *direct measurement* of electrical quantities; these may include dc and ac voltages, frequency, or resistance, and are typically found in such areas as electronic component testing, environmental studies, and quality analysis work. (b) Signals originating from *transducers*, such as strain gages and thermocouples (see Chapter 11).

Instrumentation systems can be categorized into two major classes: analog systems and digital systems. *Analog systems* deal with measurement information in analog form. An analog signal may be defined as a continuous function, such as a plot of voltage versus time, or displacement versus pressure. *Digital systems* handle information in digital form. A digital quantity may consist of a number of discrete and discontinuous pulses whose time relationship contains information about the magnitude or the nature of the quantity.

An *analog data acquisition system* typically consists of some or all of the following elements:

(a) *Transducers* for translating physical parameters into electrical signals.

(b) *Signal conditioners* for amplifying, modifying, or selecting certain portions of these signals.

(c) *Visual display devices* for continuous monitoring of the input signals. These devices may include single- or multichannel oscilloscopes, storage oscilloscopes, panel meters, numerical displays, and so on.

(d) *Graphic recording instruments* for obtaining permanent records of the input data. These instruments include stylus-and-ink recorders to provide continuous records on paper charts, optical recording systems such as mirror galvanometer recorders, and ultraviolet recorders.

(e) *Magnetic tape instrumentation* for acquiring input data, preserving their original electrical form, and reproducing them at a later date for more detailed analysis.

A *digital data acquisition system* may include some or all of the elements shown in Fig. 12-1. The essential functional operations within a digital system include handling analog signals, making the measurement, converting and handling digital data, and internal programming and control. The function of each of the system elements of Fig. 12-1 is listed below.

(a) *Transducer.* Translates physical parameters to electrical signals acceptable by the acquisition system. Some typical parameters include temperature, pressure, acceleration, weight displacement, and velocity (see Chapter 11). Electrical quantities, such as voltage, resistance, or frequency, also may be measured directly.

(b) *Signal conditioner.* Generally includes the supporting circuitry for the transducer. This circuitry may provide excitation power, balancing circuits, and calibration elements. An example of a signal conditioner is a strain-gage bridge balance and power supply unit.

(c) *Scanner, or multiplexer.* Accepts multiple analog inputs and sequentially connects them to one measuring instrument.

(d) *Signal converter.* Translates the analog signal to a form acceptable by the analog-to-digital converter. An example of a signal converter is an amplifier

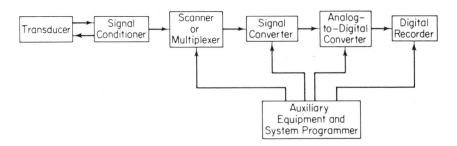

Figure 12-1   Elements of a digital data-acquisition system.

for amplifying low-level voltages generated by thermocouples or strain gages.

(e) *Analog-to-digital (A/D) converter.* Converts the analog voltage to its equivalent digital form. The output of the A/D converter may be displayed visually and is also available as voltage outputs in discrete steps for further processing or recording on a digital recorder.

(f) *Auxiliary equipment.* This section contains instruments for system programming functions and digital data processing. Typical auxiliary functions include linearizing and limit comparison. These functions may be performed by individual instruments or by a digital computer.

(g) *Digital recorder.* Records digital information on punched cards, perforated paper tape, magnetic tape, typewritten pages, or a combination of these systems. The digital recorder may be preceded by a coupling unit that translates the digital information to the proper form for entry into the particular digital recorder selected.

Data acquisition systems are used in a large and ever-increasing number of applications in a variety of industrial and scientific areas, such as the biomedical, aerospace, and telemetry industries. The type of data acquisition system, whether analog or digital, depends largely on the intended use of the recorded input data. In general, analog data systems are used when wide bandwidth is required or when lower accuracy can be tolerated. Digital systems are used when the physical process being monitored is slowly varying (narrow bandwidth) and when high accuracy and low per-channel cost is required. Digital systems range in complexity from single-channel dc voltage measuring and recording systems to sophisticated automatic multichannel systems that measure a large number of input parameters, compare against preset limits or conditions, and perform computations and decisions on the input signal. Digital data acquisition systems are in general more complex than analog systems, both in terms of the instrumentation involved and the volume and complexity of input data they can handle.

Data acquisition systems often use magnetic tape recorders, which are discussed in Sec. 12-2. Digital systems require converters to change analog voltages into discrete digital quantities or numbers. Conversely, digital information may have to be converted back into analog form, such as a voltage or a current, which can then be used as a feedback quantity controlling an industrial process. Conversion techniques are discussed in Secs. 12-4 and 12-5, while scanning or multiplexing equipment is described in Sec. 12-6. Section 12-7 briefly introduces the spatial, or shaft, encoder.

12-2 INTERFACING TRANSDUCERS TO ELECTRONIC CONTROL AND MEASURING SYSTEMS

The output voltages and currents from many transducers are very small signals. In addition to the low levels, it is often necessary to transmit the transducer output some distance to the data collection or control equipment. To compound matters even further, in an industrial environment, where there is large electrical machinery, the electrical noise present can cause serious difficulties in low-level circuits. These noises can be either radiated as an electromagnetic field or induced in the wiring of the plant as ground loops, and induced spikes on the ac power supply. Regardless of the source of noise, low-level signals must be transmitted from place to place with care.

One effective method of combating noise is to increase the strength of low-level signals before transmission through wires. This is often done with an amplifier called an instrumentation amplifier.

There are several characteristics of an instrumentation amplifier that set it apart from operational amplifiers.

(a) Instrumentation amplifiers have finite gain. An operational amplifier has a very large amount of gain, which in the ideal case is infinite. The operational amplifier is usually provided with external feedback to provide a finite gain or with other circuit elements to provide other functions, such as integrators, differentiators, filters, and so on. Other than fixed gain, the instrumentation amplifier cannot provide these functions.

(b) The instrumentation amplifier has a high-impedance differential input. The operational amplifier also has a high-impedance input. However, when the feedback elements are added around the operational amplifier, the input impedance is lowered considerably.

(c) The instrumentation amplifier has a high common-mode voltage range and a high common-mode rejection. Although op amps have common-mode rejection and voltage range, the instrumentation amplifier is superior to most op amps.

Figure 12-2 shows the block diagram of the instrumentation amplifier. After the discourse of the previous paragraphs outlining the differences between the instrumentation amplifier and the operational amplifier, the figure shows that the instrumentation amplifier is constructed of operational amplifiers.

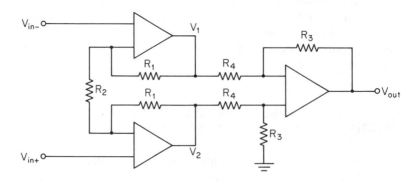

Figure 12-2 Composition of an instrumentation amplifier.

Analog and Digital Data Acquisition Systems   Chap. 12

Notice that each input to the instrumentation amplifier is the noninverting input to an operational amplifier, and thus the input impedance of the instrumentation amplifier is very high. To determine the gain of the amplifier, the usual criterion applied to operational amplifiers is applied.

The voltage at the inverting input of the lower input amplifier is V_{in+} and therefore V_1 is

$$V_1 = \left(\frac{R_1}{R_2}\right)(V_{in-} - V_{in+}) \tag{12-1}$$

In a similar fashion V_2 can be written as

$$V_2 = \left(\frac{R_1}{R_2}\right)(V_{in+} - V_{in-}) \tag{12-2}$$

The output stage is nothing more than a simple differencing amplifier, and thus the output voltage can be derived.

$$V_{out} = \left(\frac{R_3}{R_4}\right)(V_2 - V_1) = \left(\frac{2R_1R_3}{R_2R_4}\right)(V_{in+} - V_{in-}) \tag{12-3}$$

To reduce the pickup of noise voltages in the connections between the transducer and the amplifier, the leads to the transducer are kept as short as possible and the amplified signal is transmitted the required distance. There are situations where the low-level transducer must be transmitted through some length of wires. An example would be the connections to a thermocouple in a furnace where the temperature is too high to permit the introduction of electronics. This and similar situations require that the connections to the transducer be differential to prevent the introduction of noise.

Figure 12-3(a) shows a transducer connected to an instrumentation amplifier in a differential manner. Noise currents are introduced equally in both lines connecting the transducer if the wires are twisted together so that they will not separate and thus each wire will be subjected to the same induced noise voltage. The ability of an amplifier to reject signals that appear the same on both inputs of an amplifier is called the common-mode rejection. In the case of an instrumentation amplifier the common-mode rejection for low-frequency signals such as power line frequencies is as high as 100 dB.

There are methods of shielding the connecting wires from external signal pickup. One effective method is called guarding. Figure 12-3 shows a signal source connected to a differential amplifier through wires that are in close proximity to a 220-V power lead, which capacitively couples power line frequency voltages into the signal amplifier. If the amplifier is perfect, the voltage induced in each lead is the same and the capacitance and resistance paths to ground are identical. Thus the power line interference is conducted to ground with identical currents from each side of the line. If the leakage resistance or capacitance is different for one line relative to the other, as Figure 12-3(b) shows, the currents to the instrument ground are from only one side of the differential line. By the addition of a shield which is connected to one side of the signal and to the instrument case, the capacitively coupled signals from the power frequency line

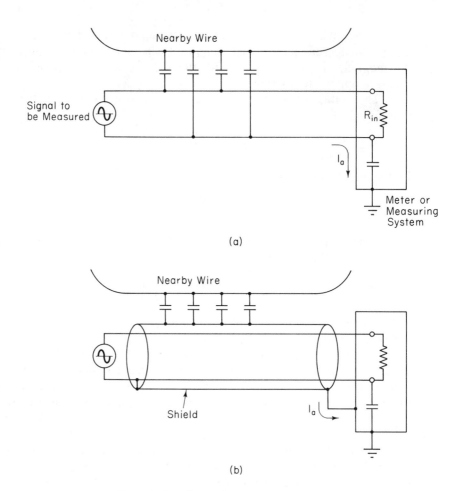

Nearby Wire

Signal to
be Measured

R_{in}

I_a

Meter or
Measuring
System

(a)

Nearby Wire

Shield

I_a

(b)

Figure 12-3 Effects of a guarded measurement.

are coupled to the shield and are safely conducted to the instrument case and to ground. The combination of the instrument case, its ground connection, and the shield extending to the signal source represents a complete shield around the entire measuring system.

There are situations where the noise environment is so severe that conventional amplifiers cannot survive the signal levels encountered. In these situations, an isolation amplifier is used to prevent the dangerously high-voltage noise signals from being conducted to the data acquisition equipment. Figure 12-4 shows the block diagram of this amplifier. The transducer is connected in a rather conventional fashion to an instrumentation amplifier. The output of this amplifier is fed to a balanced modulator which provides a bipolar square wave with an amplitude proportional to the signal level. The high-frequency square wave is called the carrier. The modulated square wave, being an ac signal with no dc level, can be

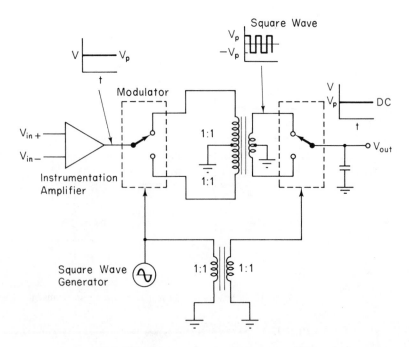

Figure 12-4 Schematic of an isolation amplifier.

coupled through a transformer to a balanced demodulator. The square-wave signal generator is transformer coupled to serve as the carrier for the demodulator,
which removes the carrier and restores the input level. After a small amount of
filtering, the output of the isolation amplifier is an accurate representation of the
input voltage.

For the sake of description, a dc input level is shown. Of course, the isolation amplifier can handle changing inputs as long as the chopping frequency used
for the balanced modulator and demodulator is sufficiently high. Many amplifiers
use chopping frequencies on the order of 25 kHz, which allows input frequencies
up to 1 kHz and higher to be present. Generally, transducer signals change slowly
and do not require fast amplifiers.

There are sources of error when transmitting a low-level signal any significant distance. One of these, as shown in Fig. 12-5, is the resistance of the interconnecting wires. As shown in the figure, three wires are used to connect an
amplifier at a distance. The power supply current flows through the power supply
wire and returns through the ground. Assume for the sake of discussion that the
wires are similar and have identical resistances as they are of equal length, as
shown in the figure. A load resistance is used to terminate the signal. Using the
topology shown in the figure, the voltage across the signal load, V_L, is

$$V_L = \frac{V_{out} + I_{supply}R \; R_L}{2R + R_L} \tag{12-4}$$

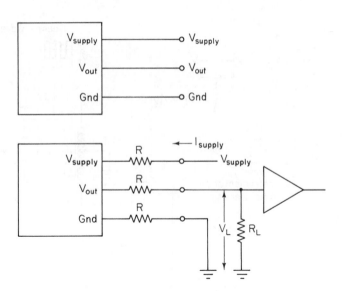

Figure 12-5 Effects of lead resistance on a remote measurement.

EXAMPLE 12-1

What is the percentage error for a remote amplifier where the resistance of the wire is 1 Ω, the load resistance is 10 kΩ, the power supply current is 50 mA, and the output of the amplifier is 1 V?

SOLUTION   Using Eq. (12-4) the load voltage is calculated to be

$$V_L = \frac{(V_{out} + I_{supply}\ R)R_L}{2R + R_L} = \frac{1 + 0.05 \times 1.0\ 10^4}{1.0002 \times 10^4}$$

$$= 1.05$$

This represents an error of about 5 per cent, which is unacceptable in most systems.

If the error were a constant, it could be compensated. However, any change in connector resistance due to corrosion or the addition of wire into the system would change the offset voltage and require a recalibration.

A method of reducing errors due to intervening resistances and the effects of the power supply currents is to use current rather than voltage for transmission. A typical current loop transmission system uses currents from 4 to 20 mA for the full-scale transmission of a specific parameter. The low-level current of 4 mA includes the power supply current, which allows for a 16-mA signal current component.

Figure 12-6 shows a voltage-to-current converter for a 4- to 20-mA current loop. Two voltage-controlled current sources are used in the converter. One current source senses the power supply current for the amplifier plus the current from the voltage-controlled current source, I_2, and sets the sum to equal 4 mA.

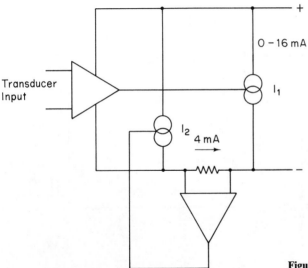

Figure 12-6   Current loop transmitter.

The second voltage-controlled current source, I_1, provides a variable current as a function of the transducer voltage as provided by the instrumentation amplifier. This second voltage-controlled current source provides from 0 to 16 mA for a total current from 4 to 20 mA.

The current source method of transmission reduces errors caused by the resistance of the interconnecting wires. There are such difficult noise environments that the transmission of analog signals of any sort even through an isolation amplifier or a current loop is difficult. For these situations, the analog signal is converted to a digital format for transmission. One simple method is to convert the analog input to a frequency using a voltage-to-frequency converter and at the receiving end convert the frequency to a voltage using a similar device.

Figure 12-7(a) shows a voltage-to-frequency converter. The converter consists of an integrator that feeds a comparator, which in turn drives a one-shot multivibrator. An electronic switch discharges the integrator via a current source.

The waveforms associated with the voltage-to-frequency converter are shown in Fig. 12-7(b). The input voltage, V_{in}, causes the integrator output to ramp in a negative direction. If the integrator output starts from some positive voltage (and it will be shown how the integrator output is forced to be positive), the output will eventually reach zero volt and the one-shot multivibrator will provide an output pulse. Using this zero voltage of the integrator as the starting point for the investigation of the voltage-to-frequency converter, the integrator output will ramp in the positive direction for a time equal to t_2. This time is set by the duration of the one-shot and causes the integrator output to achieve an output level of V_{ref}. Solving for V_{ref} in terms of t_2, the following is obtained:

$$V_{ref} = t_2 \left(\frac{I - \dfrac{V_{in}}{R}}{C} \right) \tag{12-5}$$

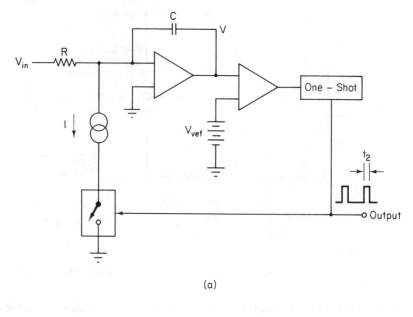

(a)

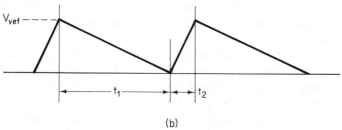

(b)

Figure 12-7 Voltage-to-frequency converter and associated waveforms.

After the one-shot is reset, the integrator output ramps down to zero volt and the one-shot is triggered again. The time taken to ramp from V_{ref} to zero volt is

$$t_1 = \frac{CV_{ref}R}{V_{in}} \tag{12-6}$$

Substituting the value of V_{ref} found from Eq. (12-5) into Eq. (12-6) yields

$$t_1 = \frac{RC}{V_{in}} t_2 \left(\frac{I - \dfrac{V_{in}}{R}}{C} \right) \tag{12-7}$$

The total time required for a complete period of the voltage-to-frequency converter is the sum of t_1 and t_2, which is

$$t = t_1 + t_2 = \frac{RC}{V_{in}} t_2 \left(\frac{I - \frac{V_{in}}{R}}{C} \right) + t_2 = \frac{\pm R t_2}{V_{in}} \qquad (12\text{-}8)$$

The frequency of the output is the reciprocal of the period and is

$$\frac{1}{t} = \frac{V_{in}}{IRt_2} \qquad (12\text{-}9)$$

As it can be seen from inspection of the relationship above, the output frequency does not involve the value of the integration capacitor. This is desirable, as precision capacitors are expensive items. The time of the one-shot is a part of the relationship, and this would imply that a precision capacitor would be required for the one-shot. For an analog one-shot this would be true. However, there are voltage-to-frequency converters where the one-shot timing is generated from a precision clock rather than an RC time constant.

The frequency is dependent on the value of R, which could easily be a precision resistor, and I, which could be a precision current source. Therefore, the voltage-to-frequency converter can be made with a reasonable amount of precision, and units with 1 per cent accuracy and better are commonplace.

The major reason for converting a voltage to a frequency is so that an analog signal can be transmitted through a communications system, whether it be wires or some more sophisticated system, where the loss through the system may either be unknown or unstable.

To implement such a system it is necessary to reconvert the frequency to an analog signal. Essentially the same elements that made up the voltage-to-frequency converter are used to create an analog signal from the varying frequency input.

The input frequency is used to trigger the one-shot as shown in Fig. 12-8. The duty of the one-shot output is

$$\frac{t_1}{t_2} = Ft_1 \qquad (12\text{-}10)$$

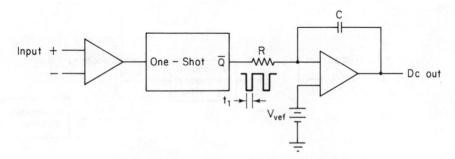

Figure 12-8   Frequency-to-voltage converter.

where t_1 = duration of the one shot

t_2 = period of the input frequency

f = input frequency

The average voltage of a pulse waveform, such as the output of the one-shot, is the peak voltage multiplied by the duty cycle, or

$$V_{avg} = V_p f t_1 \qquad (12\text{-}11)$$

This equation shows that the average value of the one-shot output is proportional to the input frequency. To convert the pulse waveform of the one-shot output to a dc level, it is necessary to integrate the one-shot output as shown in Fig. 12-8.

    With the exception of the current source and its associated electronic switch, the frequency-to-voltage converter uses the same functional elements as the voltage-to-frequency converter. Figure 12-9(a) shows a transducer connected to a voltage-to-frequency converter, through a long wire, and to a frequency converter which drives an analog meter. This system provides a remote reading of the parameter monitored by the transducer and provides an analog indication at the remote readout point. The loss of the wires connecting the transmission end to the receiving end does not affect the remote reading as long as the loss is not so great that it does not properly drive the receiving end.

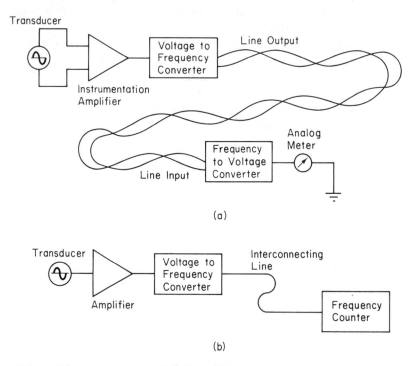

(a)

(b)

Figure 12-9 (a) Remote instrumentation system using voltage-to-frequency converter and a frequency-to-voltage converter; (b) digital readout remote instrumentation system.

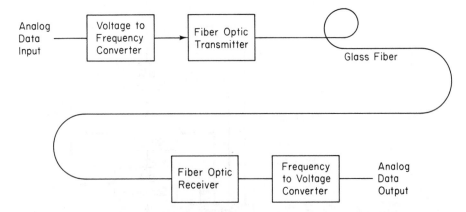

Figure 12-10   A fiber optic data transmission system.

If a remote digital readout is desired, the transmitted frequency would have to be counted and the count displayed as shown in Fig. 12-9(b).

There has been a trend in recent years to achieve the ultimate in isolation for industrial environments by the use of fiber optic transmission. This is the ultimate in isolation, as no metallic wires are used for data transmission. Power must be supplied with metallic wires but the data are transmitted through glass fiber and do not intermix with the power transmission.

Generally, data are transmitted in a digital format, although analog transmission is possible through glass fiber. The data capacity of a fiber optic transmission system is very great, usually much greater than the industrial data system requires. A typical fiber optic system is shown in Fig. 12-10. In this example, a voltage-to-frequency converter is used to convert the input analog signal to pulses, which in turn modulate a light source, either a light-emitting diode or a laser diode.

This pulsed light energy is transmitted through a glass fiber to the receiving end, where the pulsed light energy is converted to an analog signal. Further information on the techniques of data transmission through glass fibers is given in Chapter 14.

12-3 MULTIPLEXING

12-3.1 Digital-to-Analog Multiplexing

It is often necessary or desirable to combine, or *multiplex,* a number of analog signals into a single digital channel or, conversely, a single digital channel into a number of analog channels. Both digital signals and analog voltages can be multiplexed.

In digital-to-analog conversion a very common application of multiplexing is found in computer technology, where digital information, arriving sequentially from the computer, is distributed to a number of analog devices, such as an

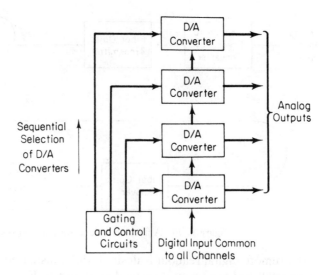

Figure 12-11 D/A multiplexer using several converters.

oscilloscope, a pen recorder, an analog tape recorder, and so on. There are two ways to accomplish multiplexing: The first method uses a *separate* D/A converter for each channel, as shown in Fig. 12-11. The second method uses one *single* D/A converter, together with a set of analog multiplexing switches and sample-and-hold circuits on each analog channel, as shown in Fig. 12-12.

In the system of Fig. 12-11 the digital information is applied simultaneously to all channels, and channel selection is made by gating clock pulses to the appro-

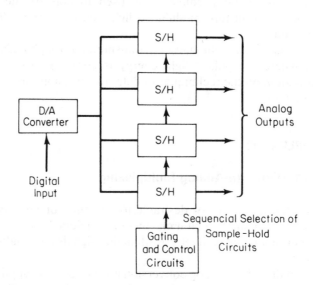

Figure 12-12 D/A multiplexer using *one* converter and several sample-and-hold circuits.

Analog and Digital Data Acquisition Systems   Chap. 12

priate output channels. One D/A converter is required per channel, so that the initial cost may be somewhat higher than the second system, but the advantage is that the analog information is available at the DAC output for an indefinite period of time (as long as the contents of the DAC flip-flop register are gated to the DAC).

The second method, illustrated in Fig. 12-12, uses only one D/A converter and is therefore slightly lower in initial cost. The multiple sample-and-hold technique, however, requires that the signal on the sample-and-hold circuits be renewed at periodic intervals (the capacitors do not hold their charge indefinitely).

12-3.2 Analog-to-Digital Multiplexing

In analog-to-digital conversion it is convenient to multiplex the analog inputs rather than the digital outputs. A possible system is given in Fig. 12-13, where switches, either solid-state or relays, are used to connect the analog inputs to a *common bus*. This bus then goes into a single A/D converter that is used for all channels.

The analog inputs are switched *sequentially* to the bus by the channel selector control circuitry. If simultaneous time samples from all channels are required, a sample-and-hold circuit may be used ahead of each multiplexer switch. In this manner, all channels would be sampled simultaneously and then switched to the converter sequentially.

It is also possible to multiplex by using a separate comparator for each analog channel. This system is shown in Fig. 12-14, where it is used with a counter-type A/D converter. The input of each comparator is connected to the output of the DAC. The other input to each comparator is connected to the separate analog input channels. Synchronization and control circuitry are required to operate the counter and sample the comparators. At the start of the multiplexing process, the counter is cleared and count pulses are applied to the counter. The D/A converter translates the counter output and provides an analog output voltage, which is fed to all the comparators. When one of the comparators

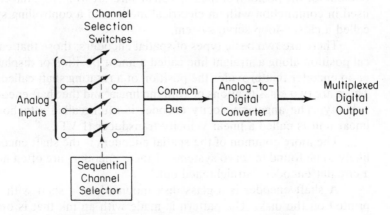

Figure 12-13 Multiplexed A/D conversion system.

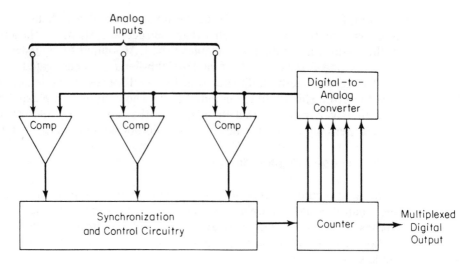

Analog
Inputs

Comp

Comp

Comp

Digital-to-
Analog
Converter

Synchronization
and Control Circuitry

Counter

Multiplexed
Digital
Output

Figure 12-14 Counter-type A/D converter with multiplexed input.

indicates that the D/A output is greater than the input voltage on that channel, the contents of the counter are read out. Counting is then resumed until the next signal is received, when the correct comparator is identified and the counter contents read out again.

12-3.3 Spatial Encoders

A spatial encoder is a mechanical converter that translates the angular position of a shaft into a digital number. It is a sort of analog-to-digital converter where the analog quantity is not an electrical signal, but mechanical. An encoder of this type finds important applications when a mechanical quantity must be entered into an electrical system, such as a remote reading wind direction indicator or a remote readout for the position of the flaps on an aircraft wing. Spatial encoders are often used in conjunction with an electrical motor and a controlling system in what is called a closed-loop servo system.

There are two basic types of spatial encoders: those that encode a mechanical position along a straight line called a linear position or displacement encoder, or an encoder that provides the position of a rotating shaft called a shaft encoder. There are two classifications of both the linear and the shaft encoder: position and velocity. The angular velocity encoder is often called a tachometer, while the linear unit is called a linear velocity transducer (LVT).

The more common of the spatial encoders is the shaft encoder, as it is most likely to be found in servo systems. Linear encoders are often nothing more than a circular encoder "straightened out."

A shaft encoder is a glass disk mounted on a shaft with a coding pattern printed on the disk. The pattern is made with an ink that is opaque to infrared light, so that optical generation of the output code is possible. Older shaft encoders used mechanical contacts using wire brushes and conductive disks. The brushes cause mechanical drag on the encoder, tend to wear out, accumulate dirt and become electrically noisy, and were damaged by rough use. The optical shaft encoders have no contacting parts, and thus brush and disk wear is nonexistent, and the effects of dirt are minimized as the dirt is not dragged along the disk by brushes.

The codes to be produced are generated by concentric rings on the disk. As an example, if the shaft position is to be determined to within 1 part in 32, and a binary output is desired, 5 binary bits would be required and thus five rings would be required, as shown in Fig. 12-15.

A source of illumination, either an incandescent lamp or a light-emitting diode (LED), is provided on one side of the disk. In the case of an incandescent lamp, because the emitting area is relatively large, one lamp may suffice for the entire disk. For the case of an LED, because the emitting area is so small, one emitter is usually provided for each bit.

The disk is placed between an illuminating source and a photodetector so that when the opaque part of the encoding disk is between the light source and the detector, a logic one or zero is encoded, while the transparent part of the disk produces the opposite logic state.

Notice the encoding pattern on the disk, as shown in the figure. The outer track is the least significant bit and makes the most changes per revolution of the encoder. This requires the finest pattern of all the tracks, and thus the least significant bit is always placed on the outside of the disk.

The output of the encoder is not always binary and Fig. 12-16 shows a binary-coded decimal encoder which is capable of resolving the rotation of a shaft to a part in 100. As the figure shows, the pattern is much finer than in the preceding example, and the least significant bit is on the outer track.

Shaft encoders with fine resolution that is able to resolve the shaft position to within 1 part in 1,024, representing a binary 10-bit converter, require very fine patterns on the encoding disk. These patterns require more precision than a printing operation can produce and are created using a photographic process. Glass is often used in shaft encoders for the encoding disk, as it is stable, does not tend to warp, and is transparent to light.

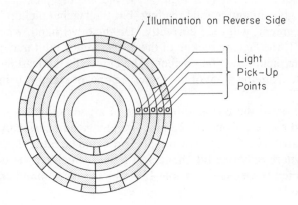

Illumination on Reverse Side

Light
Pick-Up
Points

Figure 12-15 Spatial encoder using a binary counting system.

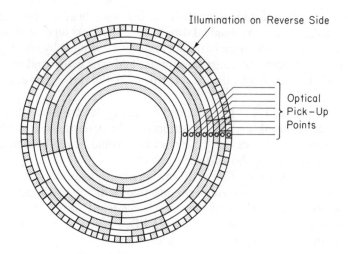

Illumination on Reverse Side

Optical
Pick–Up
Points

Figure 12-16 Binary-coded-decimal disk converter capable of a readout from 0 to 99. Since the outer commutator is divided into 100 segments, the angular position of the disk can be resolved to 3.6°.

One serious problem with shaft encoders using the disk patterns shown in Figs. 12-15 and 12-16 is the fact that when the encoder output changes from one value to another, if some of the bits change before others, there will be incorrect outputs. Although they last for only a small range of shaft position, there is no guarantee that the shaft will not assume this critical position.

One solution to this problem is to provide two pickups for all the tracks except the least significant bit, as shown in Fig. 12-17. The theory behind this type of encoder is that when the least significant bit changes from a 0 to 1, none of the other bits change state. When the next significant bit changes from a 0 to a 1, none of the higher-order bits change state. Two pickups are used for each encoding track except the least significant bit. One pickup is called the leading and the other the lagging. An encoding scheme is used to select the leading or lagging pickup of a bit relative to the state of lower-order bits.

If the least significant bit is a logic zero, the leading pickup of the next significant bit may give an ambiguous reading, but the lagging pickup, being near the center of the segment, will read correctly. On the other hand, when the LSB pickup is at a logic 1, the lagging pickup of the next significant bit may give a false output. The leading pickup will be correct. The choice of the leading or lagging pickup will be made from the state of the lesser significant pickups using the logic shown in Fig. 12-17.

Another solution to the problem is the use of a code other than the binary code. As explained earlier, if one of the bits of a binary number changes prematurely at a point of shaft rotation, the resultant error could be rather large. What is needed is a code where only one bit changes at a time. Such a code is called the Gray code or reflected binary code. Table 12-1 shows several numbers and their binary and Gray codes.

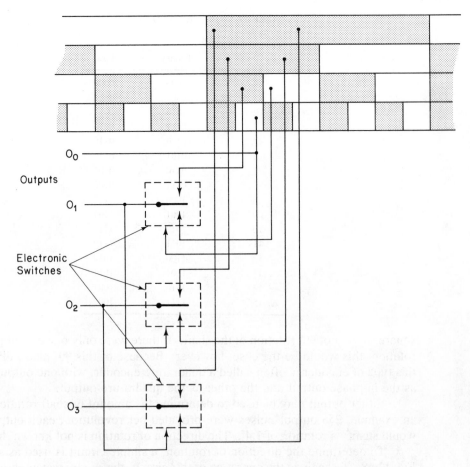

Figure 12-17 Vee brushes, or pickups, and the associated logic.

A shaft encoder providing a Gray code output would have only one bit change between subsequent values, and therefore there would be no problem with one bit changing before another. There are code converter chips for converting Gray code to binary; however, many systems can take advantage of a microprocessor to make the code conversion.

In many applications the actual position of a shaft is not an important factor, but relative motion is: that is, clockwise or counterclockwise, and the amount of rotation. A simple encoder shown in Fig. 12-18 is used for this purpose. Two outputs are available from this encoder, as shown in the figure. The resolution of the encoder (i.e., the smallest change in position discernible) is set by the number of output pulses available per revolution of the shaft. Typical shaft encoders would provide from 32 to 1,024 pulses per revolution.

Notice that the outputs from the shaft encoder are square waves for a constant rate of rotation. Notice also that the relationship between the two outputs represents a phase angle difference of 90°. This is 90° of a 360° duration of the

TABLE 12-1 BINARY AND GRAY CODES

Decimal	Binary	Gray
0	0000	0000
1	0001	0001
2	0010	0011
3	0011	0010
4	0100	0110
5	0101	0111
6	0110	0101
7	0111	0100
8	1000	1100
9	1001	1101
10	1010	1111
11	1011	1110
12	1100	1010
13	1101	1011
14	1110	1001
15	1111	1000

square wave, not 90° rotation of the shaft. If there were only one output pulse per rotation, this would be the case, however. Because of this 90° phase difference, this type of encoder is often called a quadrature encoder, with one output labeled as the in-phase output and the other as the quadrature output.

Either output may be used to determine the amount of shaft rotation. If, as an example, 256 output pulses were provided per revolution, each output pulse would signify a rotation of 1.4°. The direction of rotation is not known, however.

To determine the direction of rotation, a simple circuit is used as shown in Fig. 12-18. To analyze the operation of this circuit, the square waves shown in the

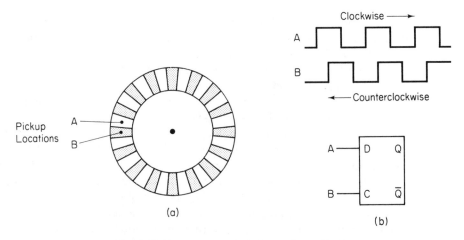

Figure 12-18 Incremental shaft encoder and logic.

Analog and Digital Data Acquisition Systems   Chap. 12

figure can be traversed in either direction from left to right for clockwise rotation or from right to left for counterclockwise rotation. The lower waveform clocks the state of the upper waveform into the D-type flip-flop. When the rotation is clockwise, the positive transitions always clock in a 1 state of the upper waveform.

In the opposite direction, the positive transitions of the lower waveform clock in a logic 0 state into the D-type flip-flop. Thus, the Q output of the D-type flip-flop indicates the direction of rotation, while positive transitions of the encoder indicate the amount of rotation.

Notice that a positive transition of the encoder in one direction becomes a negative transition in the opposite direction. If the shaft encoder is turned just beyond a positive transition and reversed in direction, the first transition encountered will be a negative transition and the rotation of the shaft will not be noticed until the second transition in the reverse direction. The area is called the dead zone and can be either a help or a hindrance, depending on the system.

REFERENCES

12-1. Bartholomew, Davis, *Electrical Measurements and Instrumentation,* chap. 7. Boston: Allyn and Bacon, Inc., 1963.

12-2. *How to Use Shaft Encoders,* Datex Division, Conrac Corporation, 1965.

12-3. *Logic Handbook,* Digital Equipment Corporation, Maynard, Mass., 1967.

12-4. *Magnetic Tape Recording Handbook,* Application Note AN-89, Hewlett-Packard Company, Palo Alto, Calif., 1967.

12-5. Ryder, John D., *Electronic Fundamentals and Applications,* 3rd ed., chaps. 14, 15. Englewood Cliffs, N.J.: Prentice-Hall, Inc., 1964.

12-6. Thomas, Harry E., and Clark, Carole A., *Handbook of Electronic Instruments and Measurement Techniques,* chap. 6. Englewood Cliffs, N.J.: Prentice-Hall, Inc., 1967.

13

Computer-Controlled Test Systems

13-1 INTRODUCTION

Perhaps one of the most valuable and powerful advances in the development of test equipment is the computer-operated test and evaluation systems sometimes referred to as automatic test equipment (ATE). The computer requirement is not extensive and thus "computers" more on the order of a calculator were used in some of the earlier ATE systems.

Three components are required for a computer-operated test system. First, computer-compatible test equipment must be available, and enough test equipment must be employed to make the necessary measurements. Second, a computer must be employed. Software must be available that will perform the desired test and present the data in the correct form. And, finally, a communications system is required that will allow the test equipment and the computer to communicate reliably.

Numerous tasks can be handled by a computer-operated test system, and it is not possible to outline all the possible tests. A few tests will be described to give a sampling of the possibilities of what can be done with such a system. The possibilities are endless.

13-2 TESTING AN AUDIO AMPLIFIER

Figure 13-1 shows a block diagram of an automatic test system to analyze an audio amplifier. This could be used in a factory as a final test from a production line or possibly by a research facility to evaluate the reliability of an amplifier relative to time or over environmental extremes.

First, the amplifier must be connected to the test system, which involves a manual operation. In some sophisticated systems, special connectors are provided so that the connecting to the automatic test system can be done with a simple operation. This would more likely occur in a commercial or military system, where automatic testing may be performed after the equipment leaves the factory as a part of a maintenance program. The audio amplifier under test may require the setting of various switches or other controls, which would involve further manual operations. Beyond these simple tasks, the testing can be fully automatic.

The computer performs three basic tasks: it supplies stimulus to the unit under test and determines the response of the unit under test to that stimulus; then the response is analyzed and the data presented in various ways.

For testing an audio amplifier, the stimulus is the application of an audio input signal. Determining the response of the amplifier involves measuring the power output or the harmonic distortion with a wattmeter or distortion analyzer. Data analysis could require calculating the amplifier gain or percentage of distortion and comparing it to the maximum or minimum allowed values.

For the example of the audio amplifier, the computer applies the ac supply voltage and measures the supply current. This checks for amplifiers with defective components that cause excessive supply current; if found, the testing is terminated immediately to prevent any damage to the unit under test or to the test system.

If the supply current is below the normal operating range, this is also an indication of a defective part in the amplifier, and the test could be terminated or continued to determine where the problem lies. Usually, a low supply current will

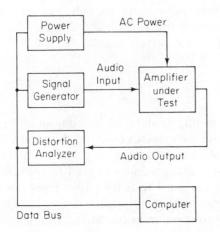

Figure 13-1 Computer-controlled measurement system for testing an audio amplifier.

not result in damage to the unit under test and the test system, so the test can continue. As an example, if a stereo amplifier is being tested and the supply current is low, one of the amplifiers could be defective and further testing would reveal which amplifier was at fault.

Assuming that the power supply current is normal, the computer-operated test system supplies an input signal to the amplifier and analyzes the output signal. To accomplish this, the computer has control over a signal generator and is able to set the frequency and amplitude of the test signal. In addition, some form of signal-analyzing tool is required to analyze the output from the amplifier to determine if the unit is performing its intended function. This requires that data be received from an analyzing instrument.

Various tests could be made on the audio amplifier. The gain of the amplifier could be determined by measuring the output power into the load. The computer sets the input signal amplitude and receives the power output from the signal-analyzing instrument. A simple calculation determines the gain of the amplifier, and this is handled by the computer. The harmonic distortion could be measured. This requires that the harmonic distortion analyzer be nulled and the results be transmitted to the computer. The system noise could be measured. In this case, the input signal must be removed, which involves controlling the signal source and reading the resultant noise level from the output-analyzing instrument.

A frequency response could be measured by varying the frequency from the signal source while measuring the output level at the output analyzer. The output level measurement is the source of the frequency response calculation. Transient response and intermodulation distortions may also be measured by controlling the signal source and the output-measuring instrument.

All the previous measurements could be repeated for different values of line voltage. Typically, a few tests are made at high and low line voltages while comprehensive tests are made at the nominal line voltage. If the unit is to operate over a wide temperature range, various measurements are repeated at different temperatures by inserting the unit under test into an environmental chamber, which would also be computer controlled.

If a signal source capable of supplying a varied range of signals is used in conjunction with a versatile distortion/wave analyzer, only two pieces of test equipment are required for this automatic test system; and although the tests are quite comprehensive, this test system is relatively simple.

13-3 TESTING A RADIO RECEIVER

As an additional example, consider testing a radio receiver using an automatic test system as shown in Fig. 13-2. As with the previous example, the receiver must be connected to the automatic test system and the necessary controls and switches placed in their proper positions. A signal generator is used to provide the RF input signal and modulation. After the usual tests for power input current, the sensitivity of the receiver may be measured by setting the output level of the RF generator while reading the signal-to-noise ratio of the receiver output. Most

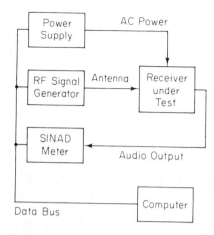

Power
Supply

AC Power

RF Signal
Generator

Antenna

Receiver
under
Test

SINAD
Meter

Audio Output

Data Bus

Computer

Figure 13-2 Computer-controlled measurement system for testing a radio receiver.

receiver sensitivity specifications require the output signal-to-noise ratio to be greater than a certain value, such as 6 or 10 dB, which can be determined by an audio signal analyzer, or a signal plus distortion to distortion (SINAD), ratio meter. In this case the level of the RF signal generator is modulated with a specific frequency, typically 1,000 or 400 Hz, at the specified deviation for frequency modulation or percentage modulation for amplitude modulation, and the level is continually lowered until the SINAD or signal-to-noise ratio is reduced to the minimum. At this point the signal generator level is stopped and the value is read by the computer. The receiver selectivity is measured by varying the frequency of the signal generator while the output is measured. An alternative is to measure the AGC voltage of the receiver, which is accomplished by bringing out a special test connection, which is easily done for a radio receiver.

If the receiver has a manually tuned dial, testing the accuracy requires that an operator set the dial while observing the accuracy. When human intervention is required, the computer can be programmed to provide prompting, such as a message on a CRT, such as "set receiver to 550 kHz, press enter when ready." This allows lower-skilled labor to accomplish sophisticated tasks but is not as desirable as a fully automatic test system where no intervention is required.

The two examples were of tests that can be made manually with conventional test equipment. The significant advantage of computer measurements is that the measurements can be made faster or less expensively because of reduction of labor costs. However, in some applications of computer-controlled measurements, the test equipment cannot be operated by human operators with the same results. One example is the simulation of complex signals. Simulating navigation signals as received by an aircraft is an excellent example. If the aircraft is receiving signals from more than one navigation aid, the aircraft position relative to each navigation aid transmitter, which is on the ground, is constantly changing, and the path of the aircraft relative to each is different. Navigation signal simulators allow the units to provide simple simulation of aircraft motion, but the computer-controlled test system is capable of accurate simulation of several navigation signals to an aircraft in flight. Navigation computers require this accurate signal simulation to measure the performance of the computer.

A growing selection of computer-controlled test equipment is being used for automatic test systems, making possible very sophisticated systems for the testing of practically any type of electronic system. When the choice of programmable test systems is large, some form of standard interface is necessary if the pool of available test equipment is to work together. The most significant interface for computer-operated test equipment is the IEEE standard 488, Digital Interface for Programmable Instrumentation (see Chapter 3 for a discussion of IEEE standards). This interface was developed around an existing interface used by several manufacturers of test equipment and is primarily suited for 8-bit microprocessors. However, use of the standard by any size or complexity of computer is not forbidden.

The IEEE 488 standard is based on the transmission of 8-bit data words with a parallel 8-bit data bus. Several status bits are used to augment the 8-bit data, but these are transmitted on separate lines. The 488 system is basically a short-distance system for test equipment mounted in a rack within a room and not intended for transmission over long distances or via telephone or other communication means. The typical 488 bus system consists of less than 15 pieces of test equipment mounted within one or two instrument racks, with the computer located within 10 feet or so.

Figure 13-3 shows a block diagram of a computer-controlled measurement system based on the IEEE standard 488 instrument bus. A *bus* is a set of interconnecting wires shared by several pieces of test equipment. As an example, the figure shows four units on the bus. The bus transmits data in and out of the test equipment. Because the bus is shared, only one unit can be transmitting the data at one time. To use the terminology of IEEE 488, one unit is a *talker* and the remaining units are *listeners*. The traffic cop in this system is called the controller and is typically the computer, although control can be passed on to other units. Usually, one unit has the ultimate authority for assigning control, and this usually is the main computer.

The test equipment used in the IEEE 488 system has a standard connector, usually on the rear panel, which allows most, if not all, of the functions of the test unit to be controlled externally if the proper signals are applied to the connector. In addition, the local control, usually the front panel controls, can be disabled to prevent inadvertent operation of the front panel during computer operation.

The interface is divided into two areas, the data bus and the control or status lines. The data are passed in 8-bit bytes to be compatible with the popular 8-bit microprocessors. Eight additional lines, called *interface signal lines,* transmit data necessary for the operation of the system but separate from measurement parameters. These lines are defined as follows:

(a) *DAV, data valid.* This signal indicates that the data on the data line are valid. When an addressed device is to supply a data word for processing, a certain amount of time delay is required for the addressed unit to obtain and

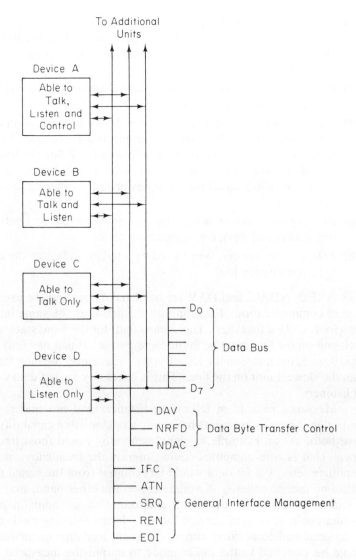

Figure 13-3 Schematic representation of the IEEE 488 instrumentation bus.

output the data on the data bus. When the data are correct, the DAV line will assume a logic zero state, indicating that the data are correct.

(b) *NRFD, not ready for data.* Although called *not* ready for data, this status line is the indication that the unit to receive the data is ready. As an example, a unit being designated as a listener would return a logic 0 for NRFD when all the internal circuits had been readied to accept the input data.

(c) *NDAC, not data accepted.* When this line goes to a logic 0, it is an indication that the data transmitted to the device have been accepted and that new data may now be supplied.

(d) *ATN, attention,* is used by the controller to specify how data on the data lines are to be used and which devices on the bus are to respond. Various messages are transmitted over the bus system in conjunction with the ATN signal.

(e) *IFC, interface clear,* is used by the controller to place the entire interface system in a quiescent state.

(f) *SRQ, service request,* is used by any device to request service and interrupt the current task. This could be used, as an example, by the line current monitor used in the audio amplifier example. When the line current exceeds a predetermined value, any test in progress will be halted and the unit shut down. An SRQ signal from the power line measuring device would initiate this action.

(g) *REN, remote enable,* is used by the controller to select between two alternative sources of device programming data.

(h) *EOI, end or identify,* when used by a talker, indicates the end of a multiple-byte communication.

NRFD, NDAC, and DAV are handshake lines and are used to transfer each byte of communication. Each unit on the line uses the same handshake line and are *wired-ANDed* together. This means that, for the logic state of the line to be 1, each unit on the line must be in the logic 1 state. If just one unit on the line is at a logic 0 state, the line is in the logic 0 state. Therefore, the interface bus is no faster than the slowest unit on the line. This is necessary so that data will be received by all listeners.

Messages pass from talkers to listeners and are under control from the controller. Some instruments have only a talk or listen capability, whereas others have both. As an example, a signal generator would most likely be only a listener. That is, the computer could program the frequency, modulation, output amplitude, etc., yet no data would be required from the signal generator since it makes no measurements. A voltmeter, on the other hand, may be only a talker, providing voltage measurements and requiring no adjustments. A frequency counter could be both talker and listener. In the listening mode the counter could be programmed to perform either a frequency or time measurement, and then it would be switched to the talker mode to output the measured data. This could also be extended to the previous example of the voltmeter. As an example, the meter was capable of measuring both ac and dc parameters, the meter would be set to the listener mode and placed in either the ac or dc mode, and then changed to the talker mode to provide the data.

13-4 INSTRUMENTS USED IN COMPUTER-CONTROLLED INSTRUMENTATION

Most test instruments require special circuits to be interfaced with a computer. In some instruments the modifications are simple, whereas in others they are significant. Generally, instruments that use any sort of mechanical device to effect a measurement, such as the precision variable capacitor or resistor used in a bridge, or a meter movement are not suited for adapting to a computer-controlled measurement system. Purely electronic means are usually available to substitute for the mechanical devices.

One of the easier instruments to interface with the IEEE 488 bus is the frequency counter shown in Fig. 13-4. Since the counter is a digital machine, the interface is a simple task. The data, which normally would appear only on a readout, are placed on the bus a digit at a time. This involves either a shift register, as shown, or a multiplexer and the necessary interface circuits to be compatible with the bus's electrical requirements. As previously mentioned, the frequency counter can be either a listener or a talker. This requires both data-generating and data-receiving circuits, as well as some method of switching between the two. Messages received from the controlling computer will be decoded and will control whether the frequency counter is a listener or a talker.

A signal generator requires a more involved procedure to make the unit computer compatible. Many signal generators, as discussed in Chapter 8, use mechanically tuned oscillators and dial plates. This effectively renders that type of signal generator unacceptable for computer control. Unless the signal generator is required to provide only one frequency at a fixed level, the mechanically

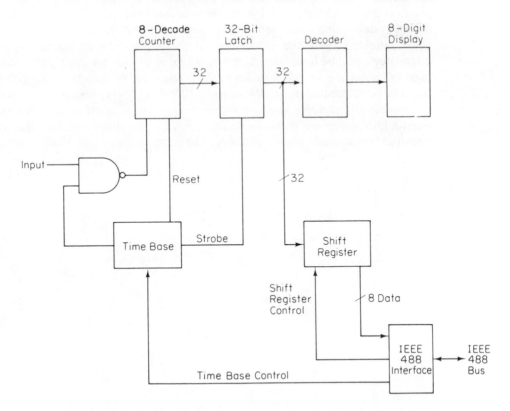

Figure 13-4 Frequency counter modified for operation with the IEEE 488 bus.

tuned generator cannot be used in a computer-operated system. Even when used for the simple single-frequency task, there would be no safeguards to determine when the frequency of the generator was moved accidentally. If a frequency counter was in the system, it would be possible to use it as a check on the frequency. However, if the counter determines that the frequency has drifted or has been changed, the only alternative is to shut down the system until the fault is corrected. In spite of the serious drawbacks, the low cost of the mechanically tuned signal generator or a special crystal-controlled generator makes it attractive for applications requiring a single-frequency source.

For more versatile systems, the signal generator would be a synthesized instrument. Figure 13-5 shows a signal generator with complete control over frequency, modulation, and signal level suitable for interfacing with test systems based on the IEEE 488 bus. This example is a synthesized generator where the frequency of the generator can be set via digital inputs from the bus. Modulation can also be easily controlled from the bus by the proper interface electronics. The output signal level from a signal generator is usually controlled via an attenuator, and it is in this area that computer control becomes somewhat difficult. As outlined in Chapter 8, there are two basic types of attenuators, piston types and switched pi sections, both requiring mechanical motion. The piston attenuator is not usually used for computer-controlled signal generators as it is difficult to position the piston attenuator with the necessary precision. Certainly, a motor-driven system using some form of displacement transducer could be used, but this could quickly negate the advantages of the piston-type attenuator. A pi-type attenuator could be implemented in two ways. Electromechanical relays could be used in lieu of the switches found in the attenuator, or the electronic equivalent could be implemented using *PIN* diodes. *PIN* diodes were discussed in Chapter 8 as variable attenuators; however, these diodes can be used as switches as well. Figure 13-6 shows an attenuator using relays rather than switches that can be computer controlled. In this example, attenuations from 0 to 15 dB can be pro-

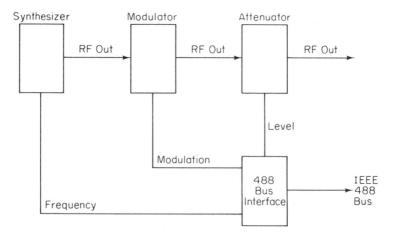

Figure 13-5 Synthesized signal generator interfaced with the IEEE 488 bus.

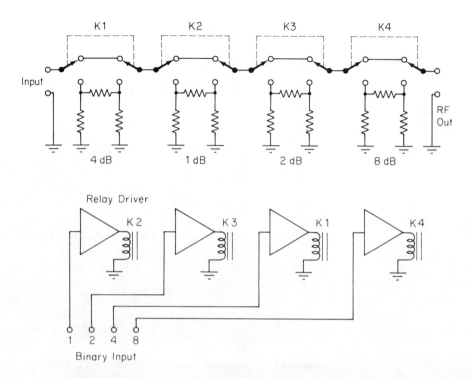

Figure 13-6 Relay-switched attenuator suitable for computer control.

grammed by purely electronic inputs. Relays, being partially mechanical devices, will have a lifetime of perhaps a few million operations. This may be an acceptable life for a laboratory attenuator but may be objectionable for a computer-operated test system where the number of operations may rapidly approach a million. When the lifetime of the relays becomes a significant problem, the attenuator of Fig. 13-7 is used. In this example, *PIN* diodes are used, which removes the mechanical element completely and improves reliability enormously. In addition to improvement in reliability, the speed of the attenuator is increased; and since the speed of the entire measurement system is dependent on the speed of the slowest device, this increase will translate into an overall increase for the entire measuring system. Figure 13-8 shows a photograph of a representative example of a completely programmable signal generator for use with the IEEE 488 bus.

Audiofrequency measurements and measurements of radio receivers usually require audiofrequency signal analysis. As described in Chapter 9, audiofrequency signal-analyzing instruments are level or voltmeters, harmonic distortion analyzers, wave analyzers, and spectrum analyzers. The latter two instruments are quite similar, being basically the same instrument. The spectrum analyzer could be used to make practically any measurement required for audiofrequency signal analysis, and a computer-interfaced analyzer is shown in Fig. 13-9. The significant differences between the spectrum analyzer shown here and the units described in Chapter 9 are the synthesized first local oscillator, the substitution of

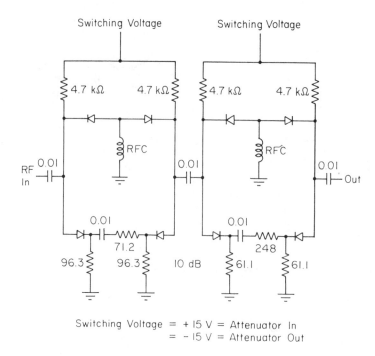

Switching Voltage = + 15 V = Attenuator In
= − 15 V = Attenuator Out

Figure 13-7 Electronically switched attenuator using PIN diodes.

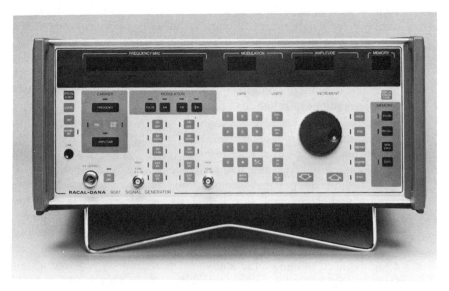

Figure 13-8 Example of a completely programmable signal generator for use on the IEEE 488 bus. (Courtesy of Racal-Dana Instruments, Inc.)

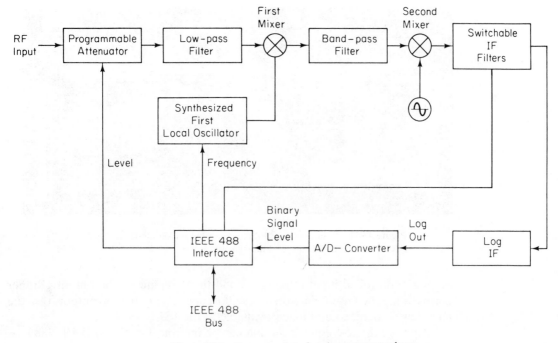

Figure 13-9 Computer-interfaced spectrum analyzer.

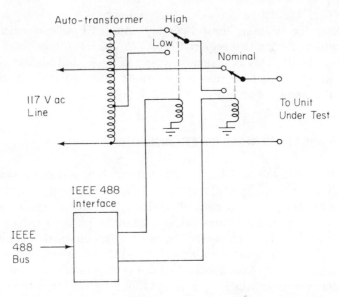

Figure 13-10 Adjustable ac power supply for automated testing using the IEEE 488 bus.

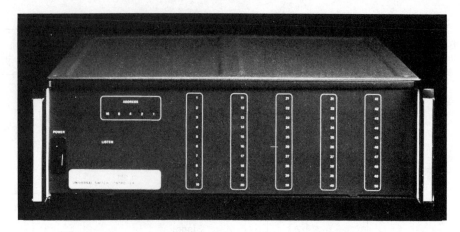

Figure 13-11 Multipurpose switching unit for use on the IEEE 488 bus. (Courtesy of Racal-Dana Instruments, Inc.)

a computer-controlled input attenuator, and the digitizing of the output. Rather than presenting the output data on an oscilloscope screen, the log output from the IF amplifier is digitized and made available to the IEEE 488 bus.

In spite of the test equipment that can be interfaced with the IEEE 488 bus, many measurement systems will require specialized test equipment that will have to be constructed for a specific task. This gives rise to specialized interface units for making the task of building unique test systems easier.

One example of a specialized computer-controlled unit is a device for supplying nominal, low, and high ac line voltages for testing line-operated equipment. The actual circuit for such a device is a simple tapped transformer with a relay for selecting the proper tap. An IEEE 488 interface unit supplying two outputs for the two required relays can be used to perform the task as shown in Fig. 13-10. This is a relatively simple task, and similar specialized computer-controlled functions may be generated with a basic IEEE 488 bus interface unit. An example of an IEEE 488 switch unit is shown in Fig. 13-11.

13-5 IEEE 488 ELECTRICAL INTERFACE

The IEEE 488 interface is intended for short distances where electrical noise is relatively low. Typical distances are less than 20-m total cable length. All instruments are placed in parallel, and it is possible to stack the connectors so that several instruments connect at a common point to reduce the amount of cable used. The data rate used by the system depends on the speed of the individual test equipment connected to the cable and the computer hardware or software. The response time of each individual test system on the bus will vary, with those using mechanical devices such as relays requiring more time to respond to an input than the purely electrical systems. There are some exceptions to this, such as a frequency synthesizer where the lock-up time could be considerable. In a test system, if frequency response is being measured, it is necessary to insure that the frequency synthesizer is locked and stable before any measurements are made.

The logic levels of the IEEE 488 bus are based on TTL levels with a logic 0 state being defined as less than 0.8 V and a logic 1 state being greater than 2.0 V. The logic driver is required to provide more than 2.4 V output at logic 1 and less than 0.5 V at logic 0. This provides an excess voltage of 0.4 V for a logic high and an excess of 0.3 V for a logic low, which is called the noise *immunity*. The logic levels are defined relative to a common ground, and it is not unusual for the grounds of individual systems to vary by a few hundred millivolts. These ground potential differences tend to be high-frequency signals where the reactance of the interconnecting cables is a causing factor rather than the resistance of the cable. It is advantageous to provide low-inductance ground cables between the units in the test system to reduce the ground noise to a minimum.

Open-collector or three-state drives may be used to drive the DAV, IFC, ATN, REN, and EOI lines. An open-collector driver must be used for the SRQ, NFRD, and NDAC lines because all units in the system are wire-ANDed for these functions. Where it is permissible, three state drivers should be used to preserve system speed.

Figure 13-12 shows an open-collector driver as it would appear connected to the system bus. Unlike the normal open-collector gate found in TTL technology, which is, as the name implies, nothing more than a bare collector, the IEEE 488 driver contains terminating resistors. These are required to provide a defined voltage on the bus when all the drivers are in a high impedance state. The open-collector output is required to sink a maximum of 48 mA while maintaining less than 0.5 V relative to ground.

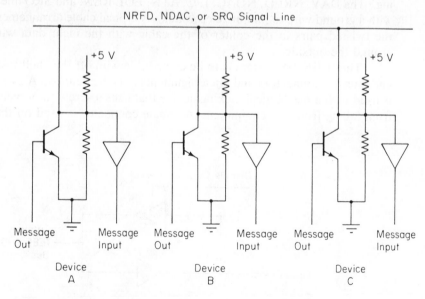

Figure 13-12 Schematic representation of an open-collector IEEE 488 bus transceiver.

The bus receiver is typically a TTL-type device, but improved results can be obtained when a Schmitt trigger input is used to prevent excess noise from being received. The receiver should also provide a diode for negative voltage clamping. Because negative transients can be generated from fast risetime signals traveling through cables, the clamp diode is required to prevent damage to the receiver and to attenuate the signal on the line. TTL gates and Schmitt trigger circuits usually have this diode.

Typically, the IEEE 488 interface element is both driver and receiver or transceiver, and it contains, for each line to be driven, an output driver, terminating resistors, a clamp diode, and a receiver.

The three-state IEEE 488 driver shown in Fig. 13-13 has three possible modes of output driver. First is logic 0, which allows up to 48 mA to pass into the device while the voltage of the driver does not exceed a logic 0. A second mode is logic 1, where current passes out of the device and into the bus to charge any cable capacitance. Unlike the open-collector driver, where the terminating resistors provide the current to charge the bus capacitance, the three-state driver can provide considerable current. Finally, a third state is a high impedance condition so that the driver does not load the line, except for the terminating resistors. In either case, the open collector or the three-state output, the terminating resistors are required so that the voltage of the line is defined when all the drivers are in the high impedance state.

The cable requirement is not difficult, but cables used for the IEEE 488 system should meet the following basic requirements. The cable should be a shielded 24-conductor cable with a minimum of 85 per cent shielding effectiveness. Sixteen of the wires are used for the eight data lines and the eight status lines, while the remaining wires are used for signal returns, grounds, and shielding. The DAV, NRFD, NDAC, IFC, ATN, EOI, REN, and SRQ lines are twisted with a ground wire to minimize crosstalk. A typical cable arrangement is to place the twisted pairs in the center of the cable with the eight data wires arranged around the outside.

The IEEE 488 system is to be compatible with off-the-shelf test equipment, and thus the connectors and pin assignments must be defined. A 24-pin connector is used with a mechanical locking device that uses a screw to prevent connectors from falling from the equipment. A female connector is used on the test equip-

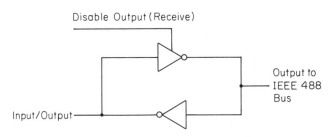

Figure 13-13   Three-state bus transceiver.

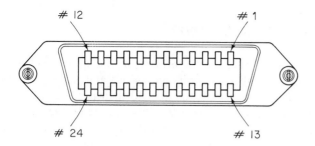

Pin	Signal	Pin	Signal
1.	Data 1	13.	Data 5
2.	Data 2	14	Data 6
3.	Data 3	15	Data 7
4.	Data 4	16.	Data 8
5.	EOI	17.	REN
6.	DAV	18.	Gnd
7.	NRFD	19.	Gnd
8.	NDAC	20.	Gnd
9.	IFC	21.	Gnd
10.	SRQ	22.	Gnd
11.	ATN	23.	Gnd
12.	Shield	24.	Logic ground

Figure 13-14 Pin assignments for the IEEE 488 instrumentation bus.

ment, while most cable assemblies provide both male and female ends so that cables can be chained from one unit to another. Figure 13-14 shows the pin assignments used for the IEEE 488 connector.

13-6 DIGITAL CONTROL DESCRIPTION

There are two basic divisions of the interface system from a functional viewpoint, device functions and interface functions. *Interface functions* are defined for the system and are the same for each test system connected to the bus. This would include status determination, clear and reset functions, etc. The *device functions* are variable and depend on the type of test equipment being addressed. As an example, a signal generator would have functions relating to signal level and modulation, which would not be applicable to a frequency counter.

Although there is a fixed set of available interface functions, the individual test equipment does not have to employ all the interface functions, just those that are useful for the type of tests made by the equipment.

There are two types of messages for the interface function, single-line and multiline messages. Only one message can exist at any time for multiline messages, whereas more than one single-line message can exist concurrently.

The single-line messages are DAV, NRFD, NDAC, IFC, ATN, SRQ, REN, and EOI and have been discussed previously. The list of multiline messages is extensive and will not be discussed.

13-7 EXAMPLE OF SIGNAL TIMING IN A
MICROPROCESSOR-BASED MEASUREMENT

At power turn-on or at a time indicated by the processor, the interface clear signal is set to a logic 1 state. This initializes all units on the bus and readies the system for operation. The microprocessor sets the devices to their predetermined states by sending the DCL or device clear message. This is a multiline message as described in the previous paragraphs. The processor sends the listen address of the power supply and follows with the data for the device. Power is applied to the device and testing commences. The processor sends the unlisten command and the power supply will no longer respond to commands or data until the listen address is sent again. The processor will send the listen address of the signal generator followed by the data for the frequency and amplitude for the signal input.

Figure 13-15 shows the timing diagram for the transfer of data from the computer to and from the addressed device. At $t = 0$, the listener is ready to receive data. If the system includes more than one piece of interfaced test equipment, this would indicate that each connected piece of equipment on the bus is ready to receive data. Provisions for unused test equipment on the bus prevent these pieces from disabling the bus. The talker then places data on the bus, and when the data are valid, the DAV line goes low, which indicates that the data are good. The NRFD signal goes low from the listener, which indicates that no further data can be accepted until the present are dispensed with. Once the listener

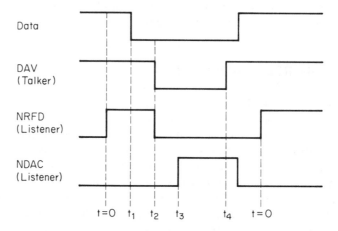

t_0 Listener indicates it is ready to accept data.

t_1 Data are applied to the line.

t_2 Talker indicates that data are valid.

t_3 Data are accepted by listener.

t_4 Talker indicates that data are no longer valid and can be changed.

Figure 13-15   Timing relationship of signals on the IEEE 488 bus.

unit has entered the data and no longer requires the data on the bus, the NDAC signal goes to a logic 1, which indicates that the data have been accepted.

The data may be used for a variety of purposes within the unit, and after the data have been applied, the listener unit will set the NRFD line low, which indicates that new data may be sent. It is this time delay that mostly affects the speed of the data bus. If the data were used to set an attenuator where electromechanical relays were used, the time delay could be several milliseconds before further data could be accepted.

The operation of the IEEE 488 bus is completely asynchronous, which requires one task to be completed before another task can commence. The data rate can extend from as slow as necessary to as high as 1 Mb/s, which is the specified maximum.

REFERENCES

13-1. IEEE, *IEEE Standard Digital Interface for Programmable Instrumentation* (number 488). New York: Institute of Electrical and Electronics Engineers, Inc., 1978.

13-2. Leventhal, Lance A., *Introduction to Microprocessors: Software, Hardware, Programming.* Englewood Cliffs, N.J.: Prentice-Hall, Inc., 1978.

13-3. Short, Kenneth L., *Microprocessors and Programmed Logic.* Englewood Cliffs, N.J.: Prentice-Hall, Inc., 1980.

PROBLEMS

13-1. What are the three requirements of an automatic test system?

13-2. What is the IEEE 488 bus system?

13-3. What limits the data rate in the IEEE 488 system?

13-4. How are data transmitted over the IEEE 488 bus system?

13-5. What are the handshake signals in the IEEE 488 bus system?

13-6. What are some of the functions/controls of test equipment that are not easily adapted to computer control? How are some of these functions made controllable?

13-7. What kinds of equipment can be talkers only? What types can be listeners only? What types can be talkers and listeners?

13-8. What prevents two units from talking simultaneously?

13-9. What is the recommended maximum cable length for the IEEE 488 system?

13-10. Why is it preferred to use the three-state driver for the IEEE 488 bus driver?

14

Fiber Optics Measurements

14-1 INTRODUCTION

The introduction of fiber optics into the mainstream of communications electronics has ushered in a new dimension of measurements. The new requirements for the measurement of light parameters are not the types usually encountered before the advent of fiber optics communications. Light measurement previous to fiber optics would involve the intensity of light in a room for the purpose of evaluating illumination, or sophisticated parameters associated with physics research. One of the few light measurements made in association with electronics would be light intensity and spectrum measurements, used for the evaluation of CRT displays for computer terminals and other readout devices. In the physics lab, scientists would require more sophisticated light measurements, sometimes involving short pulses of light and light spectra beyond the range of human visibility.

Fiber optics communications use light energy not visible to the human eye in the infrared region of the spectrum. The concept of spectrum should be understood by the student who is aware of the material in Chapter 9. There are some differences between the discussion of electrical spectrum and light spectrum. For example, wavelength, rather than frequency, is used to describe the relationship of light energy within a spectrum. This is because the instruments used to investigate optical spectrum measure wavelength rather than frequency. In the very early days of radio, however, wavelength was also used to describe radio signals. The wavelengths involved with fiber optics communications are very small compared with the wavelengths associated with the radio spectrum. The dimensions of light wavelengths are too small to be visualized as compared with the very long wavelengths of radio frequencies, which can be measured in meters, centimeters, or possibly millimeters. Light wavelengths are measured in nanometers, or 10^{-9} meter. Visible light covers the range 370 to 750 nm, with the larger wavelength being visible red and the shorter wavelength being the blue of the visible spectrum. The range of wavelengths used for fiber optic transmission ranges from 800 to 1,500 nm.

Fiber optic transmission is made possible by a phenomenon called total internal reflection. Every material that is transparent to light has an index of refraction. The index of refraction is the ratio of the velocity of light in a vacuum to the velocity of light in the material. When light strikes a boundary between two materials of differing indices of refraction, the path of the light ray is altered in two ways. First the light is reflected, which means that the light energy that is reflected does not enter the material at the other side of the boundary. The remaining light energy does enter the material, but the path of the light ray is altered. This is shown in Fig. 14-1. If the material of the other side of the boundary is of a higher index of refraction, the light ray is refracted away from the boundary surface, and if the index of refraction is less, the ray is refracted toward the boundary surface. The amount of refraction, or the deviation angle from the straight-line path, is given by

$$\frac{n_2}{n_1} = \frac{\cos \theta_1}{\cos \theta_2} \tag{14-1}$$

where n_1 = index of refraction of the first material

n_2 = index of refraction of the second material

θ_1 = angle between the surface and the incident ray

θ_2 = angle between the surface and the refracted ray

(In many physics texts the angles measured in the development of the equations for refraction are measured relative to the normal of the boundary surface. Because of the nature of the geometry of fiber optics, the angles in the equations above are relative to the refracting surface rather than the normal.)

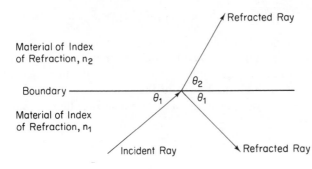

Figure 14-1 Reflected and refracted light at a boundary of differing index of refraction.

If the angle between the boundary surface and the ray were to become small, there would be a situation in which the amount of deviation of the ray from the straight-line path would cause the light energy not to enter the second material at all. This would occur when the angle θ_2 equals zero and can be expressed as

$$\theta_c = \cos^{-1} \frac{n_2}{n_1} \tag{14-2}$$

This angle, θ_c, is called the critical angle and represents a situation where all the light energy is reflected.

A narrow cylinder of glass surrounded by an outer covering of glass but with a slightly lower index of refraction is shown in Fig. 14-2. This is a simple form of glass fiber used for the transmission of information. The interior of the fiber is called the core and the outer glass layer is called the cladding. The change of the index of refraction at the boundary between the two types of glass is sudden, which is a type of glass fiber called a step index fiber. Any ray of light that is in the fiber at an angle between the centerline of the core and the light ray less than θ_c, the critical angle, is totally reflected at the boundary between the two glasses. Thus none of the energy is lost to the outside at each reflection through the length of the fiber. The angle of the cone that represents the acceptable entry angles for total internal reflection is called the cone of acceptance and can be calculated by considering the geometry of the fiber.

Consider the light rays entering a fiber as shown in Fig. 14-2. Assume that the end of the fiber is in air. Light rays entering the end of the fiber will be refracted toward the central axis of the fiber because the index of refraction of the glass is higher than air. The maximum angle, ϕ, produces an internal reflection equal to the critical angle. This relationship is

$$\sin \phi = \frac{n_1 \sin \theta_c}{n_3} = n_1 \sin \theta_c \tag{14-3}$$

where n_2 = index of refraction of the cladding

n_1 = index of refraction of the core

n_3 = index of refraction of air which is equal to 1

θ_c = critical angle

The acceptance angle is two times the result of Eq. (14-3) or

$$\theta_A = 2 \sin^{-1}(n_1 \sin \theta_c) \tag{14-4}$$

where θ_A is the acceptance angle.

Some of the light energy entering the fiber outside the cone of acceptance is lost through refraction into the cladding material with each reflection. Eventually, all the energy will be lost through the length of the fiber.

Another way of quantifying this cone of acceptance is called the numerical aperture and is given by

$$NA = \sqrt{n_2^2 - n_1^2} \tag{14-5}$$

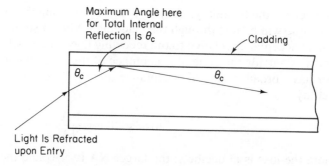

Maximum Angle here
for Total Internal
Reflection Is θ_c

Cladding

θ_c

θ_c

Light Is Refracted
upon Entry

Figure 14-2   Total internal reflection in a glass fiber.

The numerical aperture is a more convenient quantity to use when calculating the losses in a fiber optic system rather than the acceptance angle and will be explained further.

EXAMPLE 14-1

What is the acceptance angle and the numerical aperture of a fiber with a core index of refraction of 1.47 and a cladding index of refraction of 1.45?

SOLUTION   The critical angle of the fiber is

$$\theta_c = \cos^{-1}\left(\frac{1.45}{1.47}\right) = 9.46°$$

The acceptance angle is

$$\theta_A = 2\sin^{-1}(1.47\sin 9.46°) = 27.9°$$

The numerical aperture is

$$NA = \sqrt{n_1^2 - n_2^2} = \sqrt{1.47^2 - 1.45^2} = 0.242$$

Although the vast majority of the light energy in the fiber is reflected at the boundaries of the two glasses, some of the energy is lost within the fiber. Some of the light energy is simply absorbed by the glass and is called absorption loss. Some of the energy is lost by reflection from impurities or defects in the glass and is called Rayleigh scattering. These defects are microscopic and are distributed throughout the entire length of the fiber. The light is scattered in all directions; some of the energy is lost through the cladding and other energy is scattered up the fiber toward the light source.

Another type of loss occurs when the fiber is bent with a small radius. Some of the rays begin to impinge on the boundary between the core and cladding at an angle greater than the critical angle, and some of the light energy is lost into the cladding. This is called microbending loss.

When a glass fiber is coupled to a light emitter, a light detector, or another fiber, the junction between the fiber and the other entity is not perfect, and a loss

Sec. 14-1   Introduction

will occur at the boundary. Any light energy entering the fiber outside the acceptance cone will be lost through the cladding. When a source of light has a narrow cone of acceptance relative to the receiving fiber, there will be no loss, as no light energy is outside the cone of acceptance of the receiving fiber. When the source fiber has a broader cone of acceptance there will be a loss and this can be calculated by

$$\text{loss} = 20 \log \frac{NA_1}{NA_2} \qquad (14\text{-}6)$$

where the loss is in decibels; the larger NA is NA_1 and the smaller is NA_2.

EXAMPLE 14-2

How much loss will be experienced if a fiber of numerical aperture of 0.3 is the source for a fiber with a numerical aperture of 0.242?

SOLUTION Using Eq. (14-6), the light loss may be calculated. This is the energy that is lost through the cladding of the receiving fiber.

$$\text{loss} = 20 \log \frac{0.3}{0.242} = 1.87 \text{ dB}$$

As can be seen, the length of the path of the propagation of light through the fiber is dependent on the number of reflections within the fiber. If there are a large number of reflections, the distance traveled by the light energy will be greater than if the reflections are few. If a pulse of light energy were injected into a fiber, some of the energy would take the long path and some would take the short path. The velocity of propagation of the light in the glass is the same for both the long and short paths, and the energy will travel along the length of fiber at differing velocities. If a well-defined pulse of light energy were injected into the fiber, the pulse would exit a length of fiber distorted, due to the different times of arrival of the pulse energy. This phenomenon is called modal distortion and limits the useful bandwidth of a fiber.

The discussion of fiber optics to this point has treated light energy as being similar to a golf ball ricocheting through a pipe. In addition to the particle nature, light energy exhibits behavior of wave propagation. When light energy is considered strictly as a particle, there are an infinite number of paths through the optical fiber, as long as the angle of reflection is less than the critical angle. Because of the wave nature of light, the wavefronts of the reflecting light energy must combine in-phase; otherwise, there will be cancellation of the light intensity within the fiber. Because the wavelength of light is very small, there are a very large number of possible paths, or modes, that will allow the propagation of the light in phase. As the fiber diameter is reduced, the number of possible paths is reduced because of the necessity for phase coherence. If the fiber diameter is reduced to a few wavelengths, the number of modes can be reduced to one. This type of fiber is called single mode and is used to reduce the effects of modal dispersion described previously.

The index of refraction is the key to the velocity of the light in the fiber. In a vacuum, the velocity of light is always C or 3×10^8 meters per second. However, in a material, the velocity is less than the velocity in free space. The relationship between the free space velocity and the velocity in a substance of an index of refraction of n is given by

$$v = \frac{c}{n} \qquad (14\text{-}7)$$

where v is the velocity of light in the medium and n is the index of refraction of the medium.

14-2 SOURCES AND DETECTORS

The light source for a fiber optic communications system is either a light-emitting diode (LED) or a laser diode. The LED is considerably less expensive than the laser and is used for relatively slow speed or lower-frequency communications applications. The light output from the LED has a broader spectral bandwidth and emits from the diode in a larger cone than the laser. The bandwidth of the laser is very narrow, being only a few nanometers of spectrum, and the light output has a very narrow emitting cone. This makes the laser diode much easier to couple to a fiber, especially the very small diameter fibers used for single-mode applications.

Fiber optic detectors, like emitters, are usually diodes. There are two basic categories of fiber optic detectors: avalanche photodiodes and *PIN* photodiodes. In both cases, the detector diodes are reverse-biased diodes and thus have a depletion zone, or an area with no carriers available for electrical conduction. The avalanche diode uses the carrier multiplication effect of a diode that is biased to a very high potential, near where the diode would spontaneously conduct from a continuous multiplication of carriers.

As with a conventional diode, a reversed-biased dark photodiode will not support any significant amount of current. The small amount of current that does flow is due to thermal leakage currents. When the depletion zone is illuminated, however, the light photons interact with the semiconductor material and release additional carriers for conduction. If each photon could release one electron for conduction, the diode would be 100 per cent efficient. The definition of quantum efficiency is the number of electrons released for conduction divided by the number of photons times 100 per cent, or

$$QE = \frac{N_e}{N_p} (100\%) \qquad (14\text{-}8)$$

where N_e is the number of electrons released for N_p photons.

Light energy is transmitted as discrete packets of energy called photons. For monochromatic light, which describes the light used in fiber optics communications, each photon contains the same amount of energy, which is equal to

$$e = \frac{h_c}{l} \qquad (14\text{-}9)$$

where l is the wavelength of the light and h is Planck's constant. In fiber optics communications, the amount of energy contained in a photon is very small, and thus a very large number of photons is involved. It would be impossible to measure the amount of light energy or power of a single photon or to resolve a change of energy equal to a photon. Therefore, when measuring the light energy or power in a fiber optics system, the measurements appear continuous rather than quantized.

Fiber optic power is measured using a photodiode. There is a simple relationship between the incident power on the diode and the diode current and this can be derived in the following manner. The number of photons per second contained in a light source of a specific power is

$$N = \frac{pl}{hc} \qquad (14\text{-}10)$$

where N is the number of photons per second for a power level p.

The number of electrons made available for conduction is proportional to the number of photons times the quantum efficiency, and thus the number of electrons per second is

$$N(\text{QE}) = \frac{(\text{QE})pl}{hc} \qquad (14\text{-}11)$$

The actual current in amperes is the number of electrons per second times the charge per electron, or

$$I = \frac{(\text{QE})pl(1.6 \times 10^{-19})}{hc} \qquad (14\text{-}12)$$

where 1.6×10^{-19} is the charge of an electron and I is the current.

Thus, the photocurrent from a photodiode is proportional to the incident power on the diode. Included in the constant of proportionality, however, is the wavelength of the light, and thus power meters must be calibrated for a specific wavelength.

The active area of a photodiode is usually much larger than a typical fiber. In addition, the numerical aperture of a diode is essentially 1, and therefore it is safe to assume that all the light energy from a fiber is coupled to the photodiode. This is not the case with an emitter, where a large percentage of the light energy from the emitting diode may be lost in the coupling process.

EXAMPLE 14-3

How much current would be developed in a *PIN* photodiode with a quantum efficiency of 82 per cent, which is illuminated with 75 μW of 1,300-nM photons?

SOLUTION The first step in the solution of this example is to calculate the number of photons per second falling on the diode. It is safe to assume, as

---

mentioned previously, that all the energy from the fiber is coupled to the diode. Therefore, the number of photons per second is the power from the fiber divided by the energy per photon. For 1,300-nM light the energy per photon is

$$e = \frac{hc}{\lambda} = \frac{6.63 \times 10^{-34} \times 3 \times 10^8}{1.3 \times 10^{-6}} = 1.53 \times 10^{-19} \text{ joule}$$

Therefore, the number of photons per second is

$$N = \frac{75 \times 10^{-6}}{1.53 \times 10^{-19}} = 4.9 \times 10^{14}$$

The number of photons per second multiplied by the quantum efficiency, as a number rather than a percentage, results in the number of electrons per second and is

$$N(QE) = 4.02 \times 10^{14}$$

Electrical current in amperes is measured in coulombs per second rather than electrons per second. Multiplying the number of electrons per second by the charge of an electron results in the current in amperes and is

$$4.02 \times 10^{14} \times 1.6 \times 10^{-19} = 64.3 \ \mu A$$

14-3 FIBER OPTIC POWER MEASURING

The photodiode current is unique in that the current is proportional to the incident power. Usually, for a constant impedance, power is proportional to the current squared. This unusual behavior is used to an advantage in the power meter. The diode current is converted to a voltage and the resultant is displayed. Figure 14-3 shows the basic construction of a fiber optic power meter.

The diode feeds a transimpedance amplifier, which converts the diode current to a voltage while maintaining a constant voltage across the diode. The diode quantum efficiency, to a small extent, is a function of the voltage across the diode, and the transimpedance amplifier input impedance appears as a constant and low value. The current from the diode is very low if the power meter is to be sensitive. Therefore, it is desirable that a high-gain amplifier be used directly following the diode. The output voltage from a transimpedance amplifier relative to the input current is given by

$$V_{\text{out}} = R_1 I_{\text{photo}} \tag{14-13}$$

The ratio of the output voltage to the input current is called the transimpedance of the amplifier and is

$$Z_t = \frac{V_{\text{out}}}{I_{\text{photo}}} = R_1 \tag{14-14}$$

Because of the low power involved, the noise voltage from the transimpedance amplifier and the diode itself can cause unsteady readings. Therefore, imme-

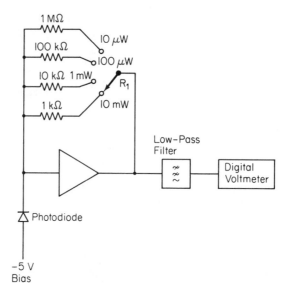

1 MΩ

100 kΩ

10 kΩ 1 mW

1 kΩ

10 μW

100 μW

R_1

10 mW

Low-Pass
Filter

Digital
Voltmeter

Photodiode

−5 V
Bias

Figure 14-3 Block diagram of an optical power meter.

diately following the transimpedance amplifier, a low-pass filter is used to remove some of the noise voltage.

The output voltage from the transimpedance amplifier, after scaling, can be used to read power directly. The range of powers encountered in a fiber optics communications system can easily span several decades. To improve the measurement resolution, the output of the transimpedance amplifier is attenuated and the power output display is divided into several ranges. One method is to scale the gain of the transimpedance amplifier in decade steps and to display the power as an exponent and significant figures. This reduces the amount of dynamic range required by the transimpedance amplifier. The switching of the gain of the transimpedance amplifier could be manual, or autoranging techniques could be used. These techniques are very similar to those used in digital voltmeters and are discussed in Chapter 6.

It is very convenient to measure fiber optics power in decibel notation such as dBm or dBμW. The fiber optic power meter shown in Fig. 14-3 can be modified to display a logarithmic power and is shown in Fig. 14-4. The output of the low-pass filter, following the transimpedance amplifier, is converted to a logarithm. The change in transimpedance amplifier gain by decade steps is equivalent to adding 10 dB to the output every time the amplifier gain is reduced by a factor of 10. To correct the output display, the equivalent of 10 dB of the output display is added to the display every time the gain of the transimpedance amplifier is reduced. The output reading, in decibels, is a composite of the fixed voltage which is a function of the gain of the amplifier plus the converted logarithmic voltage.

Often, the output display can be selected to read either dBm or dBμW. Since there is a 30-dB difference between dBμW and dBm, it is a simple matter of adding the equivalent of 30 dB to the display as shown in Fig. 14-14. A typical fiber optic power meter is shown in Fig. 14-5.

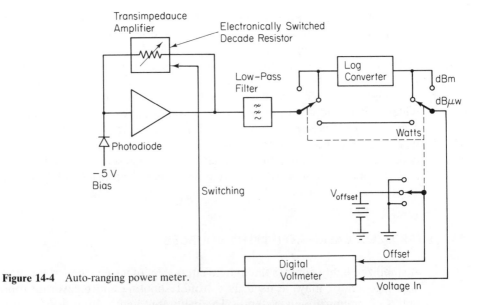

Figure 14-4 Auto-ranging power meter.

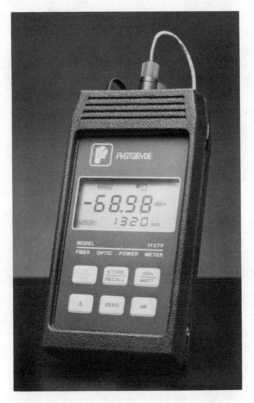

Figure 14-5 Example of an optical power meter.
(Courtesy of 3M Photodyne, Inc.)

When a fiber optic communications system is installed or when trouble must be found, light attenuation is one of the important parameters to be measured. Attenuation is usually evaluated by measuring the power of a light source before and after the attenuation to be analyzed. The loss of light energy in the device is thus determined by the difference in power levels.

Optical power is measured by two basic methods. The first method is a wideband or wide-wavelength method which provides the total optical power regardless of wavelength. The second method provides a spectral measurement where the power as a function of wavelength is displayed. The former is the more common of the measurements. Generally, only one source of energy is present in a fiber optic communications system and there is no need to separate the measured power by wavelength.

14-4 STABILIZED, CALIBRATED LIGHT SOURCES

A stabilized light source is the optical equivalent of a signal generator, and like the signal generator may be used as a troubleshooting and measuring tool in fiber optics communications systems. The calibrated light source usually uses a laser diode. However, an LED source could also be used. In some instances a white-light source consisting of a simple incandescent lamp is used as a broadband light source.

One severe problem with the semiconductor diode type of light source, particularly the laser diode, is the deterioration of light output with age. Another problem is the variation of light intensity with temperature.

A stabilized fiber optic light source uses a photodetector to sample the light output, and through feedback, sets the emitter current to provide the desired light output.

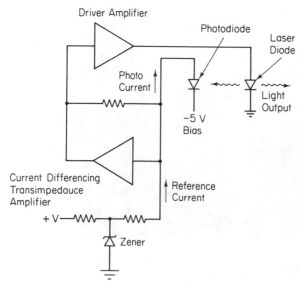

Figure 14-6 Stabilized light source.

Figure 14-6 shows a block diagram of a stabilized light source. A laser diode is the source of light and is driven by a power amplifier. A *PIN* photodiode is coupled to one of the ends of the laser diode. A laser diode can be made to emit light energy from both ends of the diode. Although the light intensity emitted from the two ends may not be the same, the proportionality of the light remains constant from either end.

The output of the sensing diode is fed to a current differencing amplifier. A reference current is also fed to the current amplifier. The diode current is, therefore, set by the feedback of the amplifier to be equal to the reference current, and thus the light output power is a constant.

The photodiode has temperature dependencies, but the magnitude of these dependencies is less than that of the laser diode. In addition, the amount of heating of the sensor diode is less because the power dissipated in the diode is small.

14-5 END-TO-END MEASUREMENT OF FIBER SYSTEM LOSS

One of the more important parameters of a fiber optic communications system is the end-to-end system loss. A simple method of measuring the end-to-end system loss is to provide a known signal at one end of the system and to measure the power available at the other end.

The stabilized fiber optic signal source provides the calibrated and known signal for the transmitting end of the fiber while a power meter measures the power received at the far end.

The significant disadvantage of this system is that the two ends of the fiber optics system may be several kilometers apart. In addition, the sources of loss are not identified. It is not known whether the loss is due to a bad connector, a break, or excessive Rayleigh scattering. It is also not known where the loss is located. When the far end of a fiber system is not accessible, there are other methods of measuring the system loss, such as the optical time-domain reflectometer, which is discussed next.

14-6 OPTICAL TIME-DOMAIN REFLECTOMETER

A very powerful tool in the maintenance and installation of a fiber optics system is the optical time-domain reflectometer. This device analyzes the reflected light energy in a fiber installation to determine the existence and location of breaks in the fiber, losses at splices and connectors, and the total loss of the system. The optical time-domain reflectometer makes use of the reflected light energy in a fiber installation. One source of reflected light power is due to the Rayleigh scattering reflections. As explained previously, Rayleigh scattering causes some of the light energy to be reflected in the return direction. This is a very small amount of light energy, but the measurement of the relative amount of reflected light due to

Rayleigh scattering can be used to measure the loss in a fiber. Larger reflections are indicators of splice losses, and very large reflections are generated by breaks in the fiber.

Figure 14-7 shows a block diagram of an optical time-domain reflectometer. A light source provides a narrow pulse of light which is coupled to the fiber to be measured. The light energy reflected is prevented from entering the transmitter by a directional coupler. The coupler also prevents the very strong light energy from the transmitter from overloading the receiver and thus maintains the sensitivity of the receiver for the weak return energy. A laser diode is usually used as the transmitter because of its high power output, a very narrow spectral bandwidth, a narrow light output beam, and its ability to be pulsed with very narrow pulses. The light receiver displays the reflected light energy as a function of time relative to the transmitted pulse.

Reflections from the fiber are measured by transmitting a very short pulse of light and measuring the reflected pulses. Very short pulses of light are required to allow the optical time-domain reflectometer to resolve distance. The velocity of propagation in a fiber is the velocity of light in a vacuum divided by the index of refraction. A typical fiber has an index of refraction of about 1.5, which when divided into the velocity of light in a vacuum results in a velocity of propagation of about 2×10^8 m/s. Thus, in 1 ns the light propagates about 20 cm in a fiber. If a pulse were broad, it would mask the reflected pulse and the system would not be able to resolve small reflections.

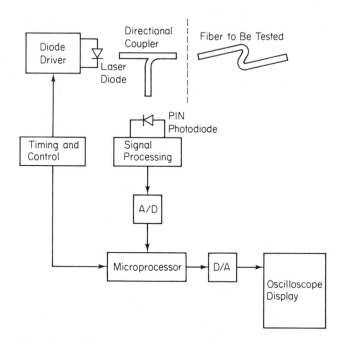

Figure 14-7 Block diagram of an optical time-domain reflectometer.

Perhaps one of the more important parameters measured by the optical time-domain reflectometer is the distance to reflection. The distance is determined by measuring the elapsed time and knowing the velocity of propagation. The time required for a reflection to arrive after the generation of the light pulse is twice the time for the light to propagate to the reflection. This is because a round trip is involved.

EXAMPLE 14-4

How much elapsed time would occur to a reflection from a break in an optical fiber at 1.4 km if the index of refraction of the core was 1.55?

SOLUTION   The velocity of propagation in the fiber is

$$v = \frac{3 \times 10^8}{1.55} = 1.94 \times 10^8 \text{ m/s}$$

The time for the light pulse to reach the break is the distance divided by the velocity of propagation in the fiber, or

$$t = \frac{1.4 \times 10^3}{1.94 \times 10^8} = 7.2 \text{ } \mu s$$

Since twice the time to reach the break is required for the reflection to arrive at the reflectometer, the total time is 14.4 μs.

A simple technique for displaying the reflected light energy would be a conventional oscilloscope triggered from the transmit pulse and displaying the photocurrent from the diode. Although this is possible in principle, there are some difficulties due to the very narrow pulses used and the extreme low level from the reflected energy. Two techniques are used to counteract these problems: sampling and averaging. The techniques of a sampling oscilloscope are outlined in Chapter 7.

A time-delay generator is triggered by the transmit pulse. This generator determines the time delay for sampling the return energy. Only that energy that returns during the sampling time will be measured. Several samples for each time slot will be accumulated where the number of samples depends on the expected strength of the return signal. The average of the return signal will be calculated and this value stored in a computer memory for display. The time-delay generator is then advanced to the next time slot and the averaging is repeated. The close-in reflections do not need as much averaging, and the number of samples per time slot will increase as the time delay is increased. The received power is converted to a logarithm, so this display will be in decibels. Actually, for every 2 dB of change of the received energy, the reflectometer will indicate a 1-dB change. This is due to the fact that any attenuation experienced by the reflected light energy will occur twice: once on the outward trip and a second time on the return trip. The optical time-domain reflectometer displays the results of the averaging on a cathode ray tube display.

To gain an insight into the amount of energy in a reflected pulse, the amount of energy per pulse for the transmitter will be investigated. A typical laser diode can provide 3 mW of power into the fiber with a risetime and a falltime of about 0.5 ns. If a pulse width, measured at the half-power points, of 1 μs is used, the resultant energy contained in the pulse is 3 picojoules (pJ). Clearly, the energy contained in a single pulse would be so minute that it would be difficult to receive any meaningful information. Therefore, a continuous train of pulses are used to make the reflected power measurement.

There is a limit to how often a pulse may be transmitted, as it is necessary to receive all the desired reflections before the next transmitted pulse is emitted. If the length of the fiber to be investigated is 10 km as an example, approximately 100 μs is required for the reflections from the end of the fiber to arrive at the sending end. Therefore, a pulse repetition rate of 10 kHz would be the maximum allowed. The total energy per second, or the average power of the transmitter, is 10,000 × 3 pJ, or 30 nW. The amount of reflected energy is considerably smaller. Only a small fraction of the emitted energy is reflected due to Rayleigh scattering. In addition, there is loss in the fiber itself.

To improve the energy received, an average of many transmissions over a period of up to 100 s is used. There are newer techniques where a scheme of varying pulse widths and computer-controlled averaging is used to reduce the average time.

The optical time domain reflectometer display must be interpreted to analyze the parameters of the fiber optic link. A typical optical time-domain reflectometer display is shown in Fig. 14-8. In this example the analysis of three sections of fiber is shown.

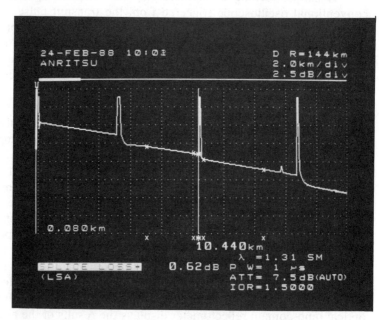

Figure 14-8 Optical time domain reflectometer display. (Courtesy Anritsu America Inc.)

Fiber Optics Measurements   Chap. 14

The display is, generally a sloping line with three discontinuities which are due to the connectors joining the fibers. The left side of the display represents the optical time-domain reflectometer end of the fiber and shows a 5 km length of fiber. The slope of the trace is 0.4 dB/km which is the loss of the first section of fiber.

There is a large peak of the trace which represents the energy reflected from the slight mismatch of the connector. This reflected energy subtracts from the energy passed through the connector and appears as a loss which appears as a sudden drop of level which is about 1 dB in the figure.

The second length of fiber is also about 5 km with a loss of 0.4 dB/km but the second connector has a loss of just over 0.5 dB.

The third length of fiber is about 6 km in length with the same 0.4 km of loss per km. Visible in this figure is a small reflection from a defect in the fiber at about 15 km from the near end.

A third connector is visible at about 16.5 km followed by more fiber. The total loss for the 20 km of the display is 10 dB which represents 8 dB of loss in the fiber, (20 km of fiber at 0.4 dB/km), plus a total of 2 dB for the three connectors.

PROBLEMS

14-1. What is the velocity of light in a glass with an index of refraction of 1.38?

14-2. How would a ray of light be deflected when entering a block of glass from air at an angle of 45° from the surface if the index of refraction of the glass is 1.6?

14-3. How would the light ray be deflected when leaving the block of glass and returning to air in Problem 14-2? Plot the ray from air to the glass and to the air.

14-4. What is the critical angle between the surface of a window pane with an index of refraction of 1.7 and air?

14-5. What is the acceptance angle of a fiber with a core index of refraction of 1.49 and a cladding index of 1.47?

14-6. What is the numerical aperture of the fiber described in Problem 14-5?

14-7. An emitting diode of numerical aperture 0.3 is coupled to a fiber of numerical aperture of 0.22. What is the loss in decibels between the coupling?

14-8. If a detector diode of aperture 0.3 is coupled to the same fiber as in Problem 14-7, what would be the loss of the coupling?

14-9. What would be the resultant photocurrent generated in a diode if 850 nm of light energy were incident on the diode of 0.1 μW? The quantum efficiency of the diode is 0.7.

14-10. A laser diode couples 50 μW of 1,300-nm light power to a 100-μm fiber. The fiber is 10 km long and has a loss of 1.2 dB/km. The numerical aperture of the fiber is 0.33 and is coupled to a diode of numerical aperture of 0.22. The quantum efficiency of the detector diode is 80 per cent. How much photocurrent is generated?

14-11. How long would be required for a reflection from a break in a fiber to return to an optical time-domain reflectometer if the distance to the break is 1.2 km? The index of refraction of the fiber is 1.33.

Appendix

ABBREVIATIONS, SYMBOLS, AND PREFIXES

The use of symbols, prefixes, and abbreviations follows the recommendations of the International Electrotechnical Commission, the American National Standards Institute, Inc., the Institute of Electrical and Electronics Engineers, and other scientific and engineering organizations. Where there is not agreement among these groups, the usage favored by the majority is chosen.

Abbreviations and Symbols

a	atto (10^{-18})		B	susceptance
A	ampere		bar	bar ($10^5 N/m^2$)
Å	angstrom		BCD	binary-coded decimal
ac	alternating current			
afc	automatic frequency control			
am	amplitude modulation		c	speed of light, centi (10^{-2})
ANSI	American National Standards Institute, Inc.		C	capacitance, coulomb
			°C	degrees Celsius (Centigrade)
APS	American Physical Society		cd	candela
ASA	Acoustical Society of America		CIF	cost, insurance, freight
ASTM	American Society for Testing and Materials		CML	current-mode logic
			COD	cash on delivery
avc	automatic volume control		cw	continuous wave
avg	average			

d	deci (10^{-1})	lb	pound
D	dissipation factor	LC	inductance-capacitance
da	deka (10)	lm	lumen
dB	decibel	log	logarithm
dBm	decibel referred to one milliwatt	lx	lux
		m	meter, milli (10^{-3})
dc	direct current	M	mega (10^6)
DCTL	direct-coupled transistor logic	max	maximum
		mbar	millibar
dia	diameter	mil	0.001 inch
DTL	diode-transistor logic	min	minimum, minute
DUT	device under test	mo	month
e	electronic charge	n	nano (10^{-9})
E	voltage	N	newton
EIA	Electronic Industries Association		
		oz	ounce
emf	electromotive force		
		p	page, parallel (as L_P), pico (10^{-12})
F	farad, Faraday	P	poise ($10^{-5}N \cdot s/m^2$)
°F	degrees Fahrenheit	PF	power factor
f	frequency, femto (10^{-15})	ppm	parts per million
fm	frequency modulation	pps	pulses per second
FOB	free on board	pk-pk	peak-to-peak
		PRF	pulse repetition frequency
G	conductance, giga (10^9)		
g	gram, gravitational constant	Q	quality factor (storage factor)
g_m	transconductance		
H	henry	R	resistance
h	hour, Planck's constant, hecto (10^2)	®	registered trademark
		rad	radian
hf	high frequency	RC	resistance-capacitance
h_f	forward current-transfer ratio	RCTL	resistor-capacitor-transistor logic
h_i	short-circuit input impedance	re	referred to
h_o	open-circuit output admittance	rf	radio frequency
		RH	relative humidity
h_r	reverse voltage-transfer ratio	rms	root-mean-square
Hz	hertz (cycle per second)	rpm	revolutions per minute
HTL	hearing threshold level	RTL	resistor-transistor logic
I	current	s	second, series (as L_s)
IC	integrated circuit	shf	super-high frequency
ID	inside diameter	sq	square
IEC	International Electro-technical Commission	sync	synchronous, synchronizing
IEEE	Institute of Electrical and Electronics Engineers	T	period, Tesla, tera (10^{12})
		t	time
if	intermediate frequency	TTL	transistor-transistor logic
in.	inch	TSA	times series analysis
ISA	Instrument Society of America	uhf	ultra-high frequency
ISO	International Standards Organization	v	velocity
		V	volt
j	$\sqrt{-1}$	VA	volt ampere
J	joule	vhf	very-high frequency
		vlf	very-low frequency
k	kilo (10^3)		
°K	degrees Kelvin	W	watt
		Wb	Weber
l	liter (10^{-3} m³)	wt	weight
L	inductance		

X	reactance		Γ	reflection coefficient
Y	admittance		Δ	increment
yr	year		δ	loss angle
Z	impedance		θ	phase angle
α	short-circuit forward current-transfer ratio (common base)		λ	wavelength
			μ	micro (10^{-6})
			Ω	ohm
β	short-circuit forward current-transfer ratio (common emitter)		$\mho$	mho
			ω	angular velocity ($2\pi f$)

Prefixes

Orders of magnitude from 10^{-18} to 10^{12} are designated by the following prefixes:

Order	Prefix	Symbol
10^{12}	tera	T
10^{9}	giga	G
10^{6}	mega	M
10^{3}	kilo	k
10^{2}	hecto	h
10	deka	da
10^{-1}	deci	d
10^{-2}	centi	c
10^{-3}	milli	m
10^{-6}	micro	μ
10^{-9}	nano	n
10^{-12}	pico	p
10^{-15}	femto	f
10^{-18}	atto	a

Selected Answers

CHAPTER 1

1-6. 1 mV
1-8. 75.0 μF $\pm$ 0.1 μF
1-10. 82 mV
1-12. (a) 147.5 Ω, (b) 0.21 Ω, (c) 0.3 Ω, (d) 0.2 Ω
1-14. (a) 36 Ω $\pm$ 1.8 Ω, 75 Ω $\pm$ 3.75 Ω, (b) 111 $\pm$ 5.55 Ω, 111 $\pm$ 5% Ω, (c) 24.32 $\pm$ 3.65 Ω
1-16. (a) 435.3 Ω, (b) 3.7%
1-18. (a) $\pm$7.55%, (b) $\pm$0.57%

CHAPTER 2

2-1. 1.5 GHz, 12,500 Hz, 0.125 μH, 346,400 V, 0.0053 A, 5,000 mH, 4.6 $\times$ 10^{-12} J, 0.0014 ms, 8.89 $\times$ 10^{-13} hr, 14 $\times$ 10^{-9} μs
2-3. 2.85 $\times$ 10^{19}
2-5. 180 cm
2-9. 35.7 m/s
2-11. 3.6 $\times$ 10^6 J
2-13. 200 V

2-15. 4.6875×10^{15}

2-17. (a) 8,930 kg/m³, (b) 557 lb/ft³

CHAPTER 3

3-6. 0.999993 Ω

3-10. 1.0190 V

CHAPTER 4

4-1. 875 Ω

4-3. 36 MΩ

4-6. 50 V and higher

4-7. (a) 0.094 mW, (b) 4.29 mW

4-9. 1.25 V

4-15. (b) 900 Ω/V

4-17. 25 W

CHAPTER 5

5-1. 0.01 Ω

5-3. 6×10^{-7}

5-5. $R = 34.3$ Ω, $L = 29$ mH

5-7. (a) $R_s = 1,000$ Ω, (b) $R_s = 250$ Ω

CHAPTER 6

6-2. 15 mV, 10,000 Ω/V

6-4. 26.6 kΩ, 2.66 kΩ, 266 Ω

6-6. 2+ pF

CHAPTER 7

7-6. 2.65×10^7 m/s

CHAPTER 8

8-1. 2.65

8-3. +35 dBm, −30 dBm, +26 dBw, 1 V, +3 dBw, −17 dBw, 0.22 μV

440

8-6. $R_1 = Z \dfrac{\sqrt{N} - 1}{\sqrt{N} + 1}$ $R_3 = \dfrac{2Z\sqrt{N}}{N - 1}$

8-15. 15.9 Hz

CHAPTER 9

9-1. 70 dB

9-5. −60 dBm

9-7. (a) Increases the third-order intercept by an amount equal to the attenuation.
   (b) Does not affect the dynamic range.
   (c) Increases the noise figure by an amount equal to the attenuation.

CHAPTER 10

10-1. 1 s

10-3. Five digits

CHAPTER 11

11-4. 694 kg/cm²

11-5. 25 μV

11-6. 2.5 × 10⁻³ mm

Index